STANLEY JAKI FOUNDATION

INTERNATIONAL CONGRESS 2024

STANLEY JAKI FOUNDATION INTERNATIONAL CONGRESS 2024

PAUL HAFFNER
JOSEPH LARACY
EDITORS

GRACEWING

First published in 2026 by
Gracewing
2 Southern Avenue
Leominster
Herefordshire HR6 0QF
United Kingdom
www.gracewing.co.uk

ISBN 978 085244 960 8

Typeset by Gracewing

Cover design by Bernardita Peña Hurtado

Cover photo: Fr Jaki receives the insignia as Honorary Member of the Pontifical Academy of Sciences from Pope St John Paul II, October 1990

CONTENTS

PREFACE

Seton Hall University's Departments of Physics and Catholic Studies, Immaculate Seminary School of Theology, Core Curriculum Program, Center for Catholic Studies, and chapter of the Society of Catholic Scientists, in collaboration with the Stanley Jaki Foundation, organized the 2024 Stanley Jaki International Congress. This conference marked the centenary of Fr. Stanley Jaki's birth on August 17, 1924, and was held on April 24, 2024, at Seton Hall University in South Orange, New Jersey. This event followed a successful inaugural 2015 Stanley Jaki International Conference. All papers were subject to double blind peer review by a distinguished group of scholars. The Program Committee invited the submission of scholarly papers in areas such as the relationship between Christian faith and the natural sciences, the relationship of physics, the philosophy of nature, and metaphysics, as well as other interdisciplinary topics related to faith and science.

Chapter 1 (Stanley Jaki: A Life Devoted to the Relationship Between Science and Faith) by Antonio Colombo provides an excellent introduction to the work of Fr. Jaki. Mr. Colombo first met Jaki in 1999. He collaborated with him from 2000 to 2009, proofreading his works in English, and translating into Italian some of them, including *Questions on Science and Religion*, *Bible and Science*, and *The Ethical Foundations of Bioethics*. Antonio Colombo's in-depth knowledge of the life and ministry of Fr. Jaki is a great contribution to this volume.

Chapter 2 (The Papers of Stanley L. Jaki: An Introduction) by Jacquelyn Deppe, Ryan Fino, and Dr. Sarah Ponichtera offers an important contribution summarizing their work thus far on Jaki's personal files. His papers were deposited in the Monsignor William Noe Field Archives and Special Collections Center at Seton Hall University in 2011, following his death in 2009. This chapter introduces Jaki's collection, the process of organizing and preserving the contents, and some highlights of the collection that have been discovered so far. This collection consists primarily of Jaki's research notes, compiled from a wide range of sources from libraries around the world. It also includes his personal correspondence with his family in Hungary and his colleagues in the United States. This chapter is a helpful resource for scholars interested in the Jaki archival materials.

Chapter 3 (Has Stanley Jaki Underestimated the Byzantine Influence on the Rise of Science?) was authored by the Rev. Dr. Paul M. Haffner, one of the leading experts on Jaki. Readers of this volume are likely well aware of Fr. Jaki's contributions to the history of science, namely that the Middle Ages represent the very cradle of science. This is in sharp contrast to the backdrop of Enlightenment ideology and propaganda that the Middle Ages were devoid of technological progress and, more specifically, that in the Eastern Roman Empire superstition and mysticism ruled and was devoid of invention. In fact, science and progress were more than alive in the Middle Ages and, even before that, in the Eastern Roman Empire. Fr. Haffner's recent research has shed new light on the importance of early Byzantine science.

Chapter 4 (Jaki and Barbour: Pilgrims on the Science and Religion Journey) by the Rev. Dr. Joseph R. Laracy discusses the striking parallels between Stanley L. Jaki and Ian G. Barbour. Despite very comparable scientific backgrounds in experimental physics, a shared commitment to philosophical realism, and similar research paths exploring the relationship of Christian theology and natural science, there is surprisingly little evidence of interaction between them. This chapter elucidates the similarities and differences in their scientific, philosophical, and theological research. It clarifies their epistemological, metaphysical, and doctrinal commitments and how these may have contributed to minimal scholarly interaction.

Chapter 5 (The Relevance of Jaki) by Dr. Peter Floriani acknowledges the vastness of Fr. Jaki's scholarly work. He argues how the breadth and depth of Jaki's research often places him beyond the grasp of the common man, the common Catholic, the common scientist, and even the common scholar. Hence, it is necessary to set forth his relevance to invite more readers to benefit from his extensive writings. Dr. Floriani explains how Fr. Jaki brought a diligent scholarly approach to the study of history, science, and theology for the glory of God and the benefit of humanity.

Chapter 6 (Chesterton, Undset, and Jaki: Three Catholic Intellectual Giants) by Geir Hasnes discusses three intellectual giants and defenders of the Catholic faith: G. K. Chesterton, Sigrid Undset, and Fr. Jaki. The chapter reveals how the relationship between Chesterton and Sigrid Undset goes much deeper than is generally known. Fr. Jaki deeply admired both and wrote a monograph on each of them. Although very different both in personality, lifestyle, and writing style, the three are united to each other through their common Catholic faith.

Chapter 7 (Gödel's Theorems in Physics Half a Century after Stanley Jaki) was written by Gergely Bognár. In 1966 Fr. Jaki first suggested in *The Relevance of Physics* that Gödel's theorems represent a limit to cognition. Bognár describes how in Jaki's later works he elaborated the details of the program he presented in *The Relevance of Physics*, but the fundamental idea remained essentially unchanged. Bognár argues that in the context of Gödel's theorems, the name of Jaki has been considered a "benchmark" to this day, despite the fact that much has happened in both physics and mathematics in the past half century.

Chapter 8 (AI, International Relations, and Catholicism) by the Rev. Dr. Brian K. Muzás engages salient issues at the intersection of human security, Pope Francis' *Laudato Si'*, and Jaki's *Brain, Mind and Computers.* Although written to address the global issues of climate and the environment, *Laudato Si'* can also address AI issues via human security's notions of freedom from want, freedom from fear, and freedom to flourish. The point of contact between the thought of Francis and Jaki is the former's call for an "adequate anthropology" to defend human dignity against technocracy and the latter's ability to provide intellectually rigorous material for that purpose. The two men share complementary frameworks that take human dignity, moral reasoning, and societal well-being seriously in the face of technological advancement. Jaki's focus on the limitations of AI and the human person relates well to Pope Francis's call for responsible innovation with justice and solidarity. Moreover, their common insights spotlight the need to consider the implications of AI not simply for technological progress but for humanity as individuals and overall.

Chapter 9 (Conciliatory Christian Humanism and the Confinement of Naturalism) by the Rev. Dr. Zachary Mabee highlights two of the core commitments of prominent varieties of philosophical naturalism and scientism, particularly their methodological or epistemological deference to the sciences as well as their anti-supernaturalist ontology. He stresses that such varieties of naturalism and scientism are indeed philosophical—that is, that they are meta-scientific assessments of the sciences and their character, aims, and contributions. They also, however, tend to be markedly anti-humanistic. Drawing especially upon Iris Murdoch and Stanley Jaki, Mabee argues that such philosophical assessments manifest their own fundamental reliance on words and philosophical language, more than on the quantitative language towards which the sciences gravitate. He emphasizes with Murdoch and Jaki that we must engage philosophically with the sciences to take stock of and advocate for their place or role within our human lives and societies. Fr. Mabee concludes that a robust and open-minded Christian humanism fosters such conciliatory engagement—philosophically and even theologically—in a way that even expansive forms of naturalism, by their own principles, undermine.

Chapter 10 (God is Still There: An Analysis of Werner Heisenberg's Reconciliation of Quantum Mechanics with Aristotelian Hylomorphism) by Dr. Stacy Trasancos focuses on an error made by Werner Heisenberg when he tried to reconcile quantum mechanics with Aristotelian hylomorphism. She argues how this error has had enormous consequences for the contemporary mentality. The error is this: Heisenberg names energy as the fundamental cause for all change in the world. He follows Aristotle's matter-form distinction but, in an effort to avoid any mention of the spiritual or of God in his philosophical assessment, he turns the distinction into matter-force to matter-energy back to matter-form, but with a changed meaning of form to refer to structure instead of being. These ideas can be salvaged by following Aristotle's Unmoved Mover argument into Thomas Aquinas's First and Second Way arguments for the existence of God. This allows one to present an interpretation of quantum mechanics consistent with Aristotelian-Thomistic natural theology and the Christian theology of creation.

Chapter 11 (Searching for Truth in Determinism: A Call for Compassion, Mercy, and Love) was written by Dr. Samantha Mattheiss. She explains how determinism is gaining popularity as neuroscientific evidence increasingly points to the immense influence of external factors on human behavior. Scholars such as Robert Sapolsky assert that our choices are completely determined by our environment, neurobiology, and interaction between the two. While much of Sapolsky's argument is based on measurable evidence, other conclusions are drawn about unmeasurable constructs, thereby contradicting Father Jaki's delineation of science and religion. In line with Jaki's metaphysical realism, Dr. Mattheiss argues how objective, scientific evidence points to a vast influence of biological and environmental factors on human behavior, thought, and emotion in support of human freedom.

Chapter 12 (Nineteenth Century American Catholic Approaches to Darwinian Evolution) by Dr. Jeffrey Morrow and Dr. Biff Rocha discusses how American Catholics in the nineteenth century struggled with Charles Darwin's theory of evolution. It seemed that Darwinian evolution offered a counter story to the biblical narrative of Genesis, thus raising the question: How should a Catholic respond? The nineteenth century saw the formation of two dominant responses to Darwin. One response in general rejected evolution. A second response generally embraced the modern ethos and welcomed evolutionary theory. Perhaps the most famous of the second approach is Fr. John Zahm, a priest of the Congregation of the Holy Cross and a professor at the University of Notre Dame, who wrote *Evolution and Dogma* in 1896, wherein he adopted Darwin's theory of evolution and attempted to adapt Catholic theology accordingly. Their chapter highlights a third approach, which although not dominant, was significant. This third approach was that of Fr. Januarius De Concilio, the first professor of Theology at Seton Hall College and Seminary and drafter of the *Baltimore Catechism.* De Concilio attempted to negotiate the more dominant polarized positions.

Chapter 13 (Nature and Scripture: Galileo's Proposed Biblical Hermeneutic in the Context of Early Modern Biblical Interpretation) was also written by Dr. Morrow. Here he explores the intersection of Galileo's scientific work with his theological work, namely his biblical hermeneutic or proposed method of biblical interpretation. He situates Galileo's hermeneutic within the context of early modern biblical hermeneutics. In order to do so, Morrow focuses on Galileo's famous 1615 "Letter to the Grand Duchess Christina," which is arguably Galileo's most important work pertaining to biblical exegesis. This chapter offers a detailed and fascinating look at Galileo's discussion of biblical hermeneutics.

Chapter 14 (Pythagorean Argument of the Grand Design) by Dr. Alexey Burov and Dr. Alexei Tsvelik presents a variant of the teleological argument for the universe. They suggest a list of intelligent properties of the fundamental physical laws and demonstrate the conflicting requirements they must satisfy to be discoverable. They conclude that we live in a very special Pythagorean universe, whose laws belong to a seemingly narrow minimax of complexity, i.e., they are complicated enough to make fine tuning for life possible, and, at the same time, they are simple enough to be discoverable for the emerging living thinking beings. Their argument is based on the history of physics, where they point out a special "Pythagorean faith," a form of Christian Platonism, which inspired the pioneers of mathematical physics to look for the mathematically elegant laws of nature. Since this faith was justified more than its enthusiasts could expect, Burov and Tsvelik argue that it deserves to be accepted as an important addendum to the working hypothesis of the Grand Design.

Chapter 15 (All Things were Made through Him: Some Thoughts about The Teaching of St. Thomas Aquinas on the Relationship between Matter and God) was written by Dr. Jordan Miller. He engages the at times paradoxical relationship between matter and God, addressing questions such as

- whether in creating material beings out of nothing, God is producing something that is utterly different from what He is, something totally unlike Himself,
- whether a material being has similarity to God, not in spite of its materiality, but because of its materiality, and
- whether in creating matter, God is imparting something of His own, giving something of Himself.

Fr. Jaki repeatedly emphasized that the Christian doctrine of creation out of nothing provided the "fertile soil" necessary for science to take root and grow. Whereas pantheism regards the natural forces of the universe as deities that must be worshipped and propitiated, Christianity proclaims that these natural forces are beautiful and very good, but never to be mistaken for the Creator who brings them into being.

Chapter 16 (Evolution, Teleology, and Purpose: Transhumanism and the Natural and Supernatural Ends of Being) by Dr. Braden Molhoek engages the issues of purpose and *telos* in natural processes such as evolution. It explores differing views on the question of teleology in evolution, including Fr. Jaki's critique of materialist Darwinism that sought to remove discussions of purpose from nature. Even though contemporary science seeks to present an understanding of evolution that is not teleological, transhumanism as a movement, believes that humans are capable of taking control of evolution and should actively use technological means to improve the human condition. While not placing teleology or purpose into evolutionary science, transhumanism does appear to make a philosophical and perhaps even theological argument for purpose. Finally, Molhoek examines the question of the natural and supernatural ends of humans and potential posthuman species, and how they would differ.

Chapter 17 (Can What Cannot be Measured Exactly Happen Exactly?) by Dr. John Long critically engages the Copenhagen interpretation of quantum mechanics, which is often stated in such a way to assert that quantum theory "overthrows" the philosophical principle of cause and effect. This claim has found its way into numerous popular works on modern physics, among them Stephen Hawking's *A Brief History of Time*. After presenting some of the basics of quantum mechanics, this chapter presents Jaki's criticism and historical analysis of this interpretation. According to Jaki, this interpretation arose from a gross error in logic. Jaki's writings abound in commentary on how ever since its formulation by Heisenberg, quantum theory has been interpreted (or mis-interpreted) in a way inconsistent with both common sense, sound philosophy, and right reason. Long's analysis highlights the need for a realist philosophy when interpreting scientific advances.

Chapter 18 (Theology as a Science and the Dialogue with Other Sciences from a Thomistic Perspective) by the Rev. João Paulo Santiago, CSJ provides some reflections on theology as a science and suggests how a dialogue with other sciences is possible. This study is based on Aquinas' conception of *sacra doctrina* as science and, consequently, theology. The chapter shows how a Thomistic understanding of theology as a science provides a solid foundation for a dialogue with other disciplines. It concludes by drawing some conclusions for both believers and non-believers, as well as creates possibilities for a better understanding of the interactions between Catholic theology and science. The respect for autonomy and boundaries of every field, as Catholic theology professes, helps those who do not have faith to avoid any form of scientism.

Chapter 19 (The Mystery of Creation, Various Ontologies, and Effective Communication) by the Rev. Brian Humphrey argues for the importance of ontology. Fr. Humphrey notes that when theologians or scientists with different ways of considering "being" seek constructive dialogue, they can easily miscommunicate. Denying ontology makes for impossible conversations, and differing ontologies make for difficult conversations. Fr. Jaki spoke about the need to admit ontology as a prerequisite for effective communication. In this chapter, he agrees with Jaki and builds on his thought by surveying classical and contemporary notions of ontology, and by offering examples and grounds for the permissibility of various ontologies. He does this within the context of a comparative study between the Thomist and Balthasarian schools of thought and claims that, in this case, communication may be difficult between the two not because one or the other denies ontology, or even because their ontologies may vary, but because their ontologies can be interpreted as either "closed" or "confused." In the end, the very fact that opinions differ about which ontological categories best describe "being" points to the mysterious nature of creation and the need for a properly relational ontology that can bridge various schools of thought.

Chapter 20 (A Comparison of Mathematical Realism in the Works of Stanley Jaki and Plato) was written by two precocious undergraduate students, Collin Doyle and Iain Morton. Throughout the world, there are many undergraduate students interested in the theology and science field. They bring to the subject innovative and fresh perspectives sometimes ignored or missed by more established scholars. Most of these people have yet to make their mark in the science and religion field. The Stanley Jaki Foundation is resolved to tap this valuable source of information, observation, analysis, and critical study. Mathematical realism is a concept that has developed over time. Doyle and Morton point out that Fr. Jaki has had a significant impact on the modern understanding of realism in general, and that his impact on the field of metaphysics, epistemology, and mathematics, remains exceedingly relevant. Although some aspects of Jaki's work regarding mathematical realism are difficult to classify as either exclusively toward one position or the other, Doyle and Morton argue that by his sympathy towards the indispensability argument, as well as his own personal religious beliefs, his voice was one that supported the concept of mathematical realism.

ACKNOWLEDGEMENTS

The publication of this volume was made possible through the generous contributions of

SETON HALL UNIVERSITY'S
Center for Catholic Studies
Department of Physics
Department of the Core Curriculum
Immaculate Conception Seminary School of Theology
College of Arts and Sciences
Office of the Provost

The editors are very grateful for the editorial assistance of Blake Harrsch.

CONFERENCE COMMITTEE

Rev. Gerald Buonopane, Ph.D.
Minister, Priest Community
Senior Lecturer, Department of Chemistry and Biochemistry
Seton Hall University

Sulie Chang, Ph.D.
Professor of Biological Sciences
Director, Institute of NeuroImmune Pharmacology (INIP)
Director, Academy for Nature and Nurture (ANN): Interdisciplinary Approaches to Resilience

Antonio G. Colombo, M.S.
Secretary, Stanley Jaki Foundation
Computer Engineer (Retired)

Rev. Paul M. Haffner, S.T.D.
General Conference Chair
President, Stanley Jaki Foundation
Invited Professor, Faculty of Theology Pontifical Gregorian University and Adjunct Professor, Department of Catholic Studies
Seton Hall University

Rev. Joseph R. Laracy, S.T.D.
Conference Program Committee Chair
Treasurer, Stanley Jaki Foundation
Assistant Professor, Department of Systematic Theology and Adjunct Professor, Department of Mathematics and Computer Science
Seton Hall University

Jose L. Lopez, Ph.D.
Professor, Department of Physics
Director, Office of Grants and Research Services
Seton Hall University

Ines A. Murzaku, Ph.D.
Professor, Department of Religion
Director, Program in Catholic Studies
Seton Hall University

M. Alper Sahiner, Ph.D.
Professor and Chair, Department of Physics
Seton Hall University

Stacy Trasancos, Ph.D.
Adjunct Professor, Department of Catholic Studies
Seton Hall University

PEER REVIEWERS

The Program Committee is very grateful to the scholars who refereed papers. Their scholarly expertise ensures the academic quality of the conference proceedings. Manuscripts were subject to double blind peer review.

Justin Anderson (Ph.D. Katholieke Universiteit Leuven)
Professor and Chair of Moral Theology, Seton Hall University

Rev. Michael Barone (M.Div. Mount Saint Mary's Seminary, M.A. New York University)
Cemetery Chaplain, Archdiocese of Newark

Stephen M. Barr (Ph.D. Princeton University)
Professor Emeritus of Physics, University of Delaware and Founding President, Society of Catholic Scientists

Rev. Giovanni Pietro Basile, S.J. (Ph.D. and Habil. Phil. Ludwig-Maximilians Universität Munich)
Associate Professor of Philosophy, Boston College

Patricia Blanchette (Ph.D. Stanford University)
McMahon-Hank Professor of Philosophy, University of Notre Dame

Gregory Burton (Ph.D. University of Connecticut)
Professor of Psychology, Seton Hall University

Sulie Chang (Ph.D. Ohio State University)
Professor of Biological Sciences, Director of the Institute of NeuroImmune Pharmacology (INIP)
Director of the Academy for Nature and Nurture,

(ANN): Interdisciplinary Approaches to Resilience
Seton Hall University

Steven Clark (B.S. Colorado School of Mines)
Doctoral Student in Physics, Rutgers University

Antonio Colombo (M.S. Politecnico di Milano)
Computer Engineer (retired)

Maureen Condic (Ph.D. University of California, Berkeley)
Associate Professor of Neurobiology and Anatomy, University of Utah School of Medicine

Rev. Gabriel B. Costa (Ph.D. Stevens Institute of Technology)
Professor of Mathematics, United States Military Academy at West Point

Rev. Thomas Davenport, O.P. (Ph.D. California Institute of Technology)
Thomistic Institute Project for Science and Doctoral Student in the Faculty of Philosophy, Pontifical University of St. Thomas

Edward B. Davis (Ph.D. Indiana University)
Professor Emeritus of the History of Science, Messiah University

Alan Delozier (D.Litt. Drew University)
Humanities and Outreach Librarian, Seton Hall University

Nancy Enright (Ph.D. Drew University)
Professor of English and Director of the University Core Curriculum, Seton Hall University

Timothy Fortin (Ph.D. Pontifical University of the Holy Cross)
Associate Professor and Chair of Philosophical Theology, Seton Hall University

Philippe Gagnon (Ph.D. Dominican University College in Ottawa, Ph.D. Université Laval)
Chair of Science, Technology, and Faith, Université Catholique de Lille

Jean De Groot (Ph.D. Harvard University)
Emerita Ordinary Professor of Philosophy, Catholic University of America

Rev. Msgr. Thomas G. Guarino (S.T.D. Catholic University of America)
Professor Emeritus of Systematic Theology
Seton Hall University

Rev. Paul M. Haffner (S.T.D. Pontifical Gregorian University)
Visiting Professor, Pontifical Gregorian University and Adjunct Professor, Seton Hall University

Mark Harris (Ph.D. University of Cambridge)
Andreas Idreos Professor of Science and Religion and Director of the Ian Ramsey Centre for Science and Religion in the Faculty of Theology and Religion
Oxford University

Geir Hasnes (M.Sc. Norwegian University of Science and Technology)
Associate Professor of Engineering Cybernetics, Norwegian University of Science and Technology

Richard F. Hassing (Ph.D. Cornell University)
Research Associate Professor of Philosophy, Catholic University of America

Rev. Peter A. Heasley (S.T.D. Pontifical Gregorian University)
Pastor, The Parish of Corpus Christi and Notre Dame, NY, NY and Adjunct Professor of Scripture St. Joseph's Seminary, Dunwoodie

Jeremy Horne (Ph.D. Florida State University)
President Emeritus Southwest Area Division of the American Association for the Advancement of Science

Leon Horsten (Ph.D. Katholieke Universiteit Leuven)
Professor of Philosophy, Universität Konstanz

Kenneth J. Howell (Ph.D. Indiana University, Ph.D. Lancaster University)
Academic Director of the Eucharist Project and President Pontifical Studies Foundation

Rev. Sony Joseph (Ph.D. University of Illinois)
Doctoral Student in Theology, Pontifical University of St. Thomas Aquinas

Very Rev. John Kartje (Ph.D. University of Chicago, S.T.D. Catholic University of America)
Rector and President University of St. Mary of the Lake

Daniel Kuebler (Ph.D. University of California, Berkeley)
Professor of Biology, Franciscan University of Steubenville

Robert Kugelmann (Ph.D. University of Dallas)
Distinguished Emeritus Professor of Psychology, University of Dallas

Gregory LaNave (Ph.D. Catholic University of America)
Ordinary Professor of Dogmatic Theology, Dominican House of Studies, DC

Carlo Lancellotti (Ph.D. University of Virginia)
Professor of Mathematics and Physics, College of Staten Island, CUNY

Elaine Landry (Ph.D. University of Western Ontario)
Professor and Chair of Philosophy, University of California, Davis

Rev. Roger Landry (S.T.L. John Paul II Institute for the Study of Marriage and Family, Lateran University)
Catholic Chaplain, Columbia University and the Thomas Merton Institute for Catholic Life

John T. Laracy (Ph.D. John Paul II Institute for Studies on Marriage and Family, DC)
Teacher of Theology, Don Bosco Preparatory High School, NJ

Rev. Joseph R. Laracy (S.T.D. Pontifical Gregorian University), Assistant Professor of Systematic Theology, Seton Hall University

Rev. Msgr. Richard M. Liddy (Ph.D. Pontifical Gregorian University)
Professor Emeritus of Catholic Studies and Religion
Seton Hall University

David Liggins (Ph.D. University of Sheffield)
Senior Lecturer in Philosophy, University of Manchester

John Long (Ph.D. Monash University)
The Institute for the Study of Christianity in an Age of Science and Technology

Rev. Zachary Mabee (Ph.D. University of Reading)
Instructor of Philosophy, Sacred Heart Major Seminary, Detroit

Rev. Thomas K. Macdonald (S.T.D. Pontifical Gregorian University)
Vice Rector, St. John's Seminary, Boston

Carol Marlowe (MLS Rutgers University)
Librarian (Retired)

Thomas J. Marlowe (Ph.D. Rutgers University, Ph.D. Rutgers University)
Professor Emeritus of Mathematics and Computer Science, Seton Hall University

Kieran McNulty (Ph.D. City University of New York)
Professor of Anthropology, University of Minnesota

Roland Millare (S.T.D. Pontifical Faculty of Theology at the University of Saint Mary of the Lake)
Vice President of Curriculum and Director of Clergy Initiatives, St. John Paul II Foundation

Jeffrey Morrow (Ph.D. University of Dayton)
Professor of Undergraduate Theology, Seton Hall University

Rev. Brian K. Muzás (Ph.D. University of Texas at Austin)
Assistant Professor of Diplomacy and International Relations, Seton Hall University

Rev. James P. Platania (S.S.L. Pontifical Biblical Institute)
Doctoral Student, Pontifical Gregorian University

J. Brian Pitts (Ph.D. University of Notre Dame, Ph.D. University of Texas at Austin)
Senior Research Fellow, University of Lincoln and The Faraday Institute for Science and Religion

Dermot Quinn (D.Phil. Oxford University)
Professor of History, Seton Hall University

Lawrence M. Principe (Ph.D. Johns Hopkins University, Ph.D. Indiana University)
Drew Professor of the Humanities and Director of the Singleton Center for the Study of Premodern Europe
Department of the History of Science and Technology
Johns Hopkins University

Rev. Thomas P. Quinn (M.Div. and M.A. Seton Hall University)
Pastor, Holy Trinity Church, Hackensack, NJ
and Adjunct Professor of Catholic Studies, Seton Hall University

Rev. John Ranieri (Ph.D. Boston College)
Professor of Philosophy, Seton Hall University

Rev. Ignacio de Ribera Martin, D.C.J.M. (Ph.D. Catholic University of America)
Associate Dean of the School of Philosophy, Catholic University of America

Robert J. Russell (Ph.D. University of California, Santa Cruz)
Professor Emeritus of Theology and Science, Graduate Theological Union and Founder of the Center for Theology and the Natural Sciences (CTNS)

Rev. Thomas P. Shubeck (Ph.D. Purdue University)
Director of Psychological Services at St. Andrew's Hall College Seminary
Adjunct Professor of Pastoral Theology, Seton Hall University

John Slattery (Ph.D. University of Notre Dame)
Director of the Grefenstette Center for Ethics in Science, Technology, and Law, Duquesne University

Jeff Speaks (Ph.D. Princeton University)
Professor of Philosophy, University of Notre Dame

Anthony L. Troha (Ph.D. University of California, Davis)
Faculty Associate in Physics and Physics Laboratory Coordinator, Seton Hall University

Rev. Pawel Tomczyk (Ph.D. Catholic University of America)
Assistant Professor of Moral Theology, Seton Hall University

Christopher Timpson (D.Phil. Oxford University)
Professor of Philosophy of Physics, Oxford University

Stacy Trasancos (Ph.D. Pennsylvania State University)
Adjunct Professor of Catholic Studies, Seton Hall University

Jodok Troy (Ph.D. University of Innsbruck)
Assistant Professor of Political Science, University of Innsbruck

Victor Velarde (M.D. University of Navarra, Ph.D. University of Madrid)
Professor of Philosophical Theology, Seton Hall University

AUTHOR BIOGRAPHIES

Gergely Bognár (M.A. Eötvös Lóránd University Faculty of Science, M.A. Pázmány Péter Catholic University Faculty of Arts, M.A. Pázmány Péter Catholic University Faculty of Religious Studies)
Mr. Bognár obtained degrees in physics (MA) at ELTE-TTK, philosophy (MA) at PPKE-BTK, and Catholic religious education (MA) at PPKE-HTK. He works in Hungary in the Religion and Science Research Group of the István Széchenyi University in Győr. He also teaches physics and philosophy at Miklós Révai High School. His main field of research is the question of the boundaries between physics, philosophy, and theology.

Alexey Burov (Ph.D. Budker Institute of Nuclear Physics, Novosibirsk)
Dr. Burov is a Senior Scientist at Fermi National Accelerator Laboratory, USA, with research focused on the beam physics. He published around 200 scientific articles and about 50 journal articles on various philosophical topics, in Russian and English languages. He is a Fellow and an Outstanding Referee of the American Physical Society, a laureate of the FQXi prize for his philosophical concept of the Pythagorean Universe. Dr. Burov is the author and the host of a weekly radio show "Science and Religion" at Russian-speaking Radio NVC, Chicago.

Antonio Colombo (M.S. Politecnico di Milano)
Antonio Colombo worked as an IT specialist for all his professional life, the first six years for the Agusta Helicopters and then for the Amdahl computer company, dealing with mainframe operating systems (MVS and

Unix), system and application programming, providing customer support to various companies in Italy and abroad, including FIAT, Citroen, IVECO, CREDEM, SIA, IPSOA, Lloyd Adriatico, and the Joint Research Center of the European Community in Ispra, Italy. He met Father Jaki in 1999, and collaborated with him from 2000 to 2009, proofreading his works in English, and translating into Italian some of them, including *Questions on Science and Religion*, *Bible and Science*, and *The Ethical Foundations of Bioethics*. Colombo belongs to the ecclesial movement Communion and Liberation.

Jacquelyn Deppe (M.I. Rutgers University)
Jacquelyn Deppe is the Technical Services Archivist in the Archives and Special Collections Center at Seton Hall University. She is focused on processing the collections and cataloging rare books to facilitate better access.

Collin Doyle
Collin Doyle is a second-year undergraduate Physics student who has done research under the New Jersey Space Grant Consortium, as well as in the Archives and Special Collections here at Seton Hall University. His work with *Mathematica*, a computer algebra system, will be featured as a part of the Time Machines Project at the Petersheim Academic Exposition. Collin currently works in Seton Hall's Academic Resource Center as a Physics Tutor-in-Residence.

Peter J. Floriani (Ph.D. Rensselaer Polytechnic Institute)
Dr. Floriani has designed, implemented, and maintained software for several different companies over some 30 years of employment, ranging from Numerical Control and related tasks to tools for software development, and

a system for local ad insertion for Cable Television. His doctoral work assisted molecular biologists by finding rRNA sequences of specified uniqueness for use in their hybridization experiments. He has written over 70 books, including some 35 on computing, science and religion, Chesterton, Jaki, and related topics. He is a student and supporter of Jaki's intellectual work.

Ryan M. Fino (M.B.A. Seton Hall University)
Ryan Fino is the technology engineer for the library at Seton Hall University. His work focuses on researching and implementing technology solutions that advance the goals of the Walsh Library. He also holds a CompTIA A+ certification as well as being Certified in Cybersecurity by the $(ISC)^2$.

Rev. Paul M. Haffner (S.T.D. Pontifical Gregorian University)
Father Paul Haffner studied physics at Corpus Christi College, Oxford. In 1987, he defended and published his doctorate at the Gregorian University, under the title *Christian Faith in God the Creator in relation to Modern Science according to the works of S. L. Jaki.*
Since 1987, he has been a lecturer at the Pontifical Gregorian University, from 2014 until the present adjunct professor at Seton Hall University in Catholic Studies, from 2001-2022 at the Duquesne University Italian campus in Rome and from 2023 until the present at Providence College Italian campus. He is president of the Stanley Jaki Foundation since 13 April 2010, representative to the Holy See of Gracewing Publishing since 1 January 2012, and member of the Pontificia Academia Mariana Internationalis since 8 December 2012. He is author of over 30 books and 150 articles on philosophical and theological themes, and many of his

books have been translated into other languages, including Italian, Spanish, Portuguese and Russian.

Geir Hasnes (M.Sc. Norwegian University of Science and Technology)
Geir Hasnes is an associate professor of engineering cybernetics at the Norwegian University of Science and Technology (NTNU) in Trondheim, Norway. His research is particularly dealing with the philosophical and conceptual basis for the description languages and processes of systems development, in short, with innovation. He authored a recent three-volume bibliography on G. K. Chesterton (2022) and received the American Chesterton Society's Chesterton Lifetime Award in 2006. He also assisted Fr. Jaki during his research on Sigrid Undset.

Rev. Brian Humphrey (S.T.L. Pontifical University of the Holy Cross)
Born in Elyria, Ohio, the Rev. Humphrey is a priest of the Archdiocese of Los Angeles. Before being ordained, Fr. Humphrey received an A.S. in Recording Engineering from Full Sail University in Florida and moved to Los Angeles where he worked for fifteen years in the music business as a producer and engineer. He received his B.A. in Theology from Cloverdale College in Indiana and his M. Div. and M.A. from St. John's Seminary in California. He received his S.T.L. in Dogmatic Theology from Santa Croce in Rome and is currently a doctoral student in the same specialization. His research focuses on Trinitarian theology as well as theological anthropology and wisdom in Augustine, Thomas, and Hans Urs von Balthasar.

Rev. Joseph R. Laracy (S.T.D. Pontifical Gregorian University)
Father Laracy is a priest of the Archdiocese of Newark and assistant professor of systematic theology at Seton Hall University. He also teaches in the Department of Mathematics and Computer Science, the Program in Catholic Studies, the Core Curriculum, and the University Honors Program. He earned his doctorate from the Pontifical Gregorian University in Rome, his master's degree from MIT, and his bachelor's degree from the University of Illinois at Urbana-Champaign. Along with Father Haffner, he is the editor of the *Acts of the 2015 Jaki Congress*. Father Laracy is also the author of *Theology and Science in the Thought of Ian Barbour: A Thomistic Evaluation for the Catholic Doctrine of Creation* (Peter Lang, 2021).

John M. Long (Ph.D. Monash University)
Dr. Long is a senior lecturer in physics and engineering at Deakin University in Geelong, Australia. He has undergraduate qualifications in physics and philosophy from University of Michigan-Flint, and originally trained with General Motors Corporation in materials testing. His research includes engineering applications of advanced materials as well as education in engineering and science. He is co-editor-in-chief of the *International Journal on Innovations in Online Education*. A member of both the Society of Catholic Scientists and ISCAST Australia, he has studied Fr. Jaki's writings for over 30 years.

Rev. Zachary Mabee (Ph.D. University of Reading)
Fr. Mabee is Instructor of Philosophy and Director of the College Propaedeutic Stage of formation at Sacred Heart Major Seminary (Detroit, MI). He is also Pastor of St. Joseph Shrine (Brooklyn, MI). Fr. Mabee earned a

licentiate degree in fundamental theology from the Pontifical Gregorian University in Rome and a doctorate in philosophy from the University of Reading in the UK.

Samantha R. Mattheiss (Ph.D. Rutgers University)
Dr. Mattheiss is an assistant professor of psychology at Felician University. Her current research focuses on self-transcendent emotions, moral beauty, and academic flourishing amongst first-generation college students. Her work is inspired by lived experiences including serving as a Salesian Lay Missionary in Bolivia and living in Catholic communities in Philadelphia, PA. She has published in scholarly journals such as *Cerebral Cortex* and *Frontiers in Psychology*, and has presented at national conferences such as the Society of Catholic Scientists.

Jordan M. Miller (Ph.D. Fordham University)
Dr. Miller is a Core Fellow at Seton Hall University, teaching in the University's Core Curriculum. His background is in Catholic systematic theology, and in addition to teaching at Seton Hall he has also taught courses at Fordham University, St. Peter's University, and the Pontifical John Paul II Institute for Studies on Marriage and Family at the Catholic University of America.

Braden Molhoek (Ph.D. Graduate Theological Union)
Braden Molhoek is Director of the Center for Theology and the Natural Sciences and the Ian G. Barbour Assistant Professor of Theology, Science, Ethics, and Technology at the Graduate Theological Union in Berkeley California. Double majoring in Religion and Genetics at Ohio Wesleyan University led him to put together his interests in science and religion. His research interests include virtue ethics, theological anthropology,

gene editing, artificial intelligence, and the ethical and theological implications of emerging technology. He is a Fellow of the International Society for Science and Religion and is currently Program Director for two CTNS grants: "Virtuous Ai?: Cultural Evolution, Artificial Intelligence, and Virtue" funded by the John Templeton Foundation and "Science, Technology, and the Environment: Exploring Implications for Science and Religion in the Bay Area and Beyond," funded by the DoSER Program of AAAS.

Jeffrey L. Morrow (Ph.D. University of Dayton)
Dr. Morrow is a professor of theology at Immaculate Conception Seminary School of Theology at Seton Hall University. His research focuses on the history of modern biblical interpretation. He is co-editor of the Verbum Domini book series (Catholic University of America Press), a member of La Société Internationale d'Études sur Alfred Loisy, and author or co-author of 10 books including *Alfred Loisy and Modern Biblical Studies* (2019).

Iain Morton
Iain Morton is a fourth year physics and mathematics undergraduate at Seton Hall University. Deeply interested in research in theoretical physics, Iain has pursued research under NASA's New Jersey Space Grant Consortium, Texas A&M's Cyclotron Institute for the Nuclear Sciences, research in optical physics, and the California Institute of Technology's study in gravitational waves. His work has led to presentations at Seton Hall, Brookhaven National Laboratory, and Texas A&M. Iain is also interested in discussing works related to mathematical Platonism.

Rev. Brian K. Muzás (Ph.D. The University of Texas at Austin)
Father Muzás is an assistant professor of diplomacy and international relations at Seton Hall University (SHU), South Orange, NJ. His research interests include international security, defense systems, and ethics. He serves as an international peace and security expert for the Permanent Observer Mission of the Holy See to the United Nations (UN), treasurer of UN civil society's Global NGO Executive Committee (GNEC), and director of SHU's Center for UN and Global Governance Studies. Father Muzás is also the faculty adviser of both SHU's Black Diplomacy Student Organization (BDSO) and the campus chapter of the United Nations Association (UNA-SHU).

Sarah Ponichtera (Ph.D. Columbia University)
Sarah Ponichtera is the Assistant Dean for Special Collections and the Gallery at Seton Hall University. Her PhD is in Germanic Languages, with a concentration in Comparative Literature. Recent research has focused on social media archiving, archival project management, and accessibility in archives.

Biff Rocha (Ph.D. University of Dayton)
Dr. Rocha is an independent scholar in Ohio who works primarily in the area of the history of catechesis and of Catholicism in the U.S. during the 19th century. He is a member of the Academy of Catholic Theology, the American Catholic Historical Association, and the American Society of Church History. He is a co-author of *A Catholic Guide to the Old Testament* (2024) from Ascension Press.

Rev. João Paulo C. F. Santiago, CSJ
A native of Brazil, Father João Paulo is a religious of the Community of St. John (Frères de Saint-Jean). He completed his philosophical and theological studies for ordination at the École Saint-Jean, France. He is currently pursuing an MA in Theology with a concentration in Systematic Theology from Immaculate Conception Seminary School of Theology, Seton Hall University, where he also serves on the Campus Ministry Staff. His research focuses on the meaning of Christian suffering in Thomas Aquinas.

Stacy Trasancos (Ph.D. Pennsylvania State University)
Stacy Trasancos has a Ph.D. in chemistry, a M.A. in dogmatic theology, and will complete a M.A. in systematic philosophy this summer. She is the author of *Particles of Faith: A Catholic Guide to Navigating Science* and has worked the last decade to share Fr. Jaki's teaching with the next generation. Dr. Trasancos is an adjunct faculty member in the Catholic Studies program at Seton Hall University.

Alexei Tsvelik (Ph.D. Kurchatov Institute for Atomic Energy, Moscow)
Dr. Tsvelik is a Senior Scientist at Brookhaven National Laboratory, USA; his scientific research focuses on the quantum field theory. He published around 250 scientific articles and 5 books; he is a laureate of Alexander von Humboldt prize in theoretical physics, a Fellow and an Outstanding Referee of the American Physical Society.

1 STANLEY JAKI: A LIFE DEVOTED TO THE RELATIONSHIP BETWEEN SCIENCE AND FAITH

ANTONIO GIOVANNI COLOMBO

This presentation is a biography of Fr Stanley Jaki. It shows how unforeseen events in his life led to his becoming first an historian of astronomy, and later a specialist on the relation between science and faith. From his native Hungary he was sent to Rome to complete his theological studies. After being ordained a priest in Assisi and after graduating in Rome, due to the political situation in the then Communist Hungary, he was asked to go to the United States, to become a teacher of French and theology. After acquiring USA citizenship, he got a PhD in Physics and moved near Princeton, where he spent the rest of his life teaching at Seton Hall, the Catholic University of New Jersey, and writing books. He is the recipient of the 1987 Templeton Prize "for his immense contribution to bridging the gap between science and religion".

The image on the cover of the book *Questions on Science and Religion,*[1] created by Father Stanley Jaki himself, symbolizes the heart of the matter: the background is full of question marks (Questions), superimposed with an oval image of the universe, produced by the cosmic microwave background radiation images (Science). In the center of the image, where our galaxy would be, is a series of crosses, to represent the Christian faith (Religion). The book is an excellent starting point to approach the work of Father Jaki, since it engages the vast majority of the main themes about which he wrote.

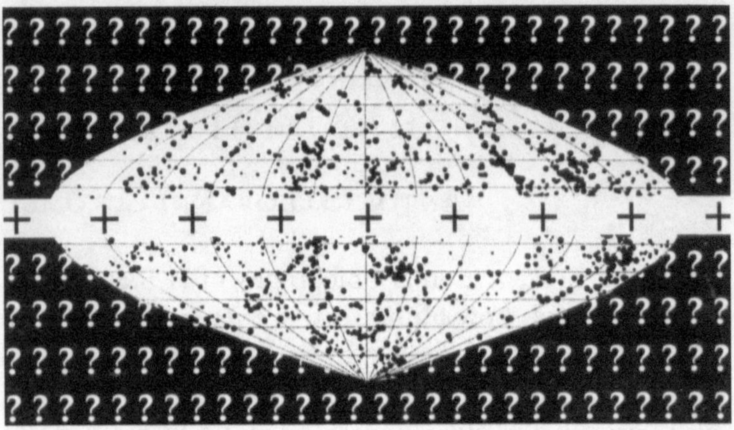

The following quote is useful in understanding the path in life taken by Father Jaki: "The circumstances in which God makes us pass are an essential, not a secondary, factor of our vocation, of the mission to which he calls us."[2] The statement reminds us that God does not intervene in our lives only miraculously, as happened to Saint Paul on the road to Damascus (Acts 9:3–6), but through a series of signs that are sent to us every day, which we are free to follow or not. These signs are what theologically define a vocation, a call. As Jesus himself put it, "it was not you who chose me, but I who chose you and appointed you to go and bear fruit that will remain" (John 15:16). In other words, it happens that our vocation comes to meet us while we are trying to do something else. Father Jaki's work shows us how he answered his vocation throughout his life.

Father Jaki was born in Győr, Hungary, on August 17, 1924. Before entering the seminary he had carried out various activities, including rowing, and acting in an amateur theater group. As a boy scout he achieved the highest rank, Eagle-Scout.

He entered the novitiate at the Benedictine Abbey in Pannonhalma in 1942. Pannonhalma is the most important Hungarian abbey; some of the oldest Hungarian manuscripts are preserved in its library. All the brothers in his family followed the same vocation to become Benedictine monks. The brothers attended a Benedictine high school, in their hometown of Győr. Father Jaki says he felt that his vocation was the priesthood even from the age of around seven years old.[3]

As a seminarian, Father Jaki was sent to Rome in 1947 to complete his studies at the Pontifical Ateneo Sant'Anselmo. He was ordained a priest in 1948, in Assisi, by the Benedictine Bishop Mgr. Giuseppe Placido Nicolini.

In December 1950 Father Jaki was granted a doctorate in theology; his thesis, written in French, was titled *Les tendances nouvelles de l'ecclésiologie* (*New Trends in Ecclesiology*).[4] It was published in 1957. A review of it was made at the time by the future Card. Ratzinger, who many years later said: "Oh, Father Jaki, your *Tendances*, which I was most pleased to review, occupies a place of honor in my library".[5] The book shows the origin of the then current trends in ecclesiology from the Council of Trento up to the first half of the 20th century, dealing also with Protestant and Orthodox authors. It was reprinted in 1963, during the Second Vatican Council. Of this book, which in a certain sense defines Jaki as a theologian, we only quote the statement taken from Saint Augustine with which Father Jaki concludes the book: "Hold then, most beloved, hold all with one mind to God the Father, and the Church our Mother."[6]

After finishing his studies, Father Jaki did not return to Hungary, due to the political situation there, in particular to the persecution against religion carried on by the local Communist regime (of 182 convents, only 6 were still active around 1950).[7] Because his thesis had been written in French, he expected to be sent to France. Instead, he was directed to go to the United States and arrived in

New York on December 21, 1950.[8] A week later he arrived at the Seminary of Saint Vincent, in Latrobe, Pennsylvania, and six months later began what was expected to be a teaching career. He taught systematic theology at the seminary and French at the adjoining college. In the courses at the seminary he drew out the omnipotence of God and His creation.[9] These theological themes have vast ramifications, affecting both philosophy and science.

Father Jaki wrote in his autobiography, "I have always been very pleased to read books and learn."[10] This is why, between 1951 and 1954, while he was already teaching, he decided to further his grasp of physics[11] and attended courses in mathematics and physics, obtaining a bachelor of science degree in physics in 1954.[12] However, on December 9, 1953, his teaching career was interrupted. The "decisive factor" was an extremely trivial event: the removal of his tonsils. While this is usually a routine operation, few days after surgery Father Jaki hemorrhaged in two spots and was almost voiceless for the next ten years. What little voice he had was insufficient for teaching, but it did not prevent him from continuing his studies and obtaining a PhD in Physics at the Fordham University in 1957.[13] Among his teachers was the 1936 Nobel Prize recipient in Physics, Victor Hess. A partnership arose between the two, and Father Jaki conducted his research under Dr. Hess' direction, presented his dissertation in 1957, and published an article signed by the two of them in the prestigious *Journal of Geophysical Research*. The thesis had as its title, *A study of the distribution of radon, thoron* (which are rare gases) *and their decay products above and below the ground.*[14] Even if the title may seem rather abstruse, radon is the second cause of lung cancer (after smoking). His work as a physicist had significant contributions to society.

In that same year, 1957, Father Jaki was ordered to move to Portola Valley, in California where with six other Hungarian Benedictines the Woodside Priory School was founded. Since his voice was still absent, Father Jaki took care of the administrative part of the school and continued his studies on the history of science.

In 1960 Father Jaki moved back to the East Coast of the United States. At Princeton University, between 1961 and 1963, he worked as Visiting Fellow on the History of Science project. There he took part to various graduate seminars in the history and philosophy of science.[15] And there he undertook the research to write his first important work, *The Relevance of Physics*.[16] Father Jaki recalls, "its idea flashed through my mind as I walked down from the steps of Princeton's Post Office on Palmer Square".[17] His doctoral thesis was based on experiments and convinced Father Jaki that exact science has to do with measurements and predictions. Accordingly, his definition of exact science became "the quantitative study of the quantitative aspects of things in motion."[18] Stated in Eddington's formulation, "the cleavage between the scientific and the extrascientific domain of experience is, I believe, not a cleavage between the concrete and the transcendental but between the metrical and the non-metrical."[19] In practical terms, this means that the competence of physics is limited to the measurable properties of everything, and in this sense science applies to everything, but only to the *measurable* side of everything. For example, the creation of the universe cannot be an object of science, but can only be dealt with by philosophy and religion, since with science we can get as close as we like to the moment of the Big Bang, but we cannot tell anything about what caused it.

In 1965, before the publication of *The Relevance of Physics*, the dean of Seton Hall University offered Father

Jaki a post as university professor, with an obligation to hold a weekly seminar, and an unwritten obligation to continue writing. Father Jaki defined his job as being *damnatus ad litteras*, i.e., condemned to write. With the permission of his superiors, Father Jaki continued to live near Princeton, commuting every week to Seton Hall, to avail himself of the ancient books at the Princeton Libraries.

The cover of *The Relevance of Physics* reads: "Physics has become the most powerful instrument at man's disposal for seeking out and revealing the hidden facts of inanimate nature. Are its methods and its insights equally relevant to other areas of human concern?" The answer, of course, is negative, and Father Jaki joked that the title should have been *The Irrelevance of Physics*. The book starts from an examination of the three models of interpretation of the universe, the organismic one of the ancient Greeks, the mechanistic one of Newton, and the contemporary one, based on mathematics. The relationship of physics with biology, metaphysics, ethics and theology are then examined.

As a kind of continuation to *The Relevance of Physics*, Father Jaki wrote *Brain, Mind and Computers*.[20] For that book he received the Lecomte de Nouy prize in 1970. Father Jaki then wrote three books on the history of astronomy, dealing respectively with Olbers' paradox, the Milky Way and the formation of planets.[21]

In the first decades of the 20th century, the French scientist and historian Pierre Duhem traced back the origin of modern mechanics to the 14th century University of Sorbonne, in Paris, specifically to the work of John Buridan, in which one can find the first formulation of the Principle of Inertia and, hence, the first of Newton's Laws of motion.[22] Duhem's main historical works are his ten volumes *Le système du monde*.[23] Father Jaki's book *Science and Creation* deals with the reasons for the stillbirths of science in each of all the main ancient cultures, and to its only birth in medieval Christian Europe.[24] I will only mention here that all ancient cultures had a cyclical view of time ("the great year") which, as a matter of fact, turns out to be "science-unfriendly." Only the Judeo-Christian culture has a linear view of time, that starts with God's creation. In the book *The Savior of Science*, Father Jaki deals with the question: "Did Christian belief have a part in the birth of science?" If this is true, then Father Jaki argues that it is appropriate to then ask "Is Jesus truly (also) the Savior of Science?" The book shows how both Judaism and Islam failed to favor the birth of science, even though they had for five centuries the monopoly of Aristotle's books. This monopoly is important because Buridan's work that contains his theory of *impetus* is a commentary on Aristotle's books on *Physics*. Christian faith worshiped Christ, the only Begotten Son of the Father, in whom everything had been created. Therefore, the universe was simply created (against the recurring temptation of pantheism), and the universe had to be rational, as an expression of a benevolent God. In the words of Father Jaki, "if the *Logos* was fully divine, its creative work [i.e., the universe] had to be the paragon of logic and order."[25] This observation has been a powerful motivation for much research over the centuries. And in this sense Jesus, seen as a guarantee of the rationality of the Universe, is the Savior of Science.

Alfred Nobel (1833–1896), the originator of the Nobel Prizes, despite being nominally a Lutheran was not particularly religious. For this reason, among the Nobel Prizes there is no prize linked to religion, even if peace prizes have occasionally been awarded to religious personalities. To fill this gap, John Marks Templeton (1912–2008), an Anglo-American businessman and philanthropist, established an award in 1972 that bears his name, "for progress in religion". Father Jaki received the award in 1987. The motivation reads: "It is above all for his immense contribution to bridging the gap between science and religion, and his making room in the midst of the most advanced modern science for deep and genuine faith, that he received the Templeton Prize."[26]

In 1990 Father Jaki was invited to join the Pontifical Academy of Sciences as an Honorary Member (namely as a member not directly involved in scientific research). All the talks of Father Jaki to the Pontifical Academy of Sciences were published posthumously in 2011 under the title *Lectures in the Vatican Gardens*.[27]

Father Jaki translated into English three philosophical works: the early work of Kant, *Universal Natural History and Theory of the Heavens*, Giordano Bruno's *The Ash Wednesday Supper*, and Lambert's *Cosmological Letters*.[28] He did so to demonstrate how ideas *a priori* could easily bring an author far from truth. Just as an example, the image of the universe that emerges from Lambert's speculations, even if contains a number of very interesting ideas, overstates the role of comets and abuses (as Kant does) of the then common conviction that all planets and comets were inhabited.

Father Jaki wrote a number of books as a theologian. We can recall the ones to defend the Papacy, the ones about the Bible, in particular about the first Chapter of the Book of Genesis and the ones about miracles.[29]

Father Jaki founded a non-profit publishing house, Real View Books, which started publishing works no longer in print, like Newman's *Difficulties of Anglicans*, and a number of books written by Father Jaki himself.[30] After Father Jaki's death, Real View Books completed the publication of works he left "ready for print" and continues to reprint his books as needed. For example, a new edition of *Science and Creation* came out in 2016. The number of items in Father Jaki's publication list is well over 700. About 100 books and booklets he wrote are available from Real View Books. The soul of the Real View Books publishing house is Dennis Musk.

Father Jaki also wrote several 32-pages booklets, which are short version of longer works or monographies. For example, *Christ and Science* is a short version of *The Savior of Science.*

A series of short books (about 80 pages each) deals with the basic topics of theology. They constitute a short compendium of apologetics. Father Jaki had in plan a book on the subject, *Apologetics in an Age of Science,* but he died before writing it. What is left in his papers is a folder (found by the present writer) with a few preparatory pages, including the titles of the chapters and a possible cover. Father Jaki had an exchange of messages about this book with John Beaumont, who wrote a paper on the matter.[31]

Father Jaki paid particular attention to converts, because they often demonstrate in their new faith a livelier intellectual acumen and apologetic than "cradle Catholics." One of Father Jaki's favorite authors was in fact a convert, Gilbert Keith Chesterton. Father Jaki dedicated several articles to him, and one book, *Chesterton, a Seer of Science,* that is, a visionary regarding science.[32] Let's remember that Chesterton was anything but tender with the scientism that dominated his era. Father Jaki's research on the conversion of Sigrid Undset was a particularly laborious undertaking, which also required the translation for the first time of some of Undset's texts from Norwegian into English. Father Jaki also visited the places where Sigrid Undset lived, while researching this book.[33]

But the convert with whom Father Jaki spent the most time is undoubtedly Saint John Henry Newman, the most illustrious convert of the nineteenth century, whom Father Jaki had quoted in his writings since his 1950 thesis. Father

Jaki dedicated several books and dozens of articles and conferences to Newman, and he also had some of Newman's works reprinted. One of his concerns was not to let Newman be transformed into the precursor of an ecumenism, very popular after the Second Vatican Council, which tended to minimize the doctrinal differences with the Protestants. Perhaps Father Jaki's most significant work on the subject is the book *Newman to Converts: An Existential Ecclesiology*, the result of meticulous research carried out in Birmingham in Newman's own library. In this book, Father Jaki traces the events that led to the conversion of around twenty people from Anglicanism to Catholicism, deriving the material from the volumes of Newman's *Letters and Diaries* and from documentation of the time.[34] The advice that Newman gave can be summed up in one of the sentences at the beginning of the book: "Not many days go by without my receiving letters from strangers, young and old, men and women, regarding the Catholic religion. I reply to them that it is the one and only true and certain religion."[35] A short version of the book appeared before the book itself, titled *The One True Fold: Newman and His Converts*.[36]

Among Father Jaki's interventions there are articles for magazines or conferences at universities. Many of these texts have been collected in book form, with the first essay giving the book its title, followed by "*and Other Essays*."[37]

Regarding Father Jaki and his vocation, since he was a child he wanted to become a priest, and this came true. Without the ruthless persecution carried out by communism in Hungary, he would never have been sent to America. Without the consequences of a banal removal of his tonsils he would never have gone from teaching to a career as a writer. Many other events in his life as a writer (and maybe in our own life) depended on seemingly random (or providential) encounters. Father Jaki liked the Portuguese proverb "God writes straight along (our) crooked lines".[38] Finally, it seems appropriate to quote a further few lines from his intellectual autobiography, *A Mind's Matter*.[39] This book is also very useful for understanding Father Jaki's intellectual itinerary:

> [Before seeing things *sub species aeternitatis*] one must feel satisfied with the fact that *militia est vita hominis super terram* (on this earth man's life is a military service) (Job 7:1), to recall from the Vulgate a sober reflection of the much tried Job. Our soldiering must go on, with the phrase "Only a few good marines are needed" in our focus, a phrase which summarizes a truly existential theology. The phrase also conveys the theological equivalent of a mere private, which is the rank of a useless servant, who merely does his duty. May this remain the matter foremost in my mind.

Another biblical phrase that Father Jaki loved to quote is: "Even to the death fight for truth, and the LORD your God will battle for you" (Sir 4:28 NAB).

Father Jaki's website (http://www.sljaki.com) contains much information about Father Jaki, and an updated list of his publications. A group of Father Jaki's friends is known as the "Stanley Brigade," a name chosen by Father Jaki himself. The group is informal. To join it, one can write an email to stanley.brigade@gmail.com.

Notes

1. Stanley Ladislas Jaki, *Questions on Science and Religion* (Pinckney, MI: Real View Books, 2004).
2. Luigi Giussani, *L'uomo e il suo destino. In cammino* (Genova: Marietti, 1999), 63. Luigi Giussani (1922–2005) was an Italian Catholic priest, theologian, educator, founder of the international Catholic movement Communion and Liberation. The quoted statement was said at the beginning of an International Assembly of Leaders of Communion and Liberation (La Thuile, Italy, August 17, 1997).
3. Stanley Ladislas Jaki, *A Mind's Matter: An Intellectual Autobiography* (Grand Rapids, MI: Wm. B. Eerdmans, 2002), 18.
4. Stanley Ladislas Jaki, *Les tendances nouvelles de l'ecclésiologie* (Rome: Casa Editrice Herder, 1957). The book has been translated into English by Alan Aversa and Antonio Colombo: *New Trends in Ecclesiology* (New Hope, KY: Real View Books, 2021).
5. See Stanley Ladislas Jaki, *Justification as argued by Newman* (Port Huron, MI: Real View Books, 2007), 269, n. 7.
6. Augustine of Hippo, *Enarrationes in Psalmis*, LXXXVIII, Sermo 2, n. 14. In the Latin original: *Tenete ergo, carissimi, tenete omnes unanimiter Deum patrem, et matrem Ecclesiam.* At page 264 in Father Jaki's book quoted in note 4.
7. See Jan Mikrut (ed.), *La Chiesa cattolica e il comunismo in Europa centro-orientale e in Unione Sovietica* (San Pietro in Cariano (Verona): Gabrielli Editori, 2016), 583.
8. Jaki, *A Mind's Matter...*, 20.
9. "*De Deo uno, De Deo creante, De homine et gratia, De novissimis*", see Paul Haffner, *Creation and Scientific Creativity: A Study in the Thought of S. L. Jaki* (Leominster: Gracewing, 2009), 13.
10. Jaki, *A Mind's Matter...*, 18.
11. Jaki, *A Mind's Matter...*, 21.
12. Haffner, *Creation and Scientific Creativity...*, 14.
13. Jaki, *A Mind's Matter...*, 27.
14. Stanley Ladislas Jaki, Victor Franz Hess, "A Study of the Distribution of Radon, Thoron, and Their Decay Products

above and below the Ground," *Journal of Geophysical Research* 63 (1958), 373–390.

15. Jaki, *A Mind's Matter...*, 28. Cfr. Haffner, *Creation and Scientific Creativity...*, 15–16.
16. Stanley Ladislas Jaki, *The Relevance of Physics* (Chicago: The University of Chicago Press, 1966).
17. Jaki, *A Mind's Matter...*, 28.
18. See Stanley Ladislas Jaki, *The Drama of Quantities* (Port Huron, MI: Real View Books, 2005), 13. See also his *Darwin's Designs* (Port Huron, MI: Real View Books, 2006), 5.
19. Arthur Eddington, *The Nature of the Physical World* (New York: The Macmillan Company; Cambridge, England: The University Press, 1929), 275.
20. Stanley Ladislas Jaki, *Brain, Mind and Computers* (New York: Herder and Herder, 1969).
21. Stanley Ladislas Jaki, *The Paradox of Olbers' Paradox: A Case History of Scientific Thought* (New York: Herder and Herder, 1969). Idem, *The Milky Way: An Elusive Road for Science* (New York: Science History Publications; Newton Abbot: David and Charles, 1972). Idem, *Planets and Planetarians: A History of Theories of the Origin of Planetary Systems* (Edinburgh: Scottish Academic Press; New York: The Halstead Press of John Wiley Inc., 1978).
22. See *Magistri Johannis Buridan Quaestiones octavi libri physicorum*, 12th question. The English translation of this text can be found in Marshall Clagett, *The Science of Mechanics in the Middle Ages* (Madison: The University of Wisconsin Press; London: Oxford University Press, 1959), 536.
23. Pierre Duhem, *Le système du monde: Histoire des doctrines cosmologiques de Platon à Copernic*, 10 Vol. (Paris: Hermann, 1913–1959).
24. Stanley Ladislas Jaki, *Science and Creation: From Eternal Cycles to an Oscillating Universe* (New York: Science History Publications; Edinburgh: Scottish Academic Press, 1974).
25. Stanley Ladislas Jaki, *The Savior of Science* (Grand Rapids, MI: W.B. Eerdmans, 2000), 83. The first edition of the book came out in 1988.
26. Templeton, *1987 Templeton Prize* (Nassau, Bahamas: Lismore Press, 1987), 4.

27. Stanley Ladislas Jaki, *Lectures in the Vatican Gardens* (New Hope, KY: Real View Books, 2011).
28. Immanuel Kant, *Universal Natural History and Theory of the Heavens*, transl. Stanley Ladislas Jaki (Edinburgh: Scottish Academic Press, 1981). Giordano Bruno, *The Ash Wednesday Supper*, transl. Stanley Ladislas Jaki (The Hague/Paris: Mouton, 1975). Johann Heinrich Lambert, *Cosmological Letters on the Arrangement of the World-Edifice*, transl. Stanley Ladislas Jaki (New York: Science History Publications; Edinburgh: Scottish Academic Press, 1976).
29. Stanley Ladislas Jaki, *And on This Rock: The Witness of One Land and Two Covenants* (Notre Dame, IN: Ave Maria Press, 1978). Idem, *The Keys of the Kingdom: A Tool's Witness to Truth* (Chicago: The Franciscan Herald Press, 1986). Idem, *Miracles and Physics* (Front Royal, VA: Christendom Press, 1989). Idem, *Genesis 1 Through the Ages* (London: Thomas More Press; New York: Wethersfield Institute, 1992). Idem, *Bible and Science* (Front Royal, VA: Christendom Press, 1996). Idem, *God and the Sun at Fatima* (Fraser, MI: Real View Books, 1999). Idem, *The Drama of Guadalupe* (New Hope, KY: Real View Books, 2009).
30. John Henry Newman, *Anglican Difficulties. (Certain Difficulties Felt by Anglicans in Catholic Teaching)* (Fraser, MI, Real View Books, 1994).
31. See John Beaumont, "Fr. Stanley Jaki on Apologetics: The Book That Never Was," in Lucía Guerra Menéndez, Rafael Pascual, Antonio Colombo (eds.), *Proceedings of the Summer Course: Science and Faith in Stanley Jaki* (Rome: Ateneo Pontificio *Regina Apostolorum*; Rome: IF Press; Madrid: CEU Universidad San Pablo, 2017), 153–165. John Beaumont collaborated with Father Jaki in the last six years of Father Jaki's life.
32. Stanley Ladislas Jaki, *Chesterton: A Seer of Science* (Urbana/Chicago: University of Illinois Press, 1986).
33. Stanley Ladislas Jaki, *Sigrid Undset's Quest for Truth* (Port Huron, MI: Real View Books, 2007).
34. Stanley Ladislas Jaki, *Newman to Converts: An Existential Ecclesiology* (Pinckney, MI: Real View Books, 2001).
35. Jaki, *Newman to Converts...*, vii.
36. Stanley Ladislas Jaki, *The One True Fold: Newman and his Converts* (Royal Oak, MI: Real View Books, 1998).

37. We will list here only the collection of essays appeared in this century. Stanley Ladislas Jaki, *The Limits of a Limitless Science and Other Essays* (Wilmington, DE: Intercollegiate Studies Institute, 2000). Idem, *The Gist of Catholicism and Other Essays* (Pinckney, MI: Real View Books, 2001). Idem, *Numbers Decide and Other Essays* (Pinckney, MI: Real View Books, 2003). Idem, *A Late Awakening and Other Essays* (Port Huron, MI: Real View Books, 2006). Idem, *Uncodified Conspiracy and Other Essays* (New Hope, KY: Real View Books, 2014).
38. Jaki, *A Mind's Matter...*, 223.
39. Jaki, *A Mind's Matter...*, 258.

2 THE PAPERS OF STANLEY L. JAKI: AN INTRODUCTION

Jacquelyn Deppe, M.I., Ryan Fino, M.B.A., and Sarah Ponichtera, Ph.D.

Introduction

Father Stanley Ladislas Jaki, O.S.B., (Jáki Szaniszló László, 1925–2009) is remembered as a transformative thinker in the field of the history of science.[1] His writings explore the boundary between science and religion, portraying them as reinforcing each other rather than being at odds.[2] Hired by Seton Hall University in 1965, he remained at the University for 40 years helping to shape the religious and scientific understanding of multiple generations. Ultimately, his work led him to achieving the rank of Distinguished Professor of Physics. He also served as a visiting professor and lecturer at universities across the globe, giving invited lectures including the Gifford Lectures at the University of Edinburgh, winning the Templeton Prize in 1987, and being appointed an honorary member of the Pontifical Academy of Sciences by Pope John Paul II in 2009.[3] Even while building a global reputation, Jaki maintained close relationships with New Jersey intellectuals such as R. H. Dicke, Eugene Wigner, and James I. McCord. His papers provide a new window into the New Jersey-based circle of physicists that changed the course of the mid twentieth century: their relationships with one another and the institutions that sustained their work.

Preparing the Collection for Research

His papers were deposited in the Monsignor William Noe Field Archives and Special Collections Center in 2011, following his death in 2009. So far they have been unprocessed and therefore unavailable to scholars, but a current project in the archives seeks to change that. This paper will introduce Jaki's collection, the process of organizing and preserving the contents, and some highlights of the collection that have been discovered so far.

This collection consists primarily of Jaki's research notes, compiled from a wide range of sources from libraries around the world. It also includes his personal correspondence with his family in Hungary and his colleagues in the United States. Correspondence regarding his travels, talks, and publications provide the raw materials for a biography. The correspondence with his brothers, both priests in the same Hungarian order as himself, may be of particular value, as the brothers discussed Church matters in addition to family matters, providing insight into Hungarian Catholic life in that period. Draft manuscripts and proofs also comprise a part of the collection. These demonstrate that Jaki was closely involved in the publication process for his books, determining the fonts and pagination used as well as the intellectual content. It represents a particularly rich repository of data on the religious philosopher Pierre Duhem, whose work was an inspiration for several books of Jaki's. It includes mostly reproductions, but also some original pieces of Duhem's writings, which were likely purchased by Jaki, who accumulated an impressive collection of scholarly resources during his lifetime.[4]

To make this collection available for scholars, the materials will be surveyed to create a list of the contents. In the course of the survey, the archivist will seek to discover the original order of the materials—the way Jaki himself organized his papers. This organizational system in itself provides insight into the way Jaki's mind worked. Retaining original order is a core principle of the archival profession. [5] Following this principle, the Duhem materials will likely be kept separate from the general research materials, just as Jaki himself maintained them. Happenstantial misplacements, however, will be rectified in order to enable efficient research. Materials will be described in terms of the individuals involved for correspondence and photographs, topic where this information is not available, and date. Unless there is compelling reason to decide on another organizational system, materials are generally organized chronologically.

After the materials are surveyed and initially described, the archivist will decide on a collection level arrangement that will enable researchers to quickly grasp the scope and nature of the collection. For the personal papers of researchers such as Jaki, materials are typically organized by type. Groupings, or series, are created of like materials. Correspondence will likely constitute one such grouping, research materials another, manuscripts a third. Creating such an arrangement is a key part of the activity of processing a collection, and is as much of an art as a science, reflecting the archivist's experience and knowledge of the research process as well as their sensitivity to the uniqueness of the individual collection they are working with. This process is enriched by collaboration, and so the processing archivist will seek the advice of peers and supervisors when making such a consequential decision.

After an arrangement is decided upon, the materials must be physically rearranged, and this is when we also rehouse the materials in archival enclosures to preserve them for generations to come. Materials are removed from the acidic envelopes and deteriorating plastic binders in which they are typically stored in their active life, and placed into chemically neutral archival folders that will slow down the inexorable acidification of paper.[6] Photographs, which represent a unique danger due to the chemical compounds present, and a fragility of their own due to their vulnerability to damage from handling, will be placed in polypropylene transparent sleeves. These sleeves protect surrounding paper from the chemicals in the photo, and also protect the photo, allowing it to be handled and seen without removing it from its sleeve.[7] In case the case of reproductions of materials published elsewhere, the archivist will create a list of citations of all the articles and book chapters present. That way, the extensive catalogue of Jaki's research interests will be available for researchers to discover, without needing to keep boxes of photocopied materials that are widely available in libraries.

One of the biggest challenges to processing this collection fully will be preserving and describing the digital files from Dr. Jaki's two laptop computers that are found in this collection. Despite the common misperception that digital files are more reliable than paper documents, in fact they are more fragile.[8] This is because older files, particularly those created before the widespread availability of cloud storage, are dependent on a particular software and hardware environment to be interpretable, whereas paper is immediately comprehensible without special equipment. The field of digital preservation has emerged to take on the challenge of ensuring that the information produced in digital form today will be available to future generations.[9]

One step Seton Hall has taken is to implement the digital preservation system Preservica, which can accept files in a wide array of formats, capture the technical metadata of the files (down to how many pixels are in an image) and then create an access copy of the file that can be opened, interpreted, and downloaded so it can operate in modern systems.[10] This both preserves and documents the original file format for historians who may need access to those details, while rendering the content so that the general public may still have access to the knowledge contained within.

In the case of the Jaki computers, we do not want to risk changing any data about the files by attempting to operate the laptops themselves. For one thing, they are Dr. Jaki's personal computers that he needs to log into to access, and we do not have the password. Even if we were able to bypass security through Seton Hall's IT department, it would require logging in as an administrator, which would give us access to Dr. Jaki's files through a file structure. We would not be seeing the computer the way he saw it, gathering important details about how he organized his files. In addition, we would be risking changing important data by accessing it in this way. Any time a file is opened, the metadata on the file is changed—particularly the key data point of "last accessed." If we were to begin opening files to see what they contained, they would update to today's date, and the key information of what documents Dr. Jaki was working on—and in what order—would be irretrievably lost.

To prevent losing this key information, we will remove the hard drives from the laptop and connect them to the archives' digital preservation laptop. This laptop is particularly safe from information compromise, because it has been set to never overwrite information automatically, unlike a normal computer, which regularly updates information. In addition, it has had the wireless card physically disconnected and the network adapters disabled, preventing any information transfer from sources outside the computer itself. We have installed a program on this machine called Bitcurator. Bitcurator is a tool that was developed by law enforcement to access the contents of computers without changing any details, and while recording all the information it finds. It can even find hidden and deleted files, and surface records of which files were changed and deleted at which times. Now this program is used extensively within the digital preservation community, to gather information to preserve human knowledge rather than prosecute crimes.[11]

In order to preserve the knowledge of which files Dr. Jaki was working on at what times, we will connect the hard drives to the Bitcurator machine and take what is known as a disk image. This is essentially an overview of the contents of the computer, listing which files are located where and the key information about the files. We have regularly done this procedure for unidentified legacy disks that we find in the archives, to determine what is on them and preserve any important files. One important advantage of this process is that it takes what are known as checksums of each file. A checksum is a sum derived from the bits that comprise a file that is unique to the state of that file, so it will be different if anything in the file is changed. This is useful to have for files that you are going to move so that you can compare the file's checksum after the move to the checksum from before the move to see if moving the file inadvertently damaged or changed anything in the file. Bitcurator will take a disk image and record checksums so that the pristine state of Dr. Jaki's computers will be recorded before we make any interventions. When we do move files, we will use a tool called Teracopy, which compares checksums from before and after the move as part of the moving process, so that we will ensure the integrity of the files we are ingesting into Preservica.

Based on the information recorded in the disk image, we will decide which files need to be transferred over to Preservica. Just like with the paper archival collection, we will focus our search on Dr. Jaki's original writings, and prioritize capturing and preserving those materials for future generations. We will use the information from Bitcurator to create archival description for the digital materials just as we do for paper materials. We will create titles, dates, and where necessary, scope notes describing the nature of the materials. Once the digital files are ingested into Preservica, we will link the digital files with this description, so that anyone browsing the finding aid will be able to click on a link right in the finding aid to access and explore the contents of the files.

Highlights of the Collection

In conclusion, here are some of the highlights of this collection found when archives staff was planning for this processing project. With about seventy linear feet of materials, there have been some hidden gems that have been discovered through initial inspection. Some of the most intriguing parts of this collection are the boxes of correspondence containing both the letters Father Jaki composed and the ones he received back. Having both the initial letter and the response allows researchers to have a more complete picture of the conversations, and in some cases the relationship Father Jaki had with his correspondents. However, there is a set of exceptions that lacks Father Jaki's initial letters.

This set is some of the oldest correspondence that comes from his time as a doctoral student at Fordham University where he studied under Victor Franz Hess, Nobel laureate who discovered cosmic rays. The letters describe details of both their lives from Jaki's laryngitis due to a tonsillectomy in 1953 to Hess's travels and the article they co-authored together: *A study of the distribution of radon, thoron, and their decay products above and below the ground.*[12] In one Hess talks about Jaki's calculations during the study that was performed asking: "Did you ever measure this outside deposit and compare the amount with the inside deposit?"[13] Additionally, this folder of correspondence contains printed articles from Hess which contain Jaki's annotations of underlines, brackets, exclamation points, and even question marks in the margins beside the text. It is clear that the two physicists maintained a lively intellectual connection that transcended their original relationship.

Other correspondence contains discussions with publishers, praise from admirers, planning for lecture tours across the globe, and requests from scholars who asked for copies of his works. Notable names which Father Jaki includes in his autobiography *A Mind's Matter: An Intellectual Autobiography* are Robert Oppenheimer, who he dined with on occasion at the Oppenheimer home in Princeton, New Jersey; Herbert Feigl, Austrian-American philosopher; Peter E. Hodgson, a British physicist; Eugene Wigner, theoretical physicist and Nobel Prize winner; and John C. Eccles, Australian neurophysiologist, philosopher, and Nobel Prize winner who was impressed by Father Jaki's work *Brain, Mind and Computers.*[14]

It is quite clear that Father Jaki sent letters to numerous people over the course of his life with several responding back. These letters are primarily in English however, there are those that are in Hungarian, French, Spanish, and German which make it difficult to understand the nature and the relationship Jaki had with those specific correspondents. Whether they discussed research, lectures, or visits. This is the case for the oldest set of correspondence that comes from around the mid to late 1940s into 1950. The letters are all written in Hungarian, with some that are known to have come from his oldest brother, Zénó, and his youngest brother, Teodóz, and all of which have been kept separate from later letters. The archives will work with students and faculty at

Seton Hall and within the New Jersey Catholic community to describe these materials accurately.

Besides the known letters from Zénó and Teodóz, other materials that are of a personal nature contain slides showing Father Jaki's return to Győr, Hungary around 1964 and then again in 1989 according to the stamp on the slides. These slides capture the members of his family in different places, such as one set that looks to be taken during a trip the Jaki family took to a lakeside town. Among slides of Father Jaki's mother, sister, and brothers, there are ones that capture the grave of his sister Etuska, who died when she was only fifteen years old.

It is not just Father Jaki's correspondence that sheds light into this scholar's life but also the copious and meticulously kept and detailed research notes he maintained as part of his process. Viewing the note-taking system which doubled like a card catalog like system he maintained will allow scholars to understand the texts he consulted and notes he recorded when writing manuscripts and lectures across the globe. Father Jaki even maintained files in preparation of his lectures. These files mainly contain letters about his scheduled lecture, clippings pertaining to the content Father Jaki was lecturing about, and handwritten notes. Among his manuscript and lecture notes are drafts of sermons both those actually given and those that never were.

Father Jaki was a scholar of science and religion, studying past physicists and interacting with those of his own time. One such physicist was Pierre Duhem. Within Jaki's collection are correspondence to various libraries and museums to track down letters Duhem had written as research for his book on this French physicist. Amazingly his collection even contains an original letter from Pierre Duhem dated 21 September 1913 as well as autographed publications which are attached to an inscription that is addressed to another French physicist, Émile Jouguet (Jacques Charles Émile Jouguet).

Much like researchers will be able to understand who Father Jaki was as a person and a scholar from his archival records, they will also be able to see him through the selection of books that came from his library. Many of these books match up with Jaki's research interests and contain more annotations and declarations within the margins of the book where Jaki agreed or disagreed with the text. A vehement *No!* was found within the margins of *Infallibility in the Church: an Anglican-Catholic Dialogue*, specifically in Chapter 4, "Infallibility: What Newman Taught in Manning's Church" by C.S. Dessain in connection to the following statement:

> After these quotations, it is not surprising to find that Newman always thought that the definition of papal infallibility was a mistake, just as he always thought the way in which it was passed was a scandal.[15]

The *No!* was found right next to *mistake* which had been underlined as was *definition*.

It is not just scholars interested in Jaki that will benefit from analyzing these materials. Scholars whose interests lie in the history of the book and how readers interacted with their books can gain insight into annotations and marginalia of an active and incisive mind. The archives plans to undertaking this processing project in the fall of 2024, and if all goes according to plan, the collection should open to scholars by the end of 2025.

Our thanks go to Professor Anthony Troha for his invaluable insights into Jaki's life and role at Seton Hall.

Notes

1. Paul Haffner, *Creation and Scientific Creativity : A Study in the Thought of S.L. Jaki* (Leominster: Gracewing, 2009).
2. S. L. Jaki, *A Mind's Matter: An Intellectual Autobiography*, (Grand Rapids, MI: W.B. Eerdmans, 2002).
3. Bruce Weber, "The Rev. Stanley L. Jaki, Physicist and Theologian, Dies at 84," *The New York Times*, April 12, 2009, sec. New York, https://www.nytimes.com/2009/04/13/nyregion/13jaki.html.
4. Jaki's book collection was donated to Seton Hall's Walsh Library, just as his collection of personal papers was deposited in its archives, in 2011.
5. Laura A. Millar, *Archives: Principles and Practices, 2nd ed., Principles and Practice in Records Management and Archives* (London: Facet Publishing, 2017), 45.
6. Elizabeth Joffrion, Michèle Valerie Cloonan, and Society of American Archivists, *Advancing Preservation for Archives and Manuscripts*, Archival Fundamentals Series III, no. 5 (Chicago: Society of American Archivists, 2020).
7. Bertrand Lavédrine, Jean-Paul Gandolfo, John P. McElhone, *Sibylle Monod, and the Getty Conservation Institute, Photographs of the Past: Process and Preservation* (Los Angeles: Getty Conservation Institute, 2009), 101.
8. Trevor Owens, *The Theory and Craft of Digital Preservation* (Baltimore, MD: Johns Hopkins University Press, 2018), 13.
9. Jeremy Myntti and Jessalyn Zoom, eds., Digital Preservation in Libraries: Preparing for a Sustainable Future, An ALCTS Monograph (Chicago: ALA Editions, 2019).
10. https://www.preservica.com.
11. https://bitcuratorconsortium.org/.
12. S. L. Jaki and V. F. Hess, "A Study of the Distribution of Radon, Thoron, and Their Decay Products Above and Below the Ground," *Journal of Geophysical Research* 63, no. 2 (1958): 373–390.
13. Jaki Papers, Box 4, Collection Number TBC, The Monsignor William Noe Field Archives & Special Collection Center. (Note: Box number may change after processing.)
14. Jaki, *A Mind's Matter*, 159–162.

15. C.S. Dessain, "Chapter 4: Infallibility: What Newman Taught in Manning's Church" in A.M. Farrer, Robert Murray, J.C. Dickinson, C.S. Dessain, *Infallibility in the Church : an Anglican-Catholic Dialogue* (London: Longman & Todd, 1968), p. 70.

3 HAS STANLEY JAKI UNDERESTIMATED THE BYZANTINE INFLUENCE ON THE RISE OF SCIENCE?

REV. PAUL HAFFNER, S.T.D.

An interesting and important question in the great scientific development over the past century concerns how it all came about in the first place. It is true that one discovery builds upon another. For example, the invention of computers is inconceivable without the prior discovery of electricity. Many scholars would say that the early moves in the development of modern science go all the way back to the Middle Ages.[1] Fr. Stanley L. Jaki OSB discovered this fundamental historical paradigm of the history of science, namely that the Middle Ages represent the very cradle of science. This is in sharp contrast to the backdrop of Enlightenment ideology and propaganda that the Middle Ages were devoid of technological progress and, more specifically, that in the Eastern Roman Empire superstition and mysticism ruled and was devoid of invention. Though this was true in some cases, superstition is still part of our world today as is science and progress. In fact, science and progress were more than alive in the Middle Ages and, even before that, in the Eastern Roman Empire.

In justifying his thesis that Christianity played a fundamental role, Jaki focuses attention on an Alexandrian philosopher John Philoponus, the author of most fascinating reflections on the cosmos.[2] Philoponus' *De opificio mundi*, written after his conversion to Christianity, deals in an elaborate manner with the biblical account of creation. Here he countered Theodore of Mopsuestia's view that the heavenly bodies were moved by angels, implying instead that they were given a certain kinetic force by God as were other moving bodies. Philoponus agreed with Aristotle on much of his view of the physical world, "but wherever Aristotelian physics and cosmology seemed to encroach on the Creator's prerogatives or on man's proper relation to his Creator, Philoponus instinctively parted ways with Aristotle." Hence Philoponus rejected the divinity of celestial matter and eternal motion as parts of a pantheistic world-picture. Jaki regards Philoponus as a forerunner of Buridan and Oresme (and thence of Galileo and Newton) in their view of inertial motion. Philoponus stated, in his commentary on Aristotle's physics, that projectiles move through the air "not because the air keeps closing in behind them, but because they were imparted a certain 'quantity of motion.'"[3] Philoponus is a clear example of one for whom Christian belief in God the Creator beneficially influenced his view of the scientific interpretation of the created world. The patristic period was very rich in its interpretation of the scriptural idea of creation. However, Jaki points out that later ages "under the impact of the Reformation came heavily under the sway of biblical fundamentalism, a far cry from the way in which Augustine of Hippo, Gregory of Nyssa, Basil the Great, to mention only a few, interpreted the biblical notion of creation."[4]

Jaki also makes much of St Athanasius who proposed that the created world can only be understood through the God-given gift of reason, properly ordered. He who created the κόσμος (cosmos) is λόγος (logos)—Reason, Himself. According to St. Athanasius of Alexandria (c. 296–373), the order of the cosmos is a motive of credibility for Christ and His Church:

> For if the movement of the universe was irrational, and the world rolled on in random (i.e. indeterminate) fashion, one would be justified in disbelieving what we (i.e. Christians) say. But if the world is founded on reason, wisdom, and science, and is filled with orderly beauty, then it must owe its origin and order to none other than the Word of God.[5]

Jaki shows with a remarkably novel analysis of Athanasius' anti-Arian writings, that the doctrine of a fully divine Logos indeed implied that his creative work was the "the paragon of logic and order." The rationality of the cosmos is thus "rooted in belief in the strict divinity of the Logos."[6]

Therefore having cited Philoponus and Athanasius, the Alexandrian philosopher and theologian respectively as important for the rise of science, it is perhaps surprising that Jaki doesn't value Byzantine science to a greater degree. In fact, he detailed the difference between East and West as follows: "Byzantium largely skirted the issue by withdrawing its orthodoxy into a lofty supernaturalism steeped in Neoplatonism. In such a framework there was no room either for science or for natural theology."[7] D. C. Allison holds, along with Jaki, that Christianity constituted a necessary basis for science and regards the absence of science from Eastern Christendom as implying that the basis was not sufficient.[8] G. S. Hendry further elaborates on this fact noting that "the cosmic dimension of theism has been more consistently kept in view in the theology of the Eastern Church than in that of the Western Churches."[9] However, such a "cosmic dimension" in Eastern theology was lacking in that rational consideration of the consequences of belief in God the Creator which was far more developed in the West; there was no real equivalent in the East of the Western flowering of scholastic philosophy. In addition to the necessary condition of Christian theism for the rise of science, other related factors are required, such as an intellectual and cultural climate in which the consequences of Christian faith can take root over a period of time.[10]

Recent research has shed new light on the importance of Byzantine science:

> Taking into account the new data on the influence and reshaping of the multiple discourses in Byzantine thought (mainly theological and philosophical) concerning the interaction of faith and secular knowledge during its long history, modern research has enabled corrective reinterpretations, critical reassessments, and more realistic evaluations of the Byzantine world-view and its position in the history of thought in the Middle Ages. Forming the link between late antiquity and the new worldview of Christian revelation, Byzantium is no longer considered only the guardian and transmitter of the textual heritage of Hellenism; it is seen as an active filter, a denominator of dynamic feedback between East and West, as well.[11]

Costache points out that "the study of Byzantine science has become a burgeoning field, part and parcel of this global effort of retrieving the many strands of human ingenuity expressed in the form of scientific enquiry and technological innovation, broadly conceived. And while currently Byzantine science constitutes a distinct branch of the history of science, attracting the interest of Byzantinists and non-Byzantinists alike, it is by no means an emerging field."[12]

The theological appreciation of beauty lay at the basis of the impetus for investigation of nature as McGuckin explains:

> This apprehension of beauty was a matter of divine epiphany: what the Greek Fathers would also called *Doxa*, the glorification of the Logos as *fundamentum* of all being, by the only *loci* of the created order who could appreciate its meaning, namely illuminated noetic intelligences. Evagrios of Pontos, and Maximus Confessor, with their doctrines of the *logoi* (revelatory principles) hidden within the World-order are perhaps the most explicit continuators of this master theme from Origen. It became, after him, the root of both the mainframe Christian soteriology (the incarnation of the Logos within the world order for the cause of saving it) and the Church's mystical theory (union with the Logos being the soul's yearning for restored communion, and its primal *raison d'être*). Gregory the Theologian represents both themes.[13]

This new picture is now illustrated with a number of examples that offer an indication as to why the seemingly backward and stagnant era was the predecessor to the most advanced and forward-thinking society in history. First, I shall consider scientific theory. As we have seen above, Aristotle in his physics had proposed that objects which weighed more fell faster and vice-versa: this theory was finally discredited by Galileo; however, Galileo was not the first to question this theory. John Philoponus, a teacher philosopher and Christian theologian from Alexandria, thought Aristotle's theory was mistaken. In his commentary on Aristotle's physics he wrote this is a complete error as we can see through observation better than through any abstract proof if you drop two bodies a vastly different weight from the same height you will see that the difference in the time that it takes for them to fall is not at all proportional to their difference in weight it is in fact a small difference.[14] Later, the great John of Damascus in his *Exposition of the Orthodox Faith* explained in detail how an eclipse occurs despite the moon being much smaller than the Sun.[15] Symeon Seth in the 11th century in his summary of physics gave a number of explanations for the earth being a sphere, one such proof being that some stars in the north cannot be seen in the south and vice-versa: were the earth not spherical all the stars would always be visible regardless of where you were.[16] The Byzantines also excelled in chronometry; portable sundials survived such as the one from Aphrodisias in Asia Minor: it has the ability for the user to adjust the time to work in different latitudes including Rome, Constantinople and Bordeaux.[17] Another portable sundial synchronized the time with the day of the week and the month and phases of the Moon as well as latitude realignment it was about the 30 cm in diameter and 0.5 cm thick.[18]

One Byzantine astronomer who predicted eclipses together with their year, date, time of day or night, and its extent was Nikephoros Gregoras (1295–1361) who was at the same time a theologian and historian during the late Byzantine period known as the Palaiologan Renaissance. However, because of opposing the prevailing theology of his time, Gregoras was put under house arrest where he worked on more astronomical theories creating many books as well, including some histories of the Palaiologos imperial family.[19] One of the things Gregoras is best known for was when he found an accurate way to calculate the date of Easter in 1324 in which he realized that the Julian calendar miscounted the length of the year by a small fraction of a day. Once he saw this mistake and fixed the calendar, he explained his findings to the emperor Andronikos II who saw that Gregoras was right but decided to not push through with changing the calendar for it might create a split in the Eastern Churches (in which it later did with the Russian Church). The calendar system Gregoras discovered was not implemented until 1582 when Pope Gregory XIII implemented the new calendar system using the one of Gregoras but only with a few changes. This calendar system is the Gregorian calendar which we use today which would have been called the *Gregoran* calendar if it weren't implemented by the pope in 1582.[20]

A remarkable military invention was Greek fire by Telemachus of Heliopolis in the seventh century. Greek Fire which was a large flamethrower placed on ships first used in protecting their capital, Constantinople, from an Arab siege. The liquid fire from the flamethrower was made up of a chemical compound of flammable resins, sulphur, and naphtha which came possibly from the Naphtha wells in the Crimea. The formula for Greek Fire is a mystery and has remained a Byzantine state secret. This weapon could only be operated by a secret branch of the army. In the book *De Ceremoniis* by Emperor Constantine VII Porphyrogennetos (r. 913–959), he explains to his son, later Emperor Romanos II (r. 959–963), that Greek Fire is something the Byzantines should keep to themselves and not share to other nations because these other nations may copy their invention. The weapon was operated by heavy air pressure pumped into a heated sealed container to ignite the compound enabling it to release a large blast of fire to burn down ships or scare the enemy away, although Greek Fire was hard to operate as it was slow and went in its own direction. Greek Fire was also used in defending the Bosporus from Viking attacks and against the Normans in Dyrrachion from 1107–8 as mentioned by the historian Anna Komnene. Some ships used the Greek fire by making it come out of the mouths of animal sculptures such as lions. However, Greek Fire was not only used in ships but on land as well as it was recorded that some Byzantine soldiers used it against the

Turks besieging the inland city of Manzikert in Asia Minor in 1054. Another version of Greek Fire was the portable *Cheirosiphon* said to be invented by Emperor Leo VI the Wise (r. 886–912) which was used by soldiers when besieging cities; this kind of crude grenade to set fire to siege weapons was mentioned by the emperor Maurice in his *Strategikon*.[21]

A slight alteration to the custom of riding a horse may have dramatically changed the way wars were fought. Humans rode bareback or mounted horses with a simple blanket after they first domesticated the animals, thousands of years after the dawn of agriculture. The fierce Avar tribe may have brought stirrups to the West when it arrived in Byzantium in the sixth century A.D. The Byzantine Empire soon adopted the stirrup—and later the Franks embraced it as well. This important innovation allowed the rider immensely greater control in horsemanship: rider and animal became almost extensions of each other.[22]

A smart invention the Byzantines made in time of war were ship mills, first made in 536 when the Goths had besieged Rome and the Byzantine Army of Emperor Justinian I, under the general Belisarius, was on their mission to reconquer Italy. The Goths had previously destroyed the Roman aqueducts to cut the food supply for the population of Rome, and when the army of Belisarius reached Rome, they too needed food supply. Belisarius thought of a smart solution by finding the part of the Tiber River where the current was at its strongest and there he stretched two ropes putting ships between them with wheels attached on their sides. In this way, the grain was milled due to the strong current of the water, and the population of Rome was fed. When the Goths learned of this, they responded by tossing debris down the river to jam the mills, but Belisarius countered it by hanging chains down the bridges to catch the debris and pull it up.[23]

Among the more trivial but interesting Byzantine inventions were the mechanisms at the throne room or *Magnaura* of the imperial palace. The mechanisms included hydraulics behind the two golden lion sculptures flanking the throne in which its sounds and movements were operated by water pumps inside and beside the throne was a golden tree with mechanical birds. The hydraulics were not operated within the statues themselves but from curtains behind them and pipes connected to the lion statues. The mechanisms in the throne room were recorded by Liutprand, bishop of Cremona, during his visit in 949 during the reign of Constantine VII. Here, he mentions that the lions moved their tails and made sounds, the birds warbled at the bronze tree, and when he bowed down he saw the throne elevated to the ceiling on it while the emperor was seated on it. Liutprand however says he was not afraid because he had been told all about the imperial throne room in advance.[24]

Like the Romans before them, the Byzantines also built aqueducts to supply water for their capital, Constantinople. The capital however was far from a large fresh water source, which was in Thrace so they had to build 3 large aqueducts spanning 592km long, which were even longer than the lengths of Rome's 11 aqueducts combined which in total was 520km. The water brought in from the aqueducts was stored in hundreds of cisterns located around Constantinople and the notable ones that still can be seen today include the open-air Aetius Cistern, the covered Philoxenos Cistern which has 224 double columns, and of course the most famous one being the Basilica Cistern across the Hagia Sophia which can hold up to 80,000 cubic meters and has 336 columns. The large water supply in the cisterns coming in from the aqueducts was able to sustain the population of Constantinople especially during times of siege when the enemy cut the water supply. Aside from providing a sustainable amount of water supply for the population, the cisterns also provided enough water for the many fountains in the city. The Basilica Cistern was abandoned and was covered up after the fall of Constantinople to the Ottomans in 1453 but was rediscovered in the 1540's by the French scientist Pierre Gilles who sent on a mission by King Francis I of France to find ancient manuscripts. When Gilles went to the basement of a house, he found an opening leading to some water and took a small boat across the abandoned cistern.[25]

On a more romantic level, the Byzantines introduced a couple of new inventions to the West, one of them being the fork which came into Western Europe when Western princes married Byzantine princesses, thus introducing a new method for eating. Apart from the fork, it is also believed that the Byzantines introduced the musical organ to the medieval west.[26] Unlike the medieval west, the Byzantines did not really use organs in churches but for shows at the Hippodrome, for court receptions in the imperial palace, and even with soldiers during battle and for emperors touring the provinces. The organ was introduced to the west in 757 when Emperor Constantine V sent one as a gift to Pepin, the king of the Franks, this organ was named *Big Mouth with a Loud Voice*. The Byzantines were certainly pioneers in the development of the hydraulic organ.[27]

Constant and acute concerns with security—physical, material, and spiritual—were as pressing a part of daily life in medieval Byzantium, as they are in our own culture. Indeed, high illiteracy, poor communications, primitive locking technology, and the uncertainty of central authority and government institutions engendered among the Byzantines a whole spectrum of security needs—and responses—for which no precise counterparts exist in our own age. Yet, perhaps because of its very pervasiveness, and its inherent subjectivity, the concept of *security* has not been traced extensively through the material culture of Byzantium. Three distinct types of objects designed and employed expressly for the maintenance of everyday security show a well-developed technology during this period: locks and keys, sealing and stamping implements, and official weights.[28]

By medieval standards the practice of medicine existed at a high level of expertise: hospitals were to be found not only in Constantinople, but also in the provinces. Byzantine medicine had professional specializations, including obstetrics and gynaecology, ophthalmology, dermatology, and dentistry. Thanks to Byzantine pharmacy, many diseases received sophisticated treatments, and the medical tracts of Alexander of Tralles and Paul of Aegina show the variety of drugs prescribed for ailments of the chest, heart, digestive system, and other organs. Byzantine surgery was also highly advanced; listings of surgical instruments suggest specialized expertise, perhaps derived from the known instances of dissections and autopsies performed by Byzantine physicians and surgeons.[29] The *Vita Theophanis Confessoris* by Nikephoros Skeuophylax written in the tenth century, mentions a fascinating surgical operation undergone by Saint Theophanes for his kidney stones. The physicians passed instruments up Theophanes' urinary tract, ground down the stones inside the bladder, and then allowed the fragments to pass out of the body with his own urine.[30]

Following the iconoclastic crisis and the seventh ecumenical synod of 787, when a learned body of clergy of the Church assumed its leadership, medicine and the medical profession gained more respectability. The climate was prepared by the great scholar and patriarch, Photios of Constantinople. He summarized several medical authors such as Dioscurides, Galen, and Aetios of Amida. For Photios, medicine is an *episteme* to be used by all, including the poor and destitute. Medical opinion deserves respect, and people should never go against the dictates of nature. When physicians perform a surgery, they do not do it because of ill intentions but out of mercy, to heal the sick from suffering.[31] Basil's and Photios' views on medicine and the medical profession became catalysts for later generations. For Ioannes Mavropous, the eleventh-century intellectual and metropolitan of Euchaita, it is natural for a sick person to seek out a physician for his treatment. And Theophylaktos, the twelfth century Archbishop of Ochrida calls medicine a philanthropic science (*episteme*) and the physician an artist. He acknowledges the therapeutic importance of medicine but also the need for religious faith and hope. The physician makes the diagnosis of an illness and provides advice but the source of therapy is God, the source of all creation.[32]

Hospitals (*nosokomeia*), clinics (*iatreia*) or temples (*asklepieia*) dedicated to medical treatment existed in the ancient Greek world and, under the names *hospitalis, hospitalia,* in the Roman world. Byzantium built upon its Greco-Roman inheritance, but under the influence of Christianity it went further than its predecessors. As in Greek and Roman antiquity, where the buildings adjacent to the temple of Asklepios or Aesculapius, the god of healing, became an early type of hospital, hospitals in Byzantium were erected next to churches or monasteries.[33]

To conclude, modern science is the product of a long and successful transformation in Europe, over four centuries from the time of Galileo. It has given our society a method of studying nature and a new vision of the relationship between man and his Universe, in other words, "modernity." The explanation of natural phenomena in mechanical, physical or chemical terms, and their mathematical representation are the foundations of this "modernity" which contributed to the birth, in the eighteenth century, of the Industrial Age. The assumption with such an explanation is that the facts one wishes to explain are first transposed in the form of an abstract "model," the elements of which can be defined by their mutual relations and, for some of them, by a relationship protocol with experimentation.[34] Since the "Galilean revolution" science is characterised by the combination of three requirements:

1. Science is a "reality," whichever interpretation philosophers confer on the term.
2. Science seeks an "explanation," i.e. the insertion of reality that it describes in an abstract system of concepts, overflowing with singular facts brought about by experimentation.
3. Science complies with criteria of "validity" that are explicitly formulated and consensual.[35]

While the Byzantine scientific beginnings did not uniformly fulfill these criteria in a consistent manner, it did so in a way which forged the path for medieval science in East and West. For example, Saints Basil of Caesarea and Gregory of Nyssa, two fourth-century Cappadocian saints, with their ways of tackling the natural philosophy of their age paved the way for contemporary science-engaged theology.[36] A fundamental figure in understanding Eastern Christianity and its impact upon science is St Basil of Caesarea and his explanation of the book of Genesis (called the *Hexaemeron*). Written in 378 it is, historically, the text most read by the Eastern Churches about the Creation.[37] His descriptions of nature show detailed knowledge:

> Take the bee for your model, which constructs its cells without injuring any one and without interfering with the goods of others. It gathers openly wax from the flowers with its mouth, drawing in the honey scattered over them like dew, and injects it into the hollow of its cells. Thus at first honey is liquid; time thickens it and gives it its sweetness. The book of Proverbs has given the bee the most honourable and the best praise by calling her wise and industrious. How much activity she exerts in gathering this precious nourishment, by which both kings and men of low degree are brought to health! How great is the art and cunning she displays in the construction of the store houses which are destined to receive the honey! After having spread the wax like a thin membrane, she distributes it in contiguous compartments which, weak though they are, by their number and by their mass, sustain the whole edifice. Each cell in fact holds to the one next to it, and is separated by a thin partition; we thus see two or three galleries of cells built one upon the other. The bee takes care not to make one vast cavity, for it might break under the weight of the liquid, and allow it to escape. See how the discoveries of geometry are mere by-works to the wise bee! The rows of honey-comb are all hexagonal with equal sides. They do not bear on each other in straight lines, lest the supports should press on empty spaces between and give way; but the angles of the lower hexagons serve as foundations and bases to those which rise above, so as to furnish a sure support to the lower mass, and so that each cell may securely keep the liquid honey.[38]

St Gregory refers to fire—another fundamental element of ancient physics—as casting light upon the Genesis narrative of creation and upon the order of things.[39] Fire does not feature in the scriptural narrative, but Gregory believes that it should not be left out from a coherent and comprehensive world-view that integrates scriptural and scientific ideas about reality. He goes to even greater lengths to elucidate for his readers the cycles of water in nature. He introduces, more specifically, the available hydrological and meteorological information into his supposed exegesis of the waters of creation.[40] Nothing of his explanation of the cycles of water relates to the Hebrew Scriptures—apart from his seasoning the discourse with likely scriptural passages—but Gregory, ever in Basil's footsteps, is of the view that people should have as accurate an idea of the world as possible. No wonder he made consistent recourse to the natural sciences of the Hellenistic era. The outcomes of this method are revealing: by engaging the available scientific culture—which they interpreted from the viewpoint of the doctrine of creation—as a sounder foundation for their theological contemplation of reality, these Christian authors were able to ask new questions and to provide new answers to their audiences' quest for meaning.[41]

To conclude with an example which unites practical science and lived faith in worship we consider the example of Anthemius of Tralles who was both a mathematician and an architect:

> The problem of how to contrive that at any hour and season a ray of the sun, passing through a small aperture, shall fall in a given spot without moving away was solved by Anthemius. He describes the construction of an elliptical reflector with one focus at the aperture and the other at the point to which the ray is to be reflected. Both winter and equinoctial rays are considered. In his treatment, Anthemius

> incidentally mentions the construction of an ellipse by means of a loop of string drawn closely around the foci. He also uses a proposition not made explicit in the Conics: that the straight line joining the focus to the intersection of two tangents bisects the angle between the two straight lines joining the focus to the two points of contact. Another construction shows how parallel rays may be reflected to one point at the focus of a parabolic reflector.[42]

He is best known, however, for replacing the old church of *Hagia Sophia* at Constantinople in 532. The new *Hagia Sophia,* as described by Procopius, "exults in an indescribable beauty". He goes on to describe the construction:

> For it proudly reveals its mass and the harmony of its proportions, having neither excess nor deficiency, since it is both more pretentious than the buildings to which we are accustomed, and considerably more noble than those which are merely huge, and it abounds exceedingly in sunlight and in the reflection of the sun's rays from the marble. Indeed one might say that its interior is not illuminated from without by the sun, but that the radiance comes into being within it, such an abundance of light bathes this shrine. And the face itself of the church (which would be the part which faces the rising sun, that portion of the building in which they perform the mysteries in worship of God) was constructed in the following manner. A structure of masonry (*oikodomia*) is built up from the ground, not made in a straight line, but gradually curving inward on its flanks and receding at the middle, so that it forms the shape of half a circle, which those who are skilled in such matters call a half-cylinder

(*hemikylindron*); and so it rises precipitously to a height. The upper part of this structure ends in the fourth part of a sphere (*sphaira*), and above it another crescent-shaped (*menoeides*) structure rises, fitted to the adjoining parts of the building, marvellous in its grace, but by reason of the seeming insecurity of its composition altogether terrifying. For it seems somehow to float in the air on no firm basis, but to be poised aloft to the peril of those inside it. Yet actually it is braced with exceptional firmness and security.[43]

Notes

1. See, for example, E. Grant, *The Foundations of Modern Science in the Middle Ages: Their religious, institutional, and intellectual contexts. Cambridge History of Science* (Cambridge: Cambridge University Press, 1996); A. C. Crombie, *Robert Grosseteste and the origins of experimental science, 1100–1700* (Oxford: Clarendon Press, 1953); Idem, *Science, optics, and music in medieval and early modern thought* (London, Ronceverte, WV, U.S.A.: Hambledon Press, 1990); Idem, *Styles of scientific thinking in the European tradition: The history of argument and explanation especially in the mathematical and biomedical sciences and arts* (London: Duckworth, 1994); Idem; *The history of science from Augustine to Galileo* (New York: Dover Publications, 1995); Idem, *Science, art, and nature in medieval and modern thought* (London, Rio Grande, Ohio: Hambledon Press, 1996).
2. See Christian Wildberg, "John Philoponus," in *The Stanford Encyclopedia of Philosophy*, Winter 2021, 2021, https://plato.stanford.edu/archives/win2021/entries/philoponus/.
3. Stanley L. Jaki, *Science and Creation: From Eternal Cycles to an Oscillating Universe*, Rev. and enl. ed (Edinburgh: Scottish Academic Press, 1986), p.183. Jaki refers to H. Vitelli, *Joannis Philoponi in Aristotelis physicorum libros quinque posteriores commentaria*, vol. II of the edition of the Greek text (Berlin: G. Reimer, 1888) pp.681–682, 682–684 (Corollarium de inani), 639-642 (ad Lib IV, cap. viii).

4. Stanley L. Jaki, *Planets and Planetarians* (Edinburgh: Scottish Academic Press; New York: The Halstead Press of John Wiley Inc., 1978), p.39.
5. St Athanasius, *Against the Heathen*, 40.
6. Stanley L. Jaki, *The Savior of Science* (Washington, DC: Regnery Gateway, 1988), pp.76–77.
7. Stanley L. Jaki, *The Road of Science and the Ways to God: The Gifford Lectures 1975 and 1976* (Chicago/Edinburgh: University of Chicago Press/Scottish Academic Press, 1978), p.35.
8. Cf. D. C. Allison Jr., Review of *Cosmos and Creator*, in *The University Bookman* 26 (1986) p.41.
9. G. S. Hendry, "New Science, Old Apologetic" (Review of The Road of Science and the Ways to God), in *Theology Today* 36 (1979/1980) p.262.
10. Vasilios N. Makrides, "Orthodoxy Matters: Why Has a Scientific Revolution Not Taken Place in the Greek East? The Role of Orthodox Christian Traditionalism," in *Orthodox Christianity and Modern Science: Past, Present and Future*, ed. Kostas Tampakis and Haralambos Ventis (Turnhout, Belgium: Brepols Publishers, 2022), 15–44.
11. Efthymios Nicolaidis et al., "Science and Orthodox Christianity: An Overview," *Isis* 107, no. 3 (September 20, 2016): 546.
12. Doru Costache, "Scientific Interests and Technological Innovation in Byzantium: Interdisciplinary Perspectives", *De Medio Aevo*, 13:2 (2024): 255–60.
13. J. A. McGuckin, "The Beauty of the World and Its Significance in St. Gregory the Theologian", in J. Chryssavgis and B. V. Foltz (eds), *Toward an Ecology of Transfiguration* (Fordham University Press, 2013): pp. 34–45.
14. Johannes Philoponus, *On Aristotle Physics* 1.4-9, trans. Catherine Rowett, Paperback edition, *Ancient Commentators on Aristotle* (London New Delhi New York Sydney: Bloomsbury, 2014). C book 5 chapter 17 page 683
15. John of Damascus, *Exposition of the Orthodox Faith*, 21.
16. Symeon Seth, "Synopsis on Physics (Conspectus Rerum Naturalium)," in *Anecdota Atheniensia et Alia, t. II: Textes Grecs Relatifs à l'histoire Des Sciences*, Bibliothèque de La Faculté de

Philosophie et Lettres de l'Université de Liège (Liège–Paris: Droz, 1939).

17. Derek J. De Solla Price, "Portable Sundials in Antiquity, Including an Account of a New Example from Aphrodisias," *Centaurus* 14, no. 1 (December 1969): 242–66.
18. J. V. Field and M. T. Wright, "Gears from the Byzantines: A Portable Sundial with Calendrical Gearing," *Annals of Science* 42, no. 2 (March 1985): 87–138.
19. See Divna Manolova, "Discourses of Science and Philosophy in the Letters of Nikephoros Gregoras" (Budapest, Central European University, 2014), http://goya.ceu.hu/record=b1174621.
20. See Nicephorus Gregoras, *Byzantina Historia*, vol. 25–27, *Corpus Scriptorum Historiae Byzantinae* (Bonn, 1829).
21. Maurice, *Strategikon: Handbook of Byzantine Military Strategy, The Middle Ages*, trans. George T. Dennis (Philadelphia: University of Pennsylvania Press, 1984), I.2, II.9.
22. Gary Stix, "The Stirrup," *Scientific American* 301, no. 3 (September 2009): 78–78.
23. Celdran Powee, "Forgotten but Significant Byzantine Science and Technology," n.d., https://byzantium-blogger.blog/2019/06/12/forgotten-but-significant-byzantine-science-and-technology/.
24. Liudprand, *The Complete Works of Liudprand of Cremona, Medieval Texts in Translation*, trans. Paolo Squatriti (Washington, D.C: Catholic University of America Press, 2007).
25. James Crow, Jonathan Bardill, and Richard Bayliss, *The Water Supply of Byzantine Constantinople, Journal of Roman Studies Monograph*, no. 11 (London: Society for the promotion of Roman studies, 2008).
26. Celdran Powee, "Forgotten but Significant Byzantine Science and Technology," n.d., https://byzantium-blogger.blog/2019/06/12/forgotten-but-significant-byzantine-science-and-technology/.
27. Details of the surviving descriptions and iconography of the *hydraulis* are included in Jean Perrot's *The Organ; from its Invention in the Hellenistic Period to the end of the Thirteenth Century*, translated from the French by Norma Deane. (Oxford Univ. Press, 1971) . The book includes photographs of major

surviving depictions of the *hydraulis,* including later Roman iconography such as the Carthage terra cotta, and portions of the Roman organ at Aquincum. Both Peter Williams' *A New History of the Organ* (Indiana University Press, 1980) and Chapter Three of *The Organ,* by Peter Williams and Barbara Owen (The New Grove Musical Instrument Series, W. W. Norton, 1988) contain summary descriptions of the surviving evidence of the early organ.

28. See Gary Vikan and John Nesbitt, *Security in Byzantium: Locking, Sealing, and Weighing,* Dumbarton Oaks. Byzantine Collection 2 (Washington, DC: Dumbarton Oaks Center for Byzantine Studies, 1980).

29. In 1903, H. Schoene published the names of 54 'Greek-Roman' surgical instruments. For the first time, J. S. Milne (1907) attempted to attribute names to the Byzantine instruments and found their Byzantine names; and, since then, with the additions of A. Maraslis (1983) and L. J. Bliquez (1985) the list has increased to 237. Several instruments, however, have different forms, shapes and names, and some were obviously made for particular operations, thus increasing the number of the various known By zantine instruments to over 500. See most recently S. Geroulanos, "Surgery in Byzantium," M. Grünbart et al (eds.), *Material Culture and Well-Being in Byzantium (400-1453),* Vienna 2007, 129-34. Unlike Arab and Western surgical instruments which were often decorated, sometimes in excess (and this was the case until the eighteenth century), which prevented them from being thoroughly cleaned, this was not the case in Byzantium and would only become the standard after 1867, with Joseph Lister, and the development of antisepsis.

30. Theophanes, *The Chronicle of Theophanes Confessor: Byzantine and Near Eastern History, AD 284-813,* ed. Cyril A. Mango, Roger Scott, and Geoffrey Greatrex (Oxford: New York: Clarendon Press; Oxford University Press, 1997). See Theophanes the Confessor and Anastasius the Librarian, *Theophanis Chronographia,* ed. Carl de Boor (Lipsiae: B. G. Teubner, 1883), p. 23.

31. *SOPHOCLES K TOU EX OIKONOMON* (ed.), PHOTIOS, *Ta Amphilochia.* Athens, 1858, pp. 46, 252, 338.

32. A. Karpozilos, (ed), *The Letters of Ioannes Mauropous Metropolitan of Euchaita.* Thessaloniki, Association for Byzantine Research,

199, n. 32, p. 121. P. Gautier (ed), *Theophylacte d'Achrida Lettres*. Thessaloniki Association de Recherches Byzantines, 1986, n. 53, 57, 60, 78, 81, 94, 97, 99, 102, 110, 129, 133.

33. Demetrios J. Constantelos, "Medicine and Social Welfare in the Byzantine Empire," *Medicina Nei Secoli. Journal of History of Medicine and Medical Humanities* 11, no. 2 (1999): 346.
34. Stavros Lazaris, ed., "Introduction," in *A Companion to Byzantine Science, Brill's Companions to the Byzantine World*, volume 6 (Leiden; Boston: Brill, 2020), 4.
35. *Ibid.*
36. Doru Costache, "Patristic and Neopatristic Antecedents of Scientifically Engaged Theology," *St Vladimir's Theological Quarterly* 67, no. 1–2 (January 1, 2023): 125.
37. Efthymios Nicolaidis et al., "Science and Orthodox Christianity: An Overview," *Isis* 107, no. 3 (September 20, 2016): 547.
38. St. Basil, *Hexaemeron*, 8, 5.
39. See St. Gregory of Nyssa, *An Apology for the Hexaemeron* 24.37.11–38.10; 28.41.12–42.6.
40. See St. Gregory of Nyssa, *An Apology for the Hexaemeron* 33.46.12–38.51.7.
41. Costache, "Patristic and Neopatristic Antecedents of Scientifically Engaged Theology," 130.
42. "Anthemius of Tralles." *Science and Its Times: Understanding the Social Significance of Scientific Discovery. Encyclopedia.com.* (March 18, 2024). https://www.encyclopedia.com/science/encyclopedias-almanacs-transcripts-and-maps/anthemius-tralles
43. Procopius of Caesarea, *On the Buildings* (*Peri Ktismaton*), edited and translated by Henry Bronson Dewing (1882-1956), from the Loeb Classical Library edition of Procopius, published in 1919/1940,1.1.29-33.

4 JAKI AND BARBOUR: PILGRIMS ON PARALLEL PATHS OF THE SCIENCE AND RELIGION JOURNEY

REV. JOSEPH R. LARACY, S.T.D.

Introduction

Two of the most influential twentieth century scholars in the science and religion field in the United States were Stanley L. Jaki and Ian G. Barbour. Despite very similar scientific backgrounds in experimental physics, a shared commitment to philosophical realism, and parallel research paths exploring the relationship of Christian theology and natural science, there is surprisingly little evidence of interaction between them. This paper elucidates some of the similarities and differences in their approaches for relating empirical science and Christianity. It will clarify their epistemological, metaphysical, and ecclesial commitments, and how these may have contributed to minimal scholarly interaction.

The parallels between Jaki and Barbour are striking. Both trained in particle and nuclear physics in the 1950s and gradually shifted their research programs to the emerging field of science and religion. They excelled in this burgeoning discipline. In fact, Jaki was the first Catholic priest to be named a Gifford Lecturer[1] 1974–1976, and he was awarded the Templeton Prize[2] in 1987.[3] Likewise, Barbour gave the Gifford Lectures from 1989–1991 and was awarded the Templeton Prize in 1999.[4]

Jaki and Barbour also appreciated the mediating roles of history and philosophy in the science-religion interaction.[5] Both were deeply interested in the dynamic relationship of the history of science and the history of Christianity. Paul Haffner points out that as early as 1957 Jaki came to the conclusion "that the real problems concerning the relation of science and religion belonged mostly to the history and philosophy of physics."[6] Jaki contributed to the historical research initiated by Pierre Duhem that showed how science was born of Christianity.[7] Barbour's historical research explicated the various ways that science and religion have interacted throughout history, thus developing his four-fold typology of conflict, independence, dialogue, and integration.[8]

Jaki and Barbour recognized the importance of epistemology and metaphysics for science and religion. At a time when various forms of idealism, positivism, and instrumentalism were in vogue, they both argued strongly for realism in both Christian theology and natural science. Barbour's "critical realism" was not formally defined, but rather inspired by his practice of science.[9] Likewise, Jaki did not specify a systematic epistemology,[10] but he is clearly a Thomistic realist.[11] His realism is highly congruent with that of Étienne Gilson.[12] Jaki encouraged the translation of Gilson's 1935 book, *Le réalisme méthodique*,[13] and even wrote the introduction to the 1990 English language edition of *Methodical Realism*.[14] Jaki writes about how Gilson "showed that even the most refined forms of 'critical realism' advocated by some Neo-Thomists were refined versions of Kantianism."[15] Jaki also takes a Thomistic approach to metaphysics,[16] while highlighting, as Jason Mitchell puts it, the "need to recognize the limitations of the Aristotelian scientific and cosmological world-view and to recognize the consequences that derive from these limitations."[17] While Barbour appreciates Thomism,[18] he ultimately commits himself to the process metaphysics of Alfred North Whitehead[19] and process theology of Charles Hartshorne.[20]

This paper will proceed to elucidate some key biographical information on Jaki and Barbour. It will delve into their early scientific research, epistemological and metaphysical views, as well as their ecclesial identities. Finally, it will attempt to explain the potential reasons for the lack of explicit scholarly interaction between these contemporaneous American scholars.

Stanley L. Jaki

Stanley Jaki was born in 1924 in Győr, Hungary, the third of five children. Stanley, along with his two brothers, Zeno and Theodose, discerned a vocation to monastic life and took vows at the Benedictine Archabbey of Pannonhalma. His undergraduate education focused on philosophy, theology, and mathematics. Aware of his intellectual gifts, the abbot sent Br. Stanley to Rome in 1947 to continue his theological studies at the Pontifical Institute of Saint Anselm. He was ordained to the priesthood in 1948. Fr. Stanley earned the degree of *Sacræ Theologiæ Doctor* (STD), an ecclesiastical research doctorate equivalent to the *Philosophiæ Doctor* (PhD) in Theology, in 1950. His dissertation was entitled *Les tendances nouvelles de l'ecclésiologie*. This work on the theology of the Church was well received. It was published as a book in 1957 and again in 1963 (during the Second Vatican Council). Joseph Ratzinger wrote a favorable review of the book, and later told Jaki that it occupied a place of honor in his library.[21]

Unable to return to Hungary in late 1950 due to the Communist takeover, Jaki was sent to the United States and assigned to the faculty of St. Vincent's Archabbey in Latrobe, PA. He taught French to the undergraduates and systematic theology to the major seminarians, including the major dogmatic tracts: *De Deo Uno*, *De Deo Creante*, *De Homine et Gratia*, and *De Novissimis*. On December 9, 1953, Jaki suffered a life-changing event. He experienced severe complications following a tonsillectomy which began a ten-year period of very limited use of his vocal cords. This forced him to resign from teaching in February of 1954.[22]

Jaki was a life-long learner. At Latrobe College Jaki was both a faculty member and a student. He furthered his studies in mathematics and the physical sciences (as well as taking courses in American history and literature), earning an American Bachelor of Science degree in 1954. In September of that year, he began graduate studies in physics at Fordham University in New York City. Jaki earned a Master of Science degree in July of 1955 and then began doctoral research under Professor Victor F. Hess in September of 1956. This was an extraordinary opportunity for Jaki as Hess had won a Nobel Prize in Physics for the discovery of cosmic rays in 1936. Jaki developed a new method for measuring the distribution of radon and thoron at the earth-air interface. He was awarded the PhD in Physics in 1957 with the completion of his dissertation, *A Study of the Distribution of Radon, Thoron, and Their Decay Products above and below the Ground*.[23]

After earning his second doctorate, Jaki was awarded a fellowship at Princeton University. In 1965 he joined the Department of Physics at Seton Hall University. Seton Hall would be his academic home for over forty years. Jaki's scholarly interests shifted from experimental physics to the philosophy of science and the history of science, particularly their relationship to Christianity. In 1975, Seton Hall conferred upon him the rare rank of "Distinguished Professor," which is above Full Professor. Over the course of his scholarly career, he delivered lectures at over fifty major universities, including Yale, Oxford, the Pontifical Gregorian University, and others. Jaki authored over fifty books and 350 articles, published in a variety of languages including English, French, Italian, Spanish, Hungarian, Polish, Russian, and Chinese.[24]

Jaki's Early Scientific Work

Jaki's PhD research focused on the distribution of Radon, Thoron, and their decay products above and below the ground. Radon is a colorless, odorless, and radioactive noble gas. It has three naturally occurring isotopes. Radon-222 has a sufficiently long half-life (3.825 days) for it to be released from the surface of the Earth and is the immediate decay product of Radium-226. Thoron is Radon-220 (i.e., two fewer neutrons). It is a decay product of Thorium-232.[25]

Jaki and Hess published some of the key results from their research in the *Journal of Geophysical Research*. Summarizing their work, they write:

> The respective concentrations of radon, thoron, and their decay products have been measured immediately above the earth and below the ground, and a study of their correlation with meteorological conditions has been carried out. It was found that the concentration of radon and thoron in the vicinity of the earth's surface depends primarily on the dryness of the ground. Measurements at and below the ground level were carried out by an indirect method, collecting the decay products of radon and thoron on the inner surface of an aluminum cylinder charged to −600 volts. From November 16, 1956, to May 7, 1957, 239 experiments were made. It was found that the concentration of thoron below the ground seems to reach a maximum value at a depth of about 75 cm, whereas the concentration of radon shows a continuous increase within the range investigated. Comparison with other experiments shows that the cylinder method can be used for investigating the exhalation of both radon and thoron at the surface of the ground. The mean value obtained for radon is $6 \times 10-^{18}$ curie per cm^2 sec. The measurements

> presented here give the first data on the exhalation of thoron through the air-soil interface.[26]

This research shows that thoron is a non-trivial contributor to radiation at the source of the Earth. It also demonstrated that while precipitation, e.g., rain and snow, significantly interfered with the release of radioactive effluent, subsequent evaporation of the precipitation contributed to elevated levels of atmospheric radon and thoron. Soil type also affects the emission of Radon and Thoron. Anthony Troha points out that even though radon remains a serious health hazard, few studies have been conducted since Jaki's seminal work, which remains widely cited in government reports and environmental journals.[27]

Jaki's Epistemology and Metaphysics

David L. Schindler, wisely notes that "a theology which begins from within faith in the Christian God must concern itself as a matter of inner principle with the reality of nature and the nature of reality."[28] Paul Haffner sagely remarks that realism "is the metaphysical bridge which guarantees the true relation between the mind and reality and is thus the right approach to link reason with Christian belief in God."[29] The metaphysical realist acknowledges the following order of philosophical discovery:

1. Being precedes perception;
2. Perception precedes apprehension;
3. Apprehension precedes judgment; and
4. Judgment precedes reasoning.

Stanley Jaki is widely known for his metaphysical realism, inspired by the perennial insights in ontology, causality, and epistemology by Aristotle and Thomas Aquinas. Jaki is certainly a member of the twentieth century Neo-Thomist tradition, which includes a very broad range of thinkers such figures as Yves Congar, Bernard Lonergan, Karl Rahner, Karol Wojtyła, and many others who attempted to develop and apply the masterful intuitions and reasoning of Aquinas in light of contemporary philosophical and theological concerns.[30] Jaki's Thomist realism, which he views as enormously important for both Catholic theology and natural science, has been widely studied.[31]

With regard to the philosophy of science, Haffner writes, "Jaki is firmly in the realist tradition which he defends against prevailing idealist and operationist interpretations of science. Beyond all the constructs and symbolism, human understanding reaches to reality itself, albeit in way which will grow and develop as science progresses."[32] Jaki himself states:

> It is a fundamental shortcoming of science that on its exact and formal level it gives the appearance of being severed from that reality which is a vast network of events standing in a causal relation. Yet, while science may and should appear in that sense severed from reality, science becomes an illusion if that apparent severance is declared to be real.[33]

He is very critical of the post-Kantian critique of knowledge. Jaki argues,

> That even the fact, let alone the nature, of external reality, however ordinary, cannot be proven by mere logic or mathematical formulas does not make one's immediate registering of external reality a less than fully rational process. To know the existence of things is in fact the very first step in reasoning. Any critical knowledge or philosophy which does not accept this will remain a mere criticism of criticism and not a criticism of the external reality one registers, and not even one's own registering it.[34]

Jaki also emphasizes that "for Aquinas it is natural for man to be in a cognitive unity with nature."[35] He explains that in "all major advances in the exact sciences, or physics, a philosophy was used, that of moderate realism, which is also the philosophy implied in the classical proofs of the existence of God."[36] As Maurice De Wulf puts it, moderate realism "reconciles the characteristics of external objects (particularity) with those of our intellectual representations (universality), and explains why science, though made up of abstract notions, is valid for the world of reality."[37]

As mentioned earlier, Jaki's realism is highly congruent with that of Gilson. With regard to the critical problem introduced by Kant, Gilson writes,

> The critique of knowledge is essentially incompatible and irreconcilable with metaphysical realism. There is no middle ground. You must either begin as a realist with being, in which case you will have a knowledge of being, or begin as a critical idealist with knowledge, in which case you will never come in contact with being.[38]

Gilson states, "a realist theory of knowledge and a realist critique of knowledge are both possible and necessary, but neither the one nor the other is equivalent to critical realism, and they should not be called by that name."[39] He argues that the fundamental problem is the impossibility of reconciling a critical epistemology that systematically suspends belief in the world outside the mind with classical realism.[40] "Common sense spontaneously affirms that normal perception attains real objects."[41] In response to Thomists who sought to construct a critical realism, Gilson points out that "St. Thomas considers the existence of the external world to be self-evident."[42]

Jaki attempts to diagnose the origin of this "critical" problem. He observes that in post-Kantian philosophy, the "universe" is often reduced to a mere word, if it is employed at all. Jaki writes:

> The word "universe" had no place in the positivism of Auguste Comte and of J.S. Mill. The pluralistic universe of W. James's pragmatism is just as illogical as is the universe within Dewey empiricism. There is no room for the universe in existentialism and in phenomenology. Whatever else the universe may be, it is not a phenomenon. And insofar as the universe means coherence, it is irreconcilable with the radical separation among all events as postulated by Sartre. It should be no surprise that deconstructionists can at most repeat Kant's false arguments about the universe. As if to prove how far they are behind times, they are the least willing to construct a philosophy about the universe in this age of scientific cosmology.[43]

Jaki also notices that the "universe" in Berkeleyan and Hegelian idealism is reduced to a mental entity. The Dutch mathematician, Willem de Sitter, in the context of discussing the formal construction of scientific laws as the sole true reality, states that the universe is a mere mathematical formula.[44]

Jaki writes, "The recognition by the human mind of the existence of the Creator by means of created things implies a philosophical framework which makes no sense unless there is a totality of things, real, orderly and contingent, which is the notion of a universe proclaimed by the dogma of creation and claimed by science as well."[45] Elsewhere he states that "it was the Biblical doctrine of creation, especially in its Christological form that God created everything through his Son, that prompted the notion of an autonomous, though contingent set of laws of nature that can and should be investigated by science."[46]

Jason Mitchell's extensive investigation into Jaki's metaphysics identifies three significant aspects of Jaki's thought:

1. the value and limits of science in relation to philosophy;
2. the nature of a non-Thomist Thomistic metaphysics and the current tasks of Thomists with regard to science; and
3. the epistemological importance of moderate realism.[47]

Like all scholars, Jaki's philosophical views develop with time. One of his later books, *Means to Message*,[48] presents his philosophical system at the close of the twentieth century. As Mitchell puts it, the book "constitutes a clear confirmation of the Jaki's endorsement of both Thomistic metaphysics and realist epistemology."[49]

Surprisingly, and provocatively, just four years later Jaki published an article entitled "A Non-Thomist Thomism."[50] How can this be? Mitchell's thorough study clearly documents how Jaki accepts most of the major principles of Thomistic metaphysics such as

1. the analogy of being,
2. the distinction between substance and accidents,
3. the distinction between prime matter and substantial form,
4. the distinction between essence and the act of being in creatures,...[and]
5. the Thomistic definition of truth as *adæquatio rei et intellectus*.[51]

In "A Non-Thomist Thomism," Jaki offers a critique of some aspects of Thomism, particularly those emphasized in the Neo-Scholastic Thomism that he was taught as a seminarian. For example, Jaki criticizes the ordering of the Pius X's Thomistic Theses[52] and the fact that Aristotle's *Physics* is not physics in the contemporary sense of the term, i.e., a quantitative, exact science.[53] On this issue, Mitchell writes:

> On the one hand, Jaki is right in arguing that Thomistic metaphysicians need to recognize the limitations of the Aristotelian scientific and cosmological world-view and to recognize the consequences that derive from these limitations. On the other, I believe that Jaki's proposal can be improved upon by a stronger recognition of philosophy's legitimate autonomy and by a stronger appreciation of role of the philosophy of nature: namely, there are perennial (Thomistic) philosophical principles—such as act and potency, essence and esse, prime matter and substantial form—which withstand the test of time and the progress made in physics through the centuries.[54]

Ultimately, in his 2003 article, when commenting on *Means to Message* and his 2002 intellectual biography, *A Mind's Matter*, Jaki writes, "I think that my philosophy, however non-Thomist in appearance and in its procedures, is a very Thomist one."[55] *Res judicata.*

Ian G. Barbour

Ian Graeme Barbour was born on October 5, 1923 in Peking (now Beijing), China. His parents, George Brown and Dorothy Barbour were Protestant lay missionaries and professors at Yenching University. George, a Scottish Presbyterian, taught in the Department of Geology and Dorothy, an American Episcopalian, taught in the Department of Religious Education. George Barbour was a colleague and close friend of the Jesuit paleontologist, Pierre Teilhard de Chardin.[56]

The Barbour family left China in 1931 due to Ian's brother, Hugh, developing a life-threatening lung infection. Hugh received medical treatment in the United States and recovered. The rise of Maoism in China prevented the Barbour family from ever returning to China. Faculty positions, like work in general, were hard to come by during the Great Depression. George Barbour had to move the family often as he took various teaching positions. While George was teaching in London, Ian attended a prestigious Quake boarding school. After returning to the USA, he attended high school at Deerfield Academy in Massachusetts. Ian Barbour began undergraduate studies in physics at Swarthmore College just as World War II was beginning. After graduation, he registered as a conscientious objector and joined the Civilian Public Service program.[57]

Following the Allied victory against the Axis powers, Barbour earned his master's degree in physics from Duke University and met his wife, Deane Kern. The couple moved to Illinois where Ian began a PhD in physics at the University of Chicago and Deane pursued graduate studies at Chicago Theological Seminary. Ian studied nuclear physics with Enrico Fermi and quantum mechanics with Edward Teller. Barbour's own research focused on the use of photographic emulsions to study cosmic-ray mesons. His experimental apparatus included a 70-pound magnet which was brought to high altitude by a helium-filled balloon in order to deflect mesons.[58]

His first faculty position was as assistant professor of physics at Kalamazoo College. In 1953, a Ford Foundation Fellowship enabled Barbour to study at Yale Divinity School, a bastion of the Neo-Orthodoxy movement. After two years, Barbour earned the Bachelor of Divinity degree, the precursor to the contemporary Master of Divinity. He subsequently changed institutions and received a joint appointment in the Departments of Physics and Philosophy at Carleton College in Northfield, Minnesota. This would become his academic home for the rest of his life. In 1960 Carleton College created a free-standing Department of Religion and Barbour was named the chairman. He would later be honored with an endowed chair as the Winifred and Atherton Bean Professor of Science, Technology, and Society.[59] Barbour died on Christmas Eve in 2013 at the age of 90 with sixteen books and numerous scholarly articles to his name.[60]

Barbour's Early Scientific Work

Ian Barbour's first research project as a graduate student at Duke was in the area of applied physics, what would be called today "biomedical engineering." Barbour's supervisor, Hans Löwenbach, MD, was a professor of neurology and psychiatry. Löwenbach introduced Barbour to the challenge of quantitatively analyzing electroencephalograms (EEGs), i.e., the recording of the brain's electrical activity over time. Up until that point, the analysis of EEGs was limited to qualitative interpretation.[61] Barbour engineered a device to display graphically a plot of the amplitude of each component sine wave against its frequency. The resulting system allowed a neurologist to monitor a patient's brain wave pattern as well as the amplitudes of the various frequencies in the EEG. Particular brain disorders were found to have distinct patterns when analyzed in the frequency domain. Barbour's device enabled neurologists to make rapid diagnoses in potentially life-threatening situations, e.g., encephalitis.[62]

When Barbour started doctoral studies at the University of Chicago, he joined the cosmic ray research group led by Marcel Schein. The high energy particle accelerators that exist today were not yet developed, so energetic nuclear particles were studied primarily through observations of cosmic radiation. Barbour and his colleagues studied the variations in the number and kinds of nuclear particles at different altitudes within Earth's atmosphere. Their important research showed that the observed phenomena could not be explained with the existing understanding of nuclear processes and their fundamental particles.[63]

Barbour's PhD project was to develop an experimental apparatus in which two emulsions were mounted with an air gap between them in a magnetic field generated by a very strong magnet from a World War II inspired klystron.[64] The device was brought by a balloon to an altitude of 90,000 feet. By measuring the deflection of a charged particle in the magnetic field, as well as the distance it traveled, Barbour computed the electric charge and mass.[65] Where research teams at the University of Bristol and UC Berkeley had failed, Barbour's experimental setup successfully measured the mass and electric charge of a newly discovered particle, the pion, as well its decay product, the muon.[66] Commenting on this discovery, Carl M. York writes: "Not only did these measurements yield corroborating evidence of the properties of these particles, but they verified that pions are produced by the cosmic rays in equal numbers of positive and negative charges in the atmosphere."[67]

During his time on the faculty at Kalamazoo College, Barbour continued his research on cosmic rays using particle tracks through emulsion plates. He measured nuclear interactions at high energies in various elements. Before modern particle accelerators, Barbour's work provided a great deal of data, that had never been available previously, on a significant number of different elements. After his move to Carleton College, Barbour chose to participate in the International Geophysical Year 1958 by measuring the neutrons produced by cosmic rays in the atmosphere over Minnesota.[68]

Barbour's Epistemology and Metaphysics

Like Jaki, Ian Barbour is a philosophical realist. He views theories as representations of the real world. The inaugural issue of *Zygon: Journal of Religion & Science* in 1966 introduced Barbour's notion of critical realism.[69] *Issues in Science and Religion* brought this epistemology to a much larger audience. Throughout his scholarly works, Barbour invokes his approach of critical realism, although, without specifying it in a formal, rigorous way. Barbour argues that practicing scientists "usually assume realism in their work."[70] Like Jaki, he is highly critical of

1. the *positivist* position that views theories as merely summaries of data,
2. the *instrumentalist* position that sees theories as simply useful tools, and
3. the *idealist* position that reduces theories to mental structures.

Barbour writes:

> Against the positivist, the realist asserts that the real is not the observable. Against the instrumentalist, he affirms that valid concepts are true as well as useful. Against the idealist, he maintains that concepts represent the structure of events in the world. The patterns in the data are not imposed by us, but originate at least in part in objective relationships in nature. The object, not the subject, makes the predominant contribution to knowledge. Hence science is discovery and exploration, not just construction and invention.[71]

Like Jaki, Barbour also critically engages the idealist legacy of Kant. In particular, Barbour argues against the

Neo-Kantianism of Ernst Cassirer, Henry Margenau, and Carl Friedrich von Weizsäcker. For example, in *The Nature of Physical Reality,* Margenau writes, "I am perfectly willing to admit that reality does change as discovery proceeds. I can see nothing basically wrong with a real world which undergoes modifications along with the flux of experience."[72] Barbour points out the absurdity of Margenau's assertion that the neutron did not exist until it was "invented" in 1932.[73]

Barbour's "realism insists that being is prior to knowing."[74] Given his training in experimental physics, and considering the challenges associated with studying atomic and sub-atomic phenomena, Barbour suggests that "direct perceptibility" cannot be the definition of that which is real. He argues that "intelligibility" may be considered the fundamental characteristic of what is real. Barbour writes that "it is precisely the organizing power of theoretical structures which shows that they correspond to the structure of the world."[75] In Whitehead's process thought, which Barbour largely adopts, while the object is given priority over the subject, the role of the subject is not neglected because:

1. reality consists not of things but of events occurring in networks of relations which include both the knower and the known;
2. knowledge arises not from either subject or object alone but from a situation of mutual interaction; and
3. scientific language is symbolic, deriving from the subject's selective abstraction from the total situation.[76]

Barbour's critical realism views scientific theories as representations of an objective, external world. Barbour states that a theory is "valid" for a scientist if it is both true and useful. Always concerned about the accusation that his realism was "naive," Barbour clearly states that all scientific theories are incomplete and selective, i.e., they describe particular aspects of the natural world for specific purposes. Thus, with his epistemology he attempts to avoid both the errors of literalism and fictionalism in theoretical models.[77] Barbour often repeats in his works the claim that a critical realism takes models seriously, but not literally.

Unfortunately, Barbour's critical realism could be interpreted as an attempt to elevate potency over act. Barbour posits that being precedes subjective knowledge as "process" or "fluid event," not as actuality. This implies that knowing, even for God, is a momentary actualization of what is first only potentially true.[78] In addition, "being as process" can never be fully known. The mind grasps only changing parts, not the actual whole. Thus, a Kantian tendency remains in Barbour's epistemology. His epistemology suggests that being is only known through diverse models that provide momentary glimpses into what is "real," but is in fact always incomplete and fundamentally unintelligible. Barbour's philosophy of knowledge implies that the mind actualizes "reality" through a process of integrating otherwise separate and unintelligible data. Like Heraclitus of old, Barbour frames science itself as a process of becoming. If being is not one prior to being known, and only becomes one through finite knowing, all reality is reducible to the process of coming to know.[79]

Barbour's critical realism places significant emphasis on the subjective character of knowledge. He writes, "For when human interpretation is present, there is always the possibility of misinterpretation, especially through wishful thinking, which reads into experience more than is warranted. Moreover, any verbal communication requires the use of symbolic language and conceptual structures that are culturally conditioned."[80] His "epistemic humility" may seem reasonable in light of the tentative character of scientific theories. However, this attitude can become very problematic for Christian theology.

Barbour appreciates what he describes as the "rich and many-faceted" approach to Divine causality found in Neo-Thomists such as Gilson.[81] However, he ultimately commits himself to the process metaphysics of Whitehead. In his Lowell Lecture of 1925, Whitehead states that "nature is a structure of evolving processes. The reality is the process."[82] Whitehead went on to develop an entirely new metaphysics. Originally presented at the Gifford Lectures, it would take its published form in the 1929 book *Process and Reality*.

Barbour adopts, adapts, and in many ways simplifies Whitehead's exceptionally detailed and complex metaphysics. He identifies four aspects of Whitehead's system that he sees as particularly consistent with modern science and what he characterizes as its evolutionary, many-leveled view of nature:

1. the primacy of time,
2. the interconnection of events,
3. reality as an organic process, and
4. the self-creation of every entity.[83]

The primacy of time encapsulates the process view that "becoming" is more significant than "being." Barbour's process view of nature can be viewed as a dynamic network of particular "moments of experience," corresponding to Whitehead's "actual entities" or "actual occasions." Barbour slightly modifies the terminology by using the term "entities," to highlight their integration, or "events," to highlight their temporality. Barbour drastically simplifies Whitehead's complex notion of causality into three steps:

> 1. Every new entity is in part the product of efficient causation, which refers to the influence of previous entities on it. Objective "data" from the past are given to each present entity, to which it must conform, but it can do so in alternative ways.
> 2. There is thus an element of self-causation or self-creation, for an entity unifies its "data" in its own manner from its unique perspective on the universe. Every entity contributes something of its own in the way it appropriates its past, relates itself to various possibilities, and produces a novel synthesis that is not strictly deducible from its antecedents.
> 3. Thus a creative selection occurs from among alternative possibilities in terms of goals and aims, which is final causation.[84]

Barbour avers that every unique event in the universe is to be understood as a "present response (self-cause) to past entities (efficient cause) in terms of potentialities grasped (final cause)."[85]

It must be mentioned that there are many problematic aspects to process thought from a Thomistic, Catholic perspective.[86] For example, one weakness of process metaphysics is that it considers final causes as potential, not actual. It also does not account for *esse simplex*—the simple to be, the act of existing.[87] *Esse simplex* involves total stability in God. In creatures, it is the foundational stability of a nature, or essence, with the possibility of the actuality of potencies. Barbour's process metaphysics does not acknowledge God as completely perfect actuality. His process ontology implies a problematic elevation of activity over substance.[88]

Stanley Jaki observes that although the subtitle of Whitehead's famous book, *Process and Reality*, mentions the term "cosmology," it never deals with the cosmos or universe *per se*! Whitehead does not acknowledge the essential fact that the universe exists, i.e., that there is a strict totality of things. Jaki writes:

> Even the last chapter, "God and the World," is far more about a pantheistic God, who co-acts and even co-suffers with the world, than about the world itself. None of the half a dozen references in that book to relativity are to its cosmology...Whitehead gave away something of his debt to Bergson and revealed much of the self-defeating tactic whereby purpose and humanness are reclaimed through a vitalization of the cosmos.[89]

For an in-depth evaluation of Barbour's epistemology and metaphysics, see *Theology and Science in the Thought of Ian Barbour: A Thomistic Evaluation for the Catholic Doctrine of Creation*.[90]

Evaluation and Conclusion

Born just one year apart in foreign lands, both Stanley Jaki and Ian Barbour made the USA their home. Both were devout Christians, albeit one a Roman Catholic Benedictine monk and the other a married Protestant layman.[91] Both men were experimental physicists, specializing in the field of nuclear and particle physics in the 1950s. And both pivoted early in their career into the emerging field of science and religion. Both were committed to an epistemology of realism, albeit one as a Gilsonian Neo-Thomist and another as a critical realist. Both appreciated the importance of a comprehensive metaphysics to support both natural science and Christian theology, albeit one manifestly in the Thomist tradition, and another fully committed to Whiteheadian process thought.

Barbour's work has no discernible impact on Jaki's scholarship. Research to date indicates that Jaki only engaged aspects of Barbour's corpus in his book *Lord*

Gifford and His Lectures: A Centenary Retrospect. Jaki comments on Barbour's two Gifford Lectures. Barbour's first lecture series led to the book *Religion in An Age of Science.*[92] This work investigates the impact of science and its relationship to religious life. Great attention is paid to the compatibility of theology and science. *Religion in an Age of Science* would later be revised, expanded, and re-released under the title: *Religion and Science: Historical and Contemporary Issues* in 1997.[93] With regard to Barbour's first Gifford Lecture, Jaki writes:

> Fault lines are everywhere in the lectures given by Barbour, a professor of physics as well as of religion. His primary aim was to specify the main features which religion should have in an increasingly scientific age. Those features do not include at least one which is a belief in a personal Creator. Not surprisingly, the religion Barbour pleads for is markedly impersonal, however subjective. In sum it is an amorphous religion. No less amorphous is the picture of science he paints. The sum total of fashionable views on science, no matter how extensively laid out in their mostly psychological variants, is still to qualify as a definable notion of science. Cavorting in those views may tempt non-scientists to think that it is enough to gather a great deal of enthusiasm and a lot of studied vague words about science in order to become an expert on it.[94]

Barbour's second lecture series was published as *Ethics in An Age of Technology.* It explores the ethical implications of modern engineering accomplishments, reflecting on the value of these achievements in relation to potential environmental and human detriments. On this work, Jaki writes:

> "Who speaks to whom?" This question is particularly applicable to Barbour's second series of lectures dealing with ethical problems in an age of science. The appropriateness of the question has much to do with the fashion, ever spreading in the academia, to make one's position as inclusive as possible. Operative in this process is a logic which works in a manner rather similar to the procedure whereby the baby is thrown out with the bathwater. There is no trace in Barbour's lectures of the ethical questions that most radically cut into the flesh and blood of modern individuals. His lectures seem to suggest that all ethics is equivalent to ecologically admissible, that is socially and individually comfortable, uses of science and technology both in respect to the wider as well as to the strictly personal ecosphere.[95]

Richard Liddy, Professor of Catholic Studies at Seton Hall University, who had been a Priest Community colleague of Jaki for many years, recounts Jaki's "confrontational style."[96] Stephen Barr, Professor of Physics at the University of Delaware, wrote a tribute in honor of Jaki in *First Things* following his death. Barr praises Jaki throughout, but also observes that the polemical tone in some of his writings was "doubtless one reason his work has less influence among scholars than it deserves."[97]

Perhaps Barbour's identity as a Protestant, *critical* realist, process thinker constituted "three strikes" for Jaki, and Barbour was "out" (to borrow an expression from American baseball). After Jaki's death in 2009, another one of his Seton Hall Priest Community colleagues, Thomas Guarino, Professor of Systematic Theology, preached the Memorial Mass at the Chapel of the Immaculate Conception. He praised Jaki's brilliant intellect, award-winning scholarship, as well as his life of service to the Church and the Academy. With a smile on his face, Guarino genially noted that Jaki will be remembered for his "holy irascibility in service to Truth."

Barbour shared Jaki's appreciation of history and philosophy as mediating disciplines between science and theology. He was exceptionally ecumenical and eclectic (in the best sense) in his approach to research, reading scholars ranging from Albert Einstein to Réginald Garrigou-Lagrange to Karl Barth to Alfred North Whitehead. Barbour was aware of Jaki, given that on a few occasions he references *The Road of Science and the Ways to God* in a footnote. Barbour cites that book in *Religion in an Age of Science* when discussing "methodological secularism" in science.[98] He raises this issue as well in *Religion and Science: Historical and Contemporary Issues*[99] but also cites Jaki's book when discussing the "desacralization of nature."[100] In *When Science Meets Religion*, when referring to the Patristic theology of *creatio ex nihilo*, Barbour again references the same book.[101] It is surprising that Barbour never took up any of Jaki's research themes in a substantial way, whether to express agreement or disagreement. Perhaps at the end of the day, they were not of great interest to Barbour.

Postscript

On a personal note, I have benefitted greatly from reading the works of these brilliant scholars of science and religion. As a professor, I introduce the thought of these two great Christian intellectuals to my students and hope that they too will be inspired by their quest that manifested the compatibility, and possibility of mutual enrichment, between natural science with Christian theology. Stanley Jaki and Ian Barbour leave behind a tremendous intellectual legacy. One can only imagine what they might have accomplished in collaboration.

Acknowledgments

The author is very grateful for the insights and feedback from Mr. Antonio Colombo, the Rev. Dr. Gabriel Costa, Dr. John T. Laracy, the Rev. Dr. Thomas G. Guarino, the Rev. Dr. Gerald Buonopane, and Dr. Stacy Trasancos.

Notes

1. According to the will of the nineteenth century Scottish advocate and judge, Adam Lord Gifford, the lecture series is offered annually at one of four prominent Scottish universities: the University of Saint Andrews, the University of Glasgow, the University of Aberdeen, or the University of Edinburgh. The lecture series was established to "promote and diffuse the study of natural theology in the widest sense of the term— in other words, the knowledge of God." "The Gifford Lectures," The Gifford Lectures, accessed November 20, 2017, https://www.giffordlectures.org/.
2. Established in 1972 by American-born British investment fund manager, Sir John Templeton, the Templeton Prize recognizes research that has contributed to new insights about religion. Templeton set the prize above that of the Nobel Prizes in order to highlight the importance of what he called "progress in religion." "About the Templeton Prize," Templeton Foundation, accessed February 28, 2024, https://www.templetonprize.org/templeton-prize-history/.
3. Paul Haffner, *Creation and Scientific Creativity: A Study in the Thought of S. L. Jaki*, 2nd ed. (Leominster, UK: Gracewing Publishing, 2009), 15.
4. Joseph R. Laracy, *Theology and Science in the Thought of Ian Barbour: A Thomistic Evaluation for the Catholic Doctrine of Creation* (New York: Peter Lang, 2021), 27–28.
5. See Joseph R. Laracy, "History as a Key Mediating Discipline in the Science-Religion Interaction," in *Integratio*, vol. 2 (Center for Catholic Studies Faculty Summer Seminar, South Orange, NJ, 2023), 22–28.
6. Haffner, *Creation and Scientific Creativity*, 2009, 18.
7. See Stanley L. Jaki, *Science and Creation* (Lanham: University

Press of America, 1990); Stanley L. Jaki, *The Savior of Science* (Grand Rapids: Wm. B. Eerdmans Publishing Company, 2000); Stanley L. Jaki, *Scientist and Catholic: Pierre Duhem* (Front Royal, VA: Christendom Press, 2004); Stanley L. Jaki, *The Road of Science and the Ways to God*, 3rd ed. (Port Huron, MI: Real View Books, 2005); Stacy Trasancos, *Science Was Born of Christianity* (Titusville, FL: Habitation of Chimham Publishing, 2014).

8. Ian G. Barbour, *Religion in an Age of Science* (San Francisco: Harper Collins, 1990).
9. Laracy, *Theology and Science in the Thought of Ian Barbour: A Thomistic Evaluation for the Catholic Doctrine of Creation*, 85.
10. Paul Haffner, "Chesterton's Realism and the Work of Stanley L. Jaki," *The Chesterton Review* 47, no. 3/4 (Fall/Winter 2021): 378, https://doi.org/10.5840/chesterton2021473/469.
11. Hrvoje Relja, "Il realismo di S.L. Jaki: Dalla convinzione religiosa tramite il realismo moderato e la creatività scientifica fino al realismo metodico" (PhD, Rome, Ateneo Pontificio Regina Apostolorum, 2008).
12. Haffner, *Creation and Scientific Creativity*, 2009, 19.
13. Stanley L. Jaki, *A Mind's Matter: An Intellectual Autobiography* (Grand Rapids, MI: Eerdmans, 2002), 177.
14. Étienne Gilson and Stanley L. Jaki, *Methodical Realism*, trans. Philip Trower (Front Royal, VA: Christendom Press, 1990).
15. Jaki, *A Mind's Matter: An Intellectual Autobiography*, 185.
16. Stanley L. Jaki, *Is There a Universe?* (New York: Wethersfield Institute, 1993); John-Peter Pham, "The Importance of The Reality of the Universe in the Work of Stanley L. Jaki," *Faith & Reason* 21, no. 4 (December 26, 1995): 337–54; Haffner, *Creation and Scientific Creativity*, 2009, chaps. 6–7; Jason Mitchell, "Stanley L. Jaki on Thomistic Metaphysics," in *Commemorazione Di Padre Stanley L. Jaki, OSB, Nel Primo Anniversario Della Sua Morte*, ed. Rafael Pascual and Antonio Colombo (Rome: IF Press, 2014), 53–88.
17. Mitchell, "Stanley L. Jaki on Thomistic Metaphysics," 60.
18. Laracy, *Theology and Science in the Thought of Ian Barbour: A Thomistic Evaluation for the Catholic Doctrine of Creation*, 129.
19. Ian G. Barbour, *Issues in Science and Religion* (Upper Saddle River, NJ: Prentice-Hall, 1966), 128; Alfred North Whitehead, *Process and Reality* (New York: Free Press, 1969).

20. Charles Hartshorne, *The Divine Relativity: A Social Conception of God* (New Haven, CT: Yale University Press, 1948); Charles Hartshorne, *Reality as Social Process: Studies in Metaphysics and Religion* (New York: Free Press, 1953); Ian G. Barbour, *Religion and Science: Historical and Contemporary Issues* (San Francisco: Harper, 1997), 294.
21. Antonio Colombo, "From Győr to Madrid: A Biographical Sketch of Father Stanley Jaki," in *Stanley Jaki Foundation International Congress*, ed. Paul Haffner and Joseph R. Laracy (Leominster, UK: Gracewing Publishing, 2020), 193–95.
22. Colombo, 196.
23. Anthony L. Troha, "The Early Scientific Work of the Rev. Dr. Stanley L. Jaki," in *Stanley Jaki Foundation International Congress*, ed. Paul Haffner and Joseph R. Laracy (Leominster, UK: Gracewing Publishing, 2020), 14–15.
24. Haffner, *Creation and Scientific Creativity*, 2009, 15–17.
25. For more details, see Walter D. Loveland, David J. Morrissey, and Glenn T. Seaborg, *Modern Nuclear Chemistry*, 2nd ed. (Hoboken, NJ: Wiley, 2017).
26. Stanley L. Jaki and Victor F. Hess, "A Study of the Distribution of Radon, Thoron, and Their Decay Products Above and Below the Ground," *Journal of Geophysical Research (1896–1977)* 63, no. 2 (1958): 373, https://doi.org/10.1029/JZ063i002p00373.
27. Troha, "The Early Scientific Work of the Rev. Dr. Stanley L. Jaki," 22.
28. David L. Schindler, "Theology, Science, and Cosmology," *Communio* 15, no. 3 (Fall 1988): 270.
29. Paul Haffner, *The Mystery of Reason* (Leominster, UK: Gracewing Publishing, 2001), 12.
30. Thomas Joseph White, "Thomism after Vatican II," *Nova et Vetera* 12, no. 4 (2014): 185–202; Tracey Rowland, *Catholic Theology*, Doing Theology (New York: T&T Clark, 2017), 43–90.
31. Relja, "Il realismo di S.L. Jaki: Dalla convinzione religiosa tramite il realismo moderato e la creatività scientifica fino al realismo metodico"; Paul Haffner, "Creation and Scientific Creativity," in *Commemorazione Di Padre Stanley L. Jaki, OSB, Nel Primo Anniversario Della Sua Morte*, ed. Rafael Pascual and Antonio Colombo (Rome: IF Press, 2014), 79–100; Mitchell, "Stanley L.

Jaki on Thomistic Metaphysics"; Paul Robinson, *The Realist Guide to Religion and Science* (Leominster, UK: Gracewing, 2018).

32. Haffner, "Creation and Scientific Creativity," 2014, 91.
33. Jaki, *The Road of Science and the Ways to God*, 275.
34. Jaki, *Is There a Universe?*, 108–9.
35. Jaki, *The Road of Science and the Ways to God*, 37.
36. Stanley L. Jaki, "The Demarcation Line between Science and Religion," *Angelicum* 87, no. 1 (2010): 81.
37. Maurice De Wulf, "Nominalism, Realism, Conceptualism," in *The Catholic Encyclopedia* (New York: Robert Appleton Company, 1911), http://www.newadvent.org/cathen/11090c.htm.
38. Étienne Gilson, *Thomist Realism and the Critique of Knowledge*, trans. Mark A. Wauck (San Francisco: Ignatius Press, 2012), 149.
39. Gilson, 170.
40. Balthasar and Przywara highlight an additional issue. These Catholic philosophers point out that being is not first "being out there" which is then known in a second moment. Rather, being is the intelligible fullness of act. Consequently, there must be a foundational "polarity" of object and subject in being. Otherwise, a danger may arise of slipping back into the idealistic notion that objects are potency for an actualizing subject. "Objectivism" and "subjectivism" can be viewed as two sides of the same dialectical coin. See Erich Przywara, *Analogia Entis: Metaphysics*, trans. John R. Betz and David Bentley Hart (Grand Rapids, MI: Eerdmans, 2014), I.
41. Gilson, *Thomist Realism and the Critique of Knowledge*, 124.
42. Gilson, 100.
43. Stanley L. Jaki, *The Limits of a Limitless Science and Other Essays* (Wilmington, DE: ISI Books, 2000), 110.
44. Willem de Sitter, *Kosmos* (Cambridge, MA: Harvard University Press, 1932), 133–34.
45. Stanley L. Jaki, *Cosmos and Creator* (Edinburgh: Scottish Academic Press, 1980), 86.
46. Jaki, "The Demarcation Line between Science and Religion," 81.
47. Mitchell, "Stanley L. Jaki on Thomistic Metaphysics," 55.
48. Stanley L. Jaki, *Means to Message: A Treatise on Truth* (Grand

Rapids, MI: Eerdmans, 1999).

49. Mitchell, "Stanley L. Jaki on Thomistic Metaphysics," 53.
50. Stanley L. Jaki, "A Non-Thomist Thomism," *Sensus Communis* 4, no. 1 (March 2003): 75–88.
51. Mitchell, "Stanley L. Jaki on Thomistic Metaphysics," 57.
52. St. Pius X, *Postquam Sanctissimus*, 1914.
53. Stanley L. Jaki, *A Late Awakening and Other Essays* (Port Huron, MI: Real View Books, 2006), 27.
54. Mitchell, "Stanley L. Jaki on Thomistic Metaphysics," 60.
55. Jaki, "A Non-Thomist Thomism," 87.
56. Ian G. Barbour, "A Personal Odyssey," *Theology and Science* 15, no. 1 (January 2, 2017): 5, https://doi.org/10.1080/14746700.2016.1265215.
57. Barbour, 6.
58. Barbour, 6.
59. Barbour, 7.
60. William Yardley, "Ian Barbour, Who Found a Balance Between Faith and Science, Dies at 90," *The New York Times*, January 12, 2014, sec. U.S., https://www.nytimes.com/2014/01/13/us/ian-barbour-academic-who-resisted-conflicts-of-faith-and-science-dies-at-90.html.
61. Carl M. York, "Ian Barbour's Contributions as a Scientist," *Theology and Science* 15, no. 1 (January 2, 2017): 19, https://doi.org/10.1080/14746700.2016.1265228.
62. Hans Löwenbach and Ian G. Barbour, "An Automatic Device for Continuous Frequency Analysis of Electroencephalograms," *Federation Proceedings* 5, no. 1 (1946): II:65; Ian G. Barbour, "An Automatic Low Frequency Analyzer," *Review of Scientific Instruments* 18, no. 7 (July 1, 1947): 516–22, https://doi.org/10.1063/1.1740990.
63. York, "Ian Barbour's Contributions as a Scientist," 17.
64. A klystron is specialized linear-beam vacuum tube used as an amplifier for high radio frequencies. It was invented by two brothers, Russell and Sigurd Varian. For more details, see Russell H. Varian and Sigurd F. Varian, "A High Frequency Oscillator and Amplifier," *Journal of Applied Physics* 10, no. 5 (May 1939): 321–27, https://doi.org/10.1063/1.1707311.
65. York, "Ian Barbour's Contributions as a Scientist," 21.
66. Ian G. Barbour, "Magnetic Deflection of Cosmic-Ray Mesons

Using Nuclear Plates" (PhD, Chicago, University of Chicago, 1950).

67. York, "Ian Barbour's Contributions as a Scientist," 21.
68. York, 22–24.
69. Ian G. Barbour, "Commentary on Theological Resources from the Physical Sciences," *Zygon* 1, no. 1 (March 1966).
70. Barbour, *Issues in Science and Religion*, 171.
71. Barbour, 168–69.
72. Henry Margenau, *The Nature of Physical Reality*, 2nd ed. (New York: McGraw-Hill Book Company, 1950), 295.
73. Barbour, *Issues in Science and Religion*, 168.
74. Barbour, 169.
75. Barbour, 170.
76. Barbour, 171.
77. Ian G. Barbour, *Myths, Models, and Paradigms* (New York: HarperCollins College Division, 1974), 37.
78. Barbour, *Religion and Science*, 294.
79. Laracy, *Theology and Science in the Thought of Ian Barbour: A Thomistic Evaluation for the Catholic Doctrine of Creation*, 168–69.
80. Barbour, *Issues in Science and Religion*, 210–11.
81. Barbour, *Religion and Science*, 310.
82. Alfred North Whitehead, *Science and the Modern World* (New York: The Macmillan Co., 1931), 106.
83. Barbour, *Religion and Science*, 285.
84. Barbour, 286.
85. Barbour, 286.
86. Leo A. Foley, *A Critique of the Philosophy of Being of Alfred North Whitehead in the Light of Thomistic Philosophy* (Washington, DC: The Catholic University of America Press, 1946).
87. According to James Doig, "physics must suppose that there is a mobile body, and mathematics that there is quantified or discrete continuum; metaphysics precisely by studying the *esse simplex* of everything, gives to the other two sciences their objects and their principles." James C. Doig, *Aquinas on Metaphysics: A Historico-Doctrinal Study of the Commentary on the Metaphysics* (The Hague: Martinus Nijhoff, 1972), 78.
88. For more detail, see Laracy, *Theology and Science in the Thought of*

Ian Barbour: A Thomistic Evaluation for the Catholic Doctrine of Creation, 218–32.

89. Jaki, *Is There a Universe?*, 70.
90. Laracy, *Theology and Science in the Thought of Ian Barbour: A Thomistic Evaluation for the Catholic Doctrine of Creation*, chap. 4.
91. Barbour's father was a Scottish Presbyterian and his mother an Episcopalian. He ultimately became a member of the United Church of Christ. John Barbour, The Life of Ian Barbour: An Interview with His Son, John, interview by Joseph R. Laracy, Email, November 14, 2017.
92. Barbour, *Religion in an Age of Science*.
93. Barbour, *Religion and Science*.
94. Stanley L. Jaki, *Lord Gifford and His Lectures: A Centenary Retrospect* (Edinburgh: Scottish Academic Press, 1995), 47–48.
95. Jaki, 52.
96. Richard Liddy, "Jaki and Lonergan: Confrontation or Encounter?," in *Stanley Jaki International Congress*, ed. Paul Haffner and Joseph R. Laracy (Leominster, UK: Gracewing Publishing, 2020), 78.
97. Stephen M. Barr, "Science Seeking Understanding," *First Things*, June 1, 2009, https://www.firstthings.com/article/2009/06/science-seeking-understanding.
98. Barbour, *Religion in an Age of Science*, 281 n. 37.
99. Barbour, *Religion and Science*, 339 n. 26.
100. Barbour, 347 n. 37.
101. Ian G. Barbour, *When Science Meets Religion: Enemies, Strangers, or Partners?* (San Francisco: HarperOne, 2000), 185 n. 24.

5 THE RELEVANCE OF JAKI

Peter J. Floriani, Ph.D.

Introduction

Stanley L. Jaki released his first book, *The Relevance of Physics,* in 1966. From then until his death in 2009 he wrote dozens of books, booklets, pamphlets, and journal articles, mainly on the history and philosophy of Science Writ Large.[1] At the same time he never neglected his life as an intellectual Catholic, as a priest, as a follower of St. Benedict: saying the Divine Office and offering the Holy Sacrifice of the Mass.

The vastness of his work can put him beyond the grasp of the Common Man, the Common Catholic, the Common Scientist, and even the Common Scholar. Nevertheless, considering the problems of today, both in the social sphere as well as in the ecclesial and academic, this hesitancy needs to be reversed, and reversed rapidly, since Jaki's studies provide answers to those problems: indeed, pointers to veritable treasuries crammed full of riches, or perhaps, more fittingly, armories packed with useful weapons. We need to *reveal* Jaki—"to lift him up"—for that is the literal sense of the Latin verb behind this term *relevance,* and the reason for my title.

The riddle we face—we who do read and study his work, and try to implement his ideas, or at least use them as we work—the riddle is easily stated. How do we show the relevance of Jaki?

Jaki's work is a challenge against modern society for several reasons:

First: Jaki wrote so much, because he was always a teacher. He saw his degrees in the traditional sense: he held the title *Philosophiae Doctor* in two domains, and never forgot that *doctor* means "teacher." This detail is to be seen in coordination with that command he heard on the day he was ordained as a priest:

> Sacerdotem etenim oportet offerre, benedicere, praeesse, praedicare, et baptizare.[2]

That is:

> A priest's duties are to offer sacrifice, to bless, to preside, to preach, and to baptize.[3]

Hence Jaki was sent to preach,[4] and he perceived, as far too many today fail to perceive, that in preaching Jesus Christ and Him crucified[5] he was thereby *authorized* to speak on "all the treasures of wisdom and knowledge" which that same authority declared are hidden in the heart of Jesus Christ.[6] Nor did he fail to provide that preaching: in his most academic studies he never lost sight of that treasury, nor failed to reveal it in his writing.

A second challenge is that Jaki's work spans many disciplines. It is the work of a polymath, of a scholar whose intellect is *universal*. Here it is most fitting that we apply to Jaki the title he applied to Chesterton: "The Champion of the Universe."[7] Not only did Jaki keep constantly in mind the entirety of created beings, the "totality of consistently interacting things,"[8] he tied it, as Gilson did, and St. Thomas Aquinas did, to the deepest dicta of ontology: that is, to BEING Writ Very Large, founded upon Exodus 3:14:

> "I AM WHO AM."

and linking it to this famous line in the Second Book of Maccabees (7:28):

> I beseech thee, my son, look upon heaven and earth, and all that is in them, and consider that God made them out of nothing, and mankind also ...

and to the famous "Jaki verse" in the Book of Wisdom (11:21):

> ... Thou [O God] hast ordered all things by number, and measure, and weight.

and then to John 1:3

> All things were made by Him [i.e. the Word]: and without Him was made nothing that was made.

and its culmination in Colossians 2:3:

> ... Jesus Christ, in Whom are hid all the treasures of wisdom and knowledge.

Note that each specifies *all.*

Jaki examines this explicitly,[9] but the universality obtained therein, that is, the rational foundation it provides, thereby authorizing ALL of Science Writ Large, is strewn throughout his writing. You and I and other eager readers can perceive them since we are familiar with his work, his style, and his sources, but they escape the notice of the casual reader.

A third challenge is Jaki's very meticulous academic style. Here, we refer not only to his constant use of scholarly diction as found in journal articles,[10] but his rigorous, pedantic, even hyper-pedantic, citations of the work of other authors.[11]

It is only too common for students to miss Jaki's point, for readers to gloss over Jaki's careful work: to ignore details he went to great trouble to verify and then set forth—and their grief in reading Jaki may be summarized in the single word *footnote.*[12]

Certain readers will pick up a Jaki book and groan almost immediately when they see the number of footnotes Jaki uses.[13] That is rather funny, since a serious scholar rejoices in them: Jaki has carefully provided, according to the very best scholarship, references to all his sources,[14] and emphasizes this style in *A Mind's Matter*.[15]

None of these points—the vastness of Jaki's work, his universality, and his scholarly style—are insurmountable. We shall not solve the riddle by attempting to make "an American translation" of a work deemed unreadable in its original form, omitting all the hard words, the citations to obscure sources, the non-English quotes, shortening sentences, and paraphrasing, truncating, or omitting the intricate parts.

Nor can we "gut" its universality, so as to make it "politically correct." Jaki did not hesitate to mention[16] how copies of his *Science and Creation* were burned outside one of his lectures.

Alternatively, to prune out the ubiquitous Christocentric thought—or rather Christo-Marian-centric thought, of a devout writer. Nor dare we omit (as other writers have omitted) all "scholarly apparatus" which justly links a work to its authorities. Jaki's brilliance is only magnified by his humble reliance on the scholars he cites. Maybe that's why some fear his work.

Having noted these points and commented on them, what do we do?

Taking the three points in reverse order, there is little which needs to be said about Jaki's academic style. Certainly we can and should produce new studies, reviews, and cross-references to aid new readers and even dedicated scholars in using Jaki's works. Such results are what one expects from scholars, and our collaboration today, with our gracious hosts, is an important step in that work.

Secondly, the issue of universality. Jaki's careful cosmos-spanning scholarship provides definite boundaries which should result in a "good quarrelsome, manly friendship."[17] Far more difficult than the division between

Islam and Catholicism is the one dividing the Orthodox from the Roman rite. At this point some may want to go outside and burn *my* books, but it is worth my mentioning because I perceive Jaki's own tensions about it. Even to *begin* to address the matter would no doubt require a book, not a paper—but since it matters to me also, I shall point out one or two details which might otherwise escape notice.

For example, in one of his books on the Papacy Jaki cites St. John Chrysostom as saying Peter was the "mouth of the apostles," the "conductor of the apostolic choir," the "fisherman of the world who fished for the whole world," nay, the "ruler of the entire world."[18] I certainly do not mean this as a rude provocation of my Eastern brothers, for whom I wish to tender my deep respect. Rather I hope that it may stimulate the fruit Father Jaki and I desire: fitting studies of all his works, even sensitive ones. Perhaps even more powerful (and more sharp edged for some readers) is Jaki's mention of that same great theologian in his study of priestly celibacy—another book well deserving of study, no matter how critical, as such a work deserves, only providing that it is honest and scholarly.

More on the Eastern Churches is found in Jaki's *Gist of Catholicism*,[19] and in that excellent chapter 8, "The Leaven of Confidence" in his *Science and Creation*.[20] St. Basil comes up in various places, and St. Athanasius is the one most often mentioned, as he was the arch-foe of the Arians. Do not forget Jaki's important study, *Neo-Arianism as Foreseen by Newman*.[21]

My brothers, we must not lose sight of the fact that we have enemies, enemies who never cease their efforts to keep us from following our Lord. Let us seek that unity our Lord spoke of,[22] and take up, not as weapons but as tools, the universality of Jaki's works.

Thirdly, let us consider the vastness of Jaki's work, by considering an alternative classification by topic. (This is by no means any criticism of the work of previous scholars.)

1. Catholicism

It would be absurd to suggest that this scholar, who wrote with great impartiality (despite the claims of certain book-burning groups) about the beliefs of others, was not a Catholic, a man who did not neglect his faith or his vocation as an ordained priest and a spiritual son of St. Benedict. Unlike other writers of our time, even when he is critical he never fails to augment the Church by returning to reminders of what it was founded for, and Who it was founded by—and why it was founded. He reminds us of the decades-long programmatic omission of original sin and the devil[23] which he corrects by reminding us that "the four secondary consequences of original sin are: darkening of the intellect, weakening of the will, suffering and death."[24]

Now, let us consider several subclasses:

1.a The Papacy and Ecclesiology.

Jaki has two excellent books: *And On This Rock*, and *The Keys of the Kingdom*. Not only are these brilliant scriptural studies of the two great Signs of the Papacy, they provide theological, historical, and mystical insights. Just one tiny detail, a fact that I have never heard from any pulpit, nor explained in any commentary, of that rather odd declaration in Isaiah that

> ... I will lay the key of the house of David upon his shoulder: and he shall open, and none shall shut: and he shall shut, and none shall open.[25]

Jaki provides a reprinted illustration[26] of a man of Cairo, seen from behind, with what looks like several hockey sticks on his shoulder. As Jaki explains in some detail, an ancient form of key was a long stick which looked uncannily like our modern hockey stick, and it was carried, as I have seen young hockey players do, on the shoulder in just that way. Whoever held the key on his shoulder held the right to grant or forbid access to the entire edifice that key unlocked.

The same may be said about that odd little reference which flies past the man in the pew just as fast:

> And Jesus came into the quarters of Caesarea Philippi ...[27]

Who these days knows what that means? Yet Jaki shows us that mighty *rocky* background to those solemn words which gave Simon a new name.

1.b Original Sin.

Even when the writers of "Catholicism Lite" mention "sin" they leave things out, most grievously by adopting the Lennonistic heresy under the flag of "unconditional love." Thus, Jaki's booklet, *Original Sin?*, is an essential antidote. The term comes up very often in both *The Savior of Science* and (as one might expect) *The Theology of Priestly Celibacy*—and it is scattered throughout his writing. One of his most pungent warnings is summarized in this sentence:

> Thus it became a trend in the new or liberal theology to claim that original sin cannot be closely tied to anything concrete in human experience, be it ignorance, concupiscence, suffering, or death.[28]

1.c The Holy Sacrifice of the Mass and the Holy Eucharist

This is tied to the previous term (*sin*) in profound ways, and his booklet *Why the Mass?* gives a brief overview of this central and most profound element of Catholicism. Among other places, we should again mention his *The Theology of Priestly Celibacy*. There is also this very powerful paragraph which begins to smack of St. Athanasius *contra mundum*: "With respect to the Eucharist the liberal theologian emphasizes the natural externals of the sacrament as a banquet. The Eucharist is certainly a meal, even a banquet, but a sacrificial one, the very point which the liberal theologian is not going to emphasize ... The doctrine of transubstantiation is naturally too supernatural for the taste of the liberal theologian. While he cannot directly deny it, he will make much of the philosophical objections to the Aristotelian doctrine of substance, and will promote a ritual in which acts of adoration, such as kneeling before the sacrament, are eliminated."[29] What the Prayer Book of Edward the Sixth proposed, Vatican Two commanded.

1.d Celibacy

For the third time in a row we recommend that essential study, *The Theology of Priestly Celibacy*. In sum Jaki reminds us that to leave out the Mass, to leave out Calvary, in any study of that profound and mystical topic, is not simply insane. It is diabolical.

1.e. Miracles

There is a booklet called *Miracles and Physics*, and also the exhaustive study called *God and the Sun at Fatima*. It explains a very critical point, too readily denied—or overlooked (which is worse)—about what we might call "hagiokinetics," I mean "holy motion," or the underlying aspects of that Divine Motion in which miracles occur. It to be seen in examples such as the Calming of the Storm[30] and I will paraphrase his argument[31] by saying: *The miraculous character is not diminished, but rather greatly exalted, by observing that it is a matter of absolutely perfect timing—*

especially when applied to something which apparently is beyond human prediction.

Jaki justifies this elsewhere by citing this:

> ... a personal God might just as well do things slowly as quickly, especially if, like the Christian God, he were outside time.[32]

We also note how greatly Jaki's book overthrows many decades (if not centuries) of the infernally inspired "debunking" of miracles.

1.f Scripture

We have already mentioned Jaki's *And On This Rock,* and *The Keys of the Kingdom*; there is also his *Genesis One Through the Ages,* which addresses one of the Big Three topics always raised by anyone who wants to moan about Science and Religion, that is, Galileo, Evolution, and Genesis One—and perhaps today one might also add Number Four, Artificial Intelligence.

Also his booklets on *The Virgin Birth and the Birth of Science, Themes of Psalms,* Zachariah's Canticle, Mary's Canticle, the Beatitudes, and the Parable of the Good Samaritan and the book *Praying the Psalms.* But perhaps the most choice study to be found in Jaki's works is that great chapter 7, "The Beacon of the Covenant," in *Science and Creation,* which presents an unmatched coordination between Israel and Science.

1.g Prayer

If I have read Jaki's massive master list of publications correctly, it seems that he began releasing his smaller works (booklets and pamphlets) around 2000, and he continued their production until his death. These little studies give all sorts of fascinating background, history, theology, and nuggets of his own contemplation—in total, each is a treasury of material for meditation and a resource for study. They include

- The Stations of the Cross
- The Mysteries of the Holy Rosary
- The Apostles' Creed
- The Our Father
- The Hail Mary
- The Litany of St. Joseph
- The Litany of Loreto (to the Blessed Virgin Mary)
- The Litany of the Sacred Heart
- The Litany of the Holy Name

Besides his two studies of the Psalms already mentioned which illuminate the Divine Office, he also has studies of the canticles of the cardinal hours, the *Benedictus* and the *Magnificat.*

2. The Faith (or the Church) and the Relation With Science.

Here is *the* topic, the one for which so many expect to find dealt with by Jaki. Typically everyone expects to hear how he handles the Big Three (or Four) topics we mentioned earlier Galileo, Evolution, and Genesis One—and Artificial Intelligence.

There are several major starting points: *The Road of Science and the Paths to God*, *The Saviour of Science*, *Genesis One Through the Ages*, *Cosmos and Creator*, *Universe and Creed*, *God and the Cosmologists*, and a number of his essays, not only those in his *Catholic Essays*. Dr. Trasancos has made an excellent study, *Science Was Born of Christianity*, and there are other studies which examine the relation. If I were to say more on it here, I would only be recounting some of my own hopes in future writing: it is certainly the place where even the slightest work is fruitful as well as necessary. Therefore, I would rather refer you to something consistently ignored in all that surging river of whine (that's W-H-I-N-E) which we've heard about the Big Three (or Four) topics for decades. It was written by an African Bishop named Augustine and Jaki cites it in *Science and Creation.*[33]

Next is a book I cannot recommend more heartily in this era of the nonsensical fraud called "Artificial Intelligence": that is, his second work, *Brain, Mind and Computers.*[34] There Jaki presents Babbage's argument, almost two hundred years old, that his programmable device reveals not only the supreme Authority of God, as if He were the Master Programmer of the Universe, but even helps us understand miracles.[35]

But if there is one single chapter to be read beyond all others of Jaki's entire work, it ought to be the tenth chapter of his *Science and Creation,* called "The Sighting of New Horizons." I love the two earlier chapters which set the foundation: his *encomium*—that's a vocabulary word I learned from Father Jaki, it means great and respectful praise—his encomium of the Jews in chapter 7, "The Beacon of the Covenant"—which shows for all who struggle to "reconcile" Faith and Reason how ancient and deeply they are united—and a comparably rich and surprising encomium on the Fathers of the Early Church in chapter 8, "The Leaven of Confidence." The pricelessness of that book results from Jaki's citation of Buridan, a scholar writing about 1330[36] which declares what everyone calls "Newton's" First Law, about three centuries before Newton was born.

3. The Relevance of Jaki as Introducer

There is more which should be said about this scheme of ordering Jaki's works, but in order to continue revealing the relevance of Jaki, I will now look briefly at one other orthogonal classification, not of books or essays, but at a finer level of detail—what I call Jaki as Introducer. It is delightful and useful, and not what we, Jaki's disciples, may ordinarily note in reading his works. In future studies I shall present other classes of Jaki's introductions: to books, to topics, to forgotten facts and also to forgotten dead-ends, which are far more serious to forget than facts. For now, I shall mention only some of the scholars Jaki introduces us to.

Outside of the rare polymaths who have familiarity with multiple disciplines, we get a great shock in reading Jaki's books when we meet names we never heard of, and even if we did hear of them, we often find Jaki citing words we never heard those writers speaking! For example:

Maxwell

James Clerk Maxwell was one of the greatest physicists of the late 1800s. He unified the work of Faraday and Gauss and Ampère on electromagnetics into his famous four equations which even today one sees on tee-shirts. Far more important to Science Writ Large is this:

> ... one of the severest tests of a scientific mind is to discern the limits of the legitimate application of scientific methods ... there are many things in heaven and earth which, by the selection of our scientific methods, have been excluded from our philosophy.[37]

Duhem

Pierre Duhem, (1861–1916). A mathematician, a physicist, a master historian of science: a scholar, a Catholic. Beyond doubt, Jaki is Duhem's chief disciple in the order of scholarship, and we follow them. There is plenty of material I might quote, but one of the most astonishing is this:

> If a passage is quoted, it is usually because a particular idea is expressed in it in a remarkable way. But a quotation may also serve an author as an opportunity to put forward an even more remarkable idea. Indeed, the mere sight of reference to a shepherd in al-Ghazzali's phrase touched Duhem in the very core of his outlook on existence, a deeply religious core. That it was almost a part of his nature can be seen from the naturalness with which he offered his comment that centers on the Incarnate God, the very center of Christian religion: "Assuredly, no philosophy outside the influence of Christianity could make intelligible the benevolence by which the superior being desires, without compromising his own status, the good of the inferior being. No such [non-Christian] philosophy could comprehend that ... the good Shepherd loves his sheep to the point of giving his life for them."[38]

Oppenheimer

Jaki quotes J. Robert Oppenheimer, a nuclear physicist of the twentieth century, as saying:

> ... the main thing we know of fundamental particles is that none of them is fundamental.[39]

Chesterton

Gilbert Keith Chesterton was an English journalist, a prolific writer, and a convert to Catholicism. Scientists don't know him. The literati fear him and so reject him. Too many Catholics prefer reading non-Catholics (or what a friend of mine calls "Catholicism Lite") to reading Chesterton. Jaki knows better: he introduces Chesterton by often citing this writer of vast proportions. Several essays focus on him, but for a truly enlightening study see his *Chesterton, A Seer of Science*, in which we find this brilliant declaration, a succinct summary of Jaki's entire work:

> The rebuilding of this bridge between science and human nature is one of the greatest needs of mankind.[40]

Others

Father Jaki introduces us to many others—in particular I should mention Buridan, Chapman, Gilson, Cauchy, Gödel, Pascal, Descartes, Galileo, Newton, Babbage, Newman, Kelvin, Planck, and Einstein. That list could go on for pages. I should mention the names of some who seem quite irrelevant to our work, yet are dangers we must know of in order to avoid: Aristotle, Francis Bacon, Giordano Bruno, Kant, Nietzsche, Marx, Darwin, Comte, Hegel, Husserl, Heidegger, Lonergan, Russell, Whitehead, Turing, Wittgenstein ... and that list also goes on and on.

4. One Other Relevant Feature

I want to conclude with what I consider one other relevant feature, which I for my part deem "technically sweet"—since these quotes are very hard to find, yet provide such delights to the unsuspecting reader of the works of Father Jaki. I only have time for a few:

> ... even physicists put their trousers on one leg at a time.[41]

> ... if it was true ... that "the universe may be the last free lunch." McDonald's would be interested in the demonstration.[42]

> Programming today is a race between software engineers striving to build bigger and better idiot-proof programs, and the Universe trying to produce

> bigger and better idiots. So far, the Universe is winning.[43]

A quick aside about that "race": as the Pink Panther said about a certain priceless Steinway: "Not any more."

> Then there are his delightful recounting of two famous little tales, too long to insert here, but I will mention their titles: the first is usually known by its final line, "It's Turtles All the Way Down" in his *God and the Cosmologists*,[44] and the second we might call "Get Your Own Dirt" in his *Questions on Science and Religion*.[45]

Lastly, on July 26, 2004, he told me of his challenge-question (Michelangelo versus Renoir),[46] and added that he had another one to ponder:

> "Which is nicer: a peach or a hexagon?"

Indeed, it is fitting to end with mirth. In the very last sentence of his *Orthodoxy*, Chesterton speculated[47] that "mirth" was the one thing which Christ concealed during His life on earth.

Jaki never concealed it. Neither do I. I miss him greatly, but I assert, as he did, that we shall meet again, and then our mirth will overflow.[48]

Notes

1. Why this formula? Jaki uses "writ large" dozens of times to modify Nature, Reason, Truth, Process, Universe, Faith, and others, including Science. But it is easy to explain: many of his essays were originally written as lectures, and he uses this verbal indicator of his textual capital letters in order to emphasize the preceding term.
2. *Pontificale Romanum*, (New York: Benziger Brothers, 1895), 116. The paragraph begins with the words, "*Consecrandi filii dilectissimi*" ("Most beloved, consecrated sons...").
3. Bligh, *Ordination to the Priesthood* (New York: Sheed and Ward, 1956), 80.
4. See Romans 10:14:15 citing Isaiah 52:7. The Douay-Rheims translation is adopted for all Scriptural citations.
5. See 1 Corinthians 2:2.
6. See Colossians 2:2–3.
7. Jaki, *Chesterton, A Seer of Science* (Pinckney, MI: Real View Books, 2001), chapter 4. In both cases (Jaki and Chesterton) it is true in both senses: both championed the universe (that is, the cosmos) for it is the sign and sum of Creation, but both also were *universal*, that is absolutely inclusive in speaking of anything and everything. Cosmologists ought to take note. (Dozens of references could be cited, but this expansive topic deserves its own paper.)
8. Stanley L. Jaki, *The Road of Science and the Ways to God: The Gifford Lectures 1975 and 1976* (Chicago: University of Chicago Press; Edinburgh: Scottish Academic Press, 1978), 122. He uses this phrase dozens of times, starting as early as: "A brief look at twentieth-century cosmology might show that the idea of the totality of consistently interacting things is indispensable in a meaningful scientific discourse about the universe of galaxies."
9. For references, see *The Road of Science and the Ways to God: The Gifford Lectures 1975 and 1976* (Chicago: University of Chicago Press; Edinburgh: Scottish Academic Press, 1978), 278, note 84; *Cosmos and Creator* (Edinburgh: Scottish Academic Press, 1980), 56–57; and *Universe and Creed: The Père Marquette Lecture 1992* (Milwaukee, Wis.: Marquette University Press, 1992), 21.
10. In a personal communication, he revealed that he had trained himself in writing such English by reading and studying

countless journal articles for word-use, sentence structure, and related skills.

11. We could wax eloquent on why: it is evident from the entire collection of scholarly works, and in Jaki's case, like that of St. Thomas Aquinas, it provides strong evidence of his humility: "*I did not write this: here is where I found it, and I am glad to acknowledge the brilliance of this earlier author.*" Once that humility was the common trademark of all scholarly writing, but the opposite is far too common today: I recently examined a supposed "textbook" from an otherwise reputable Catholic publishing house which had neither references nor a bibliography. There is no justification for such a work.
12. Or endnotes, which I detest, since they are regularly arranged so as to make their use impossible.
13. For example, his *Science and Creation* has a total of 1344 footnotes, or an average of 89.6 per chapter. There is a single place (single to my own knowledge, but I may have missed others) where he actually had to footnote his own footnote. See Stanley L. Jaki, *Immanuel Kant: Universal Natural History and Theory of the Heavens* (Edinburgh: Scottish Academic Press, 1981), 90. On that page he has a footnote (noted by an asterisk) which itself has 14 alphabetical notes which appear after endnote 39. Therefore, I have no hesitation about doing this myself, as you will see later.
14. Or, almost all, as we will see shortly. The most frustrating case I have encountered is his mention (without citation) that "Moreover, the aging Newton did not mind wasting his time by erasing references to Descartes in his manuscripts." Jaki, *The Road of Science and the Ways to God*, 85. Likely, it appears in one of the various studies he references in that same chapter.
15. Stanley L. Jaki, *A Mind's Matter* (Grand Rapids, MI: William B. Eerdmans Publishing Company, 2002), 5–6.
16. Jaki, *A Mind's Matter*, 257.
17. G. K. Chesterton, *What's Wrong with the World*, in *Collected Works*, vol. 4 (San Francisco: Ignatius Press, 1987), 49.
18. Stanley L. Jaki, *The Keys of the Kingdom: A Tool's Witness to Truth* (Chicago: The Franciscan Herald Press, 1986), 86–87; his note 63 reads: "The following paragraphs on St. John Chrysostom are much indebted to ch. 4 'St. Chrysostom on St. Peter' in Chapman's *Studies on Early Papacy* quoted in note 51 above. His meticulous

documentation makes references unnecessary to Chrysostom's works except in connection with the longer quotations."

19. Stanley L. Jaki, "The Catholic Intellectual," in *The Gist of Catholicism and Other Essays* (Pinckney, MI: Real View Books, 2001), 27–28.
20. Stanley L. Jaki, *Science and Creation* (Edinburgh: Scottish Academic Press, 1986), 163–191.
21. Stanley L. Jaki, *Neo-Arianism as Foreseen by Newman* (Port Huron, MI: Real View Books, 2006).
22. See John 10:16.
23. Stanley L. Jaki, "Christ, Extraterrestrials, and the Devil," in *A Late Awakening and Other Essays* (Port Huron, MI: Real View Books, 2006), 103–104.
24. *Ibid.*
25. Isaiah 22:22; also relevant is Isaiah 9:6; see also *Divine Office*, "Antiphon for Vespers, December 20," in *The Liturgy of the Hours*, trans. International Commission on English in the Liturgy (New York: Catholic Book Publishing, 1975), 743.
26. Jaki, *The Keys of the Kingdom*, 219.
27. See Matthew 16:13.
28. Stanley L. Jaki, "Liberalism and Theology," in *The Gist of Catholicism and Other Essays* (Pinckney, MI: Real View Books, 2001), 75.
29. Jaki, "Liberalism and Theology," 77.
30. See Luke 8:23–25.
31. See Stanley L. Jaki, *God and the Sun at Fatima* (Fraser, MI: Real View Books, 1999), 300; see also chap. 8.
32. Jaki, *Chesterton, A Seer of Science*, 58, citing G. K. Chesterton, *Orthodoxy*, in *Collected Works*, vol. 1 (San Francisco: Ignatius Press, 1987), 238. This opens a significant topic which needs further work: though it cannot be started here I provide the terms "author" and "authority" to future scholars.
33. Jaki, *Science and Creation*, 182, citing St. Augustine, *De Genesi ad litteram*, ed. J. Zycha, in *Corpus Scriptorum Ecclesiasticorum Latinorum*, vol. 28, sec. 3, pt. 1 (Vienna: F. Tempsky, 1894), 28–29 (book 1, chap. 19).
34. The grave flaw of this work is his citing of a line from Von Neumann: it is a rather irrelevant aside, and quite false, but grave

since it is misleading: It is in part this circumstance that bars analysis, this most elaborated branch of mathematics, from the task of working out an exhaustive computer or automaton theory; and this is what relegates this task to the realm of combinatorics which, to use Von Neumann's appraisal, "is one of the technically most refractory part of mathematics." Stanley L. Jaki, *Brain, Mind, and Computers* (Washington, DC: Regnery Gateway, 1989), 58, citing John von Neumann, "The General and Logical Theory of Automata," paper presented at the Hixon Symposium, Pasadena, CA, September 20, 1948, reprinted in *The World of Mathematics*, ed. James R. Newman, vol. 4 (New York: Simon and Schuster, 1956), 2083. This is misleading. We understand only too well at a formal mathematical level ALL which any real physical computer (be it mechanical or electronic) can compute, since any such machine is FINITE and hence can compute nothing more than the set of strings specified by the Pumping Lemma for finite-state automata: A.B*.C∪D. For details, see, e.g., David Hopkin and Barbara Moss, *Automata* (New York: North-Holland, 1976), 35; and Peter J. Denning, Jack B. Dennis, and Joseph E. Qualitz, *Machines, Languages, and Computation* (Englewood Cliffs, NJ: Prentice-Hall, 1978), 214–15. This form demonstrates the fullness of the power of regular grammars in its three operators: concatenation, star-closure, and union. Hence, no machine (since it is finite) can even count. Eventually its memory is full, as shown in the famous story of the (old-style) car odometer which eventually turns to all zeroes. (Yes, there is a distinction to be made, and it is not served by glossing over it, neither by Von Neumann, by Jaki, or by me, at least not in a footnote.)

35. This is not the place to begin a discussion, but a careful reading of Turing's famous paper on what he called the "Imitation Game" (the "holy scripture" of AI enthusiasts) will reveal his antipathy to the work of his far greater predecessor in the discipline, and other, far darker details of this now popular fraud. See Alan Turing, "Computing Machinery and Intelligence," *Mind* 49 (1950): 433–60; reprinted as "Can a Machine Think?" in *The World of Mathematics*, ed. James R. Newman, vol. 4 (New York: Simon and Schuster, 1956), 2099–2123.

36. Jaki, *Science and Creation*, 233–34, citing Buridan's commentary on Aristotle's *Physics*. The full context of the discussion on projectile motion is available in English translation in M. Clagett,

The Science of Mechanics in the Middle Ages (Madison: University of Wisconsin Press, 1959), 536.

37. Stanley L. Jaki, *The Relevance of Physics* (Edinburgh: Scottish Academic Press, 1966, 1992), 382, citing W.D. Niven, ed., *Scientific Papers of James Clerk Maxwell*, vol. 1 (Cambridge: Cambridge University Press, 1890), 759.
38. Stanley L. Jaki, *Scientist and Catholic: Pierre Duhem* (Front Royal, VA: Christendom Press, 1991), 83, citing *Le Système du monde* (Paris: Hermann, 1913–59), vol. IV, 453. For context, see the excerpt in *Scientist and Catholic: Pierre Duhem*, 258–59. See also John 10:15.
39. Stanley L. Jaki, "The Relevance of Materials Science," in *Numbers Decide and Other Essays* (Pinckney, MI: Real View Books, 2003), 184.
40. Jaki, *Chesterton, A Seer of Science*, 45, citing Chesterton, "A Defence of Useful Information," 75.
41. Stanley L. Jaki, *The Limits of a Limitless Science* (Wilmington, DE: Intercollegiate Studies Institute, 2000), 230.
42. Stanley L. Jaki, *Questions on Science and Religion* (Pinckney, MI: Real View Books, 2004), 74–75.
43. Jaki, *Questions on Science and Religion*, 75–6.
44. Stanley L. Jaki, *God and the Cosmologists* (Edinburgh: Scottish Academic Press, 1989), 112.
45. Jaki, *Questions on Science and Religion*, 45–46.
46. Here it is. It is too priceless to omit: There (I am talking of the engineering university of the Air Force in Dayton) someone at the dinner table began to extol the superiority of science over the humanities. I was able to shift the discourse to the question of whether the scientific method is capable of deciding whether Michelangelo or Renoir was greater as an artist. Suddenly all those professors of engineering and physics found themselves arguing with one another. In doing so they merely proved the fallibility of their presumed artistic competence or the lack of it. Meanwhile I could go enjoying my dinner undisturbed. Jaki, *A Mind's Matter*, 26.
47. G.K. Chesterton, *Orthodoxy*, in *Collected Works of G.K. Chesterton*, vol. 1 (San Francisco: Ignatius Press, 1987), 366.
48. Cf. John 15:11.

6 CHESTERTON, UNDSET AND JAKI: THREE CATHOLIC INTELLECTUAL GIANTS

GEIR HASNES, M.S.

1. Anniversaries and centenaries of G. K. Chesterton, Sigrid Undset, and Fr. Stanley Jaki

On May 29 this year, we celebrate G. K. Chesterton's 150th anniversary, and on August 17, we celebrate Father Stanley Jaki's centenary. In between them, on June 10, we remind ourselves that 75 years has passed since Sigrid Undset passed away, and on November 1, we celebrate the centenary of her conversion.

2. The relationship between the three authors

Most Chesterton biographies do not mention Sigrid Undset, the Norwegian Nobel prize winning novelist, and if they do so, it is only in passing. Her Norwegian biographers, though, mention Chesterton casually, because she called him her favorite author and used to read aloud from his books to friends and family. She probably discovered him in 1912 when she spent six months in England after she had married. She became increasingly interested in him after that, to the point that she even translated *The Everlasting Man* into Norwegian in 1931[1].

Father Jaki discovered Chesterton in 1951, but did not take him in until 1973, when, as he wrote: 'Chesterton started acting on me as an irresistible magnet.'[2] Subsequently he wrote a monograph[3] and several articles on him. Late in life he also discovered Undset, and became intensely interested in her life and work, to the point that he wrote one of his last books on her.[4]

Why string these three authors together in this paper? The anniversaries seem to be a good enough reason to focus both on Undset's and Fr Jaki's interest in Chesterton and Fr Jaki's interest in Undset. Too little has been written on the relationship between Undset and Chesterton and almost nothing about Fr Jaki's relationship to Chesterton and Undset.

My personal interest in the relationship between those three is founded in my lifelong interest in G. K. Chesterton and Sigrid Undset. Less than two years ago, my bibliography on G. K. Chesterton was published.[5] When I attended the American Chesterton Society's annual Chesterton conference in 2006 to speak on my bibliographical work, I met with Fr Jaki. It turned out that he had recently taken an interest in my fellow Norwegian Undset, and I offered to help him if he wished to go Norway to do more research on her. He took up the offer, and I drove him twice to Undset's residence in Lillehammer, now a museum. Not only did these trips help Fr Jaki's research, but it also added him to my list of authors I need to collect all of.

3. The differences

Although Chesterton, Undset and Fr Jaki were both Catholics and authors, they were surprisingly different in several ways.

3.1 Genre

Chesterton was first and foremost a journalist, who also happened to write books. He wrote regularly for large newspapers and magazines, was immensely prolific and had about eighty books published in his lifetime and a lot of collections of unpublished material appearing later. He may have written (or rather, dictated) five—six thousand essays, more than one thousand poems, more than one hundred short stories, and most of his books were put together from these.[6] He was also a revered public speaker, and may have delivered more than one thousand public speeches, lectures, or participated in public debates, and a majority of these are recorded, several in full and the more than half in part.[7] One may best characterize Chesterton as an essayist, and his books often seem to be enlarged essays. He was no journalist in today's sense of the word; he decided himself what to write about.

Sigrid Undset was first and foremost a novelist. She had more than forty books published in her lifetime and many works of different kinds after her death. She wrote hundreds of articles, but was not by any means as prolific as Chesterton. She was a typical novelist and from her debut onwards she soon was able to live from her sales. However, she also adapted and translated several Catholic books into Norwegian.

Father Stanley Jaki was a Hungarian born Benedictine priest who at the age of fifty was employed as a professor at Seton Hall university. He had two doctorates in theology and physics and wrote mainly books and articles on science, the history of science, and theology. Many of his books consists of lectures, and there is overall an oral delivery in his texts. He had at least eighty books published in his lifetime and since then many collections have arrived.

3.2 Debut

Chesterton made his debut at 26 in 1900, Undset at 25 in 1907, and if we exclude Fr Jaki's doctoral dissertation in

1957, he made his debut at 42 years old in 1966 with *The Relevance of Physics*.[8] We know that both Chesterton and Undset wrote extensively from their teenage years onwards, and both were employed, Chesterton as a manuscript reader and ghost writer in publishing companies and Undset as a secretary, until they were able to live by their pens. Fr Jaki was an ordained priest who continued studying for several years and who was also employed as a teacher. A tonsillectomy made him concentrate further on studies and finally he found his vocation as an author on science and religion.

3.3 Use of sources and libraries

Chesterton had a remarkable memory and did not use libraries, except from when he was working on his life of Robert Browning. He also quoted from memory and was famous for making up lines in other people's poems; Chesterton insisted on quoting from memory because he viewed the effect the poem had on the reader as important, even if that meant that the poem would not come out as originally written.[9]

To the opposite, Undset and Fr Jaki both researched thoroughly everything that was treated in their books. Undset's father was an archaeologist and Undset took an interest in history from her earliest years. Before writing her medieval masterpieces in the 1920s, she was theoretically well founded both in Norwegian history and in Catholic theology. Fr Jaki, then, spent several decades diving deep into Catholic theology and science before he began publishing his studies.

Chesterton owned a huge collection of books which today has been scattered to the winds. Nobody really knows how large his library was. Already, in his late teens he must have read tremendously, and it may seem that the family had a huge library. He doesn't seem to have collected literature at all; he used to buy whichever book appealed to him at the instant, and he must have been sent a multitude of books to review during his life. A research project ought to be started on his knowledge of literature.

On the other hand, Undset built her collection of books consciously from the start, spending a lot of her income on books, while also receiving a torrent of books from publishers and friends. Her library numbered about 5,000 books at the outbreak of World War II, and after she had fled from the German occupiers, her library was hidden by friends in church crypts in the area. She brought with her about 4,000 books from the United States after the war, and at her death the library numbered 10,300 books. 1300 of them, her collection of old Norse literature, were willed to a library in Trondheim, and the rest of it stayed untouched, gathering dust for 48 years until the estate was bought and made into a museum. Her whole library has been catalogued and is available for research.[10]

Fr Jaki also built up a personal library during his life, and he also visited the libraries nearby for his work, for instance at Stanford University and especially at Princeton University.[11] Luckily, he had dedicated helpers, who could empty his flat after his death, keep his personal computer and make backup of everything, and make a catalogue of the inventory before it was sent to Seton Hall.

> When Fr Jaki died, his library was sent directly to Seton Hall. His papers and the books he wrote were sent to New Hope (Real View Books). I went through them, "built" a "physical" Publication List, which is still in New Hope, in some 12 boxes or so, with a few other items, but most of the stuff, letters, books, xeroxes, folders in preparation of books etc. (including some original Duhem

> material) were sent to Seton Hall ... Fr Jaki left all the rights to his books to Real View Books.[12]

3.4 Death and health

Chesterton died at 62, Undset at 67 and Fr Jaki at 84. When Fr Jaki had reached 62 he had 'only' published about seventeen books, and in his last three decades he wrote with a tremendous speed until his passing.

Chesterton died prematurely from smoking too much. He was a chain-smoker and did not take much care of his health, which probably was a reaction to his father being a hypochondriac. Undset also died from smoking too much; she used to write in the evenings out into the night after the children had gone to bed, and was chain-smoking while she wrote. She also felt very tired during her last years, as she had fled to the United States from the Germans in 1940, and acted as an ambassador for Norway there during World War II. Fr Jaki was in this respect the opposite of the other two, he was a healthy and dynamic person full of energy and vitality to the end.

Chesterton and Undset were thus children of their time, when people used to smoke everywhere. It is sad to think of how much more they could have written as everything they published was of such a high quality. Very few authors deserve that everything they have written is available for the public. All these three intellectual giants deserve to have their intellectual legacy taken care of.

3.5 Biographers

One may wonder how Undset may have corrected our view of Chesterton if she had written his biography, as most of his biographers either were not Catholics or they misread him or hadn't researched him well. Only two biographers were able to penetrate deep into Chesterton's mind; William Oddie[13] and Fr. Ian Ker[14], both Catholics.

Oddie, though, only covered Chesterton's life up to 1908 and the publication of *Orthodoxy*.

Now it may be unfair to ask of any biographer that he should be able to grasp the enormous mind of Chesterton. Not only is there more than enough material, but there is also a huge amount of controversy as Chesterton took an active part in the debates of the day. A biographer must choose what is important and what is not so important in the life of that person. This filtering process is based upon the worldview of the biographer.

The definitive biography of Sigrid Undset still remains to be written. Just as with Chesterton, most biographers view her life as 'before and after conversion.' Secular biographers tend to prioritize the 'before' part, while Catholic biographers tend to diminish her conservative and uncompromising Catholicism. Norway was in her time a Lutheran and largely anti-Catholic country and is now a very secular country, taking little interest in organized religion at all. Undset also took an active part in the debates of the day, although not being such a vigilant voice as Chesterton. One of the reasons is, of course, that Chesterton lived in the middle of the world of the news in those days, while Undset lived on the outskirts of a country on the outskirts of the world. Chesterton had a smooth machinery around him, from secretaries to literally hundreds of newspapers and magazines only too eager to print whatever he tossed off. Undset had no secretary, and she had her outlets, but most of her controversies were, in the more positive meaning of the word, provincial. What Norwegians were arguing against each other in those days did not reach far outside their borders.

Books on Fr Jaki have already appeared, but no definitive biography among them. The difference between Fr Jaki, on the one hand, and Chesterton and Sigrid Undset, on the other, is that the latter two built up a huge readership before they became converts, and most of their literary production appeal to people of most creeds or lack of it. Fr Jaki was exclusively Catholic in all his writings. One might argue that science is not exclusively Catholic and that Fr Jaki should appeal to anyone with an interest in the philosophy of science. One may not be so sure about that. Books about science always comes with a taint of religion, of dogmatism, of prejudice, of philosophical preferences. Some may for instance defend Giordano Bruno as an important person for science. Fr Jaki said that he had to translate a work of Bruno so that he could grasp how little Bruno knew about science. This is both a good anecdote and an indicator of how easy it is to build a controversy out of seemingly non-controvertible material.

Books about Fr Jaki tend to focus on his written material more than on his personal life. One could probably write a huge tome on Chesterton only quoting anecdotes and humorous stories about him, as he was a huge public figure, a landmark, and as an outspoken defender of the poor and attacker of the unjust, with an aptitude for epigrams. There is probably also a large number of anecdotes around Fr Jaki circulating, judging from his obituary alone, but not many of them have been written down.

While Chesterton was a public figure, speaking with everyone everywhere, Undset was not found of speaking to people she didn't know, to the point of being rude and dismissive even to fans. She was courteous to her friends, fellow authors and fellow Catholics, but did not deliver witty answers to her opponents for instance in debates the way Chesterton always crushed his opponents with some humorous and pointed remark. Fr Jaki was quick the way Chesterton was, and my impression is that he must have been very interesting to listen to in a debate, and probably quite annoying to the other point of view.

Compared to Chesterton and Undset, Fr Jaki doesn't have the same amount of biographical detail to delve into; he was largely 'condemned to sit at home and write'—and therefore, his biographers will mainly look into what he wrote, and therefore even become tempted to write their own treatises instead of writing a proper 'life' of him. Biographers of Chesterton tend to become hagiographers and therefore make up something to criticize along the way, biographers of Undset tend to misunderstand or dismiss her conservative Catholicism, and biographers of Fr Jaki must avoid being hagiographers or 'system builders'.

3.6 Societies

Chesterton has got his *Chesterton Society*, founded at his centenary in 1974, and the Chesterton Review, founded the same year. There is also a ferocious *American Society of Gilbert Keith Chesterton* arranging conferences and publishing a bi-monthly magazine and books on Chesterton every now and then, and there are smaller Chesterton societies in many other countries. Sigrid Undset has also got her society in Norway, where the State, together with the local museum, bought her estate and belongings and turned it into a separate institution, complete with a new exhibition building erected at the centenary of her debut. And the *Stanley Jaki Foundation* keeps up the legacy of Fr Jaki with books and conferences.

4. The similarities

Gilbert and Frances Chesterton married in 1901, and they had no children. His sister-in-law Ada Chesterton wrote about the couple that the marriage had not been consummated. The statement has since been thoroughly rejected by almost all Chestertonians, but there is reason to believe that there is something to it. We know that

Frances had an operation in order to have children, and that it failed. We also know that Gilbert Chesterton never spoke about his private life and any letters or papers deemed too personal were destroyed after his death. Frances also had a fragile health and was regularly ill. After they had married, Gilbert put on weight and drank too much; and there is reason to believe that if they had regular intercourse, it would not have lasted for more than a few years.

However, when Chesterton was asked about how it could be that he could write so much and know so much, he referred to the necessity of being chaste.

Sigrid Undset was fond of men, even of older men, and she had affairs even with married men before she met Anders Svarstad, whom she married after he had divorced his first wife.

During the next years, Undset began to develop an interest in Catholicism, and it dawned upon her that from a Catholic viewpoint, she was not properly married as Svarstad's first wife still was alive. She also moved to inland Norway in order to find a better environment for her work as an author, and Svarstad did not want to move with her. Then, in order to convert to Catholicism, she simply divorced Svarstad and never remarried. When asked about how it felt to live without a man, she replied that it was not any worse than living with only one man.

Undset, then, also lived a chaste life. So also with Fr Jaki, who as a priest lived in celibacy. Chastity produces incredible insights.

5. *Undset and Chesterton*

Undset's biographers used to say that Chesterton and Belloc were crucial to her conversion. Luckily, her library became available to researchers after 2007, and even better, Undset used to write the date of purchase in the books. By that, to put it short, one finds that Undset was fascinated by the saints and martyrs of the church; it is fair to assume that those were among the main reasons for her to study Catholicism, and the only book Chesterton had written on Christian themes until then was *Orthodoxy,* which she read in 1920. One of the biggest influences must have been Robert Hugh Benson; she read his novels during 1917 and 1918. She took a public standpoint for Christianity in 1919 and converted in 1922. She then translated two books by Benson in 1926 and 1928, showing an interest in Christian mysticism which has not been understood by any of her biographers.

On May 7, 1928, Sigrid Undset wrote to Chesterton from Wilton Hotel, Victoria, S.W.1:[15]

> G. K. Chesterton Esq.
>
> Dear Sir!
>
> Although I know that your time is very much occupied I take the liberty to beg an interview with you, as the Dominican fathers of Oslo wanted me to call on you. I am a catholic from Norway, over here principally to see something of the work the English Catholics do and see if I can learn something that might come in usefully for our own work at home.
>
> Please would you let me have a word, if, and when I might meet you.
>
> Yours respectfully
>
> Sigrid Undset

She enclosed her visiting card. On the letter is noted: Acknowledged / Will write later.

Then, on May 25, Sigrid Undset indeed visited Chesterton at his home in Beaconsfield[16]. We know nothing about the meeting, but we know what came out of it. In her papers, we can find a deal struck with Macmillan for rights to translating various books by Belloc, Chesterton etc. for the fee of $500, which meant books in Macmillan's series on the Catholic Church, edited by Hilaire Belloc.[17]

In December 1929, two books from the series appeared: *The Catholic Church and Conversion* by Chesterton, translated by Undset's stepdaughter Ebba Svarstad[18], and *The Catholic Church and History*, translated by Undset's friend Ellen Faaberg[19]. What is not commonly known is that Sigrid Undset paid for the publication of the two books out of her own pocket. Her Norwegian publisher may not have agreed that the translations were worthwhile. While Undset certainly wished to get more books from the series published, it began and ended with these two books.

However, Undset were not through with bringing the English Catholics to Norway. After having completed her two-volume novel describing the conversion process in 1930, she took a year off to translate Chesterton's *The Everlasting Man* in 1931. She had already written several biographies of Norwegian saints, and they all ended up in *Saga of Saints*, which was published in 1934 in Sheed and Ward's series of books on saints, where, for instance, both Chesterton, Belloc and Christopher Dawson had contributed to *The English Way* the year before.[20] She continued to write several biographies of European saints, but about 1935, the tide had waned and little more should come out of the cooperation between her and Chesterton and his co-Catholics.

6. Fr Jaki and Chesterton

What struck me most with Fr Jaki was his constant hunger for knowledge combined with the instant treatment of this knowledge when it was acquired, this wish to do something instantly with what he took an interest in. I can understand that he was seen as demanding by some, and polemical by others. I think he was just as demanding to himself; the wish to learn and to build a system out of what he learnt was a driving force in him, if not the main driving force. He was impatient with others inasmuch he was impatient with himself. For such a person, it is hard to take an interest in most people, because they don't bring anything he doesn't know and would not have thought through already.

When Fr Jaki discovered Chesterton, however, he found a person whose mind had seen the same things as he had, and perhaps put them in writing noticeably more eloquently than him. His book tells us about Chesterton's "sparkling gems", that he spoke "very attractively"; indeed, Fr Jaki had found somebody to marvel about.[21] Not only that, but as opposed to a lot of philosophers and scientists Fr Jaki admired, Chesterton also had the same foundation in the Faith and thought the same thoughts as Fr Jaki. In short, "the Universe had few finer champions than Gilbert Keith Chesterton". [22]

7. Fr Jaki and Sigrid Undset

When I first met Fr Jaki at the Chesterton conference back in 2006, he was so noticeably attentive. At the lunch table, when he heard that I came from Norway, he didn't hesitate, but straight on asked me to tell him about Sigrid Undset. There was no presentation either; he just jumped into it, and, I think to his appreciation, I just went on and on—and he drank it in. Because he told me that he had been looking into Undset and was working on a translation of a central text of hers,[23] it struck me that he ought to go to Norway to do research—and of course it would be easy for me to help him with that. It was nothing more to it than that; I wasn't sure whether he would pick up the offer, but he did, and that ended up with a book full of appreciation for yet another soulmate of his.

Indeed, he appreciated Sigrid Undset so much that when I sent an article on the relationship of the two to the *Chesterton Review*, I entitled it "Father Stanley Jaki and Sigrid Undset: A Love Story".[24] However, I think the editor found it too flippant, but I meant what I said: Just as I love Chesterton's, Undset's and Fr Jaki's writings, I think Fr Jaki found yet another person to love in the meaning of admiring her mind and actions.

In the book that followed, Fr Jaki discussed various books and articles on Undset, and dissected them in his usual polemical style, showing how they were missing a point or three. It is a delight to read, also because Fr Jaki is so eloquent himself. How better to state it than to say that "truth in its finest form was for Sigrid Undset to live in a deeply personal friendship with Christ as regulated by the only Church He founded."[25]

Conclusion. The teachers of the Church

This short essay can only scratch the surface of the relationship between these intellectual and Catholic giants. They were not much alike with regard to lifestyle and personality, but they had their Faith and their incredible ability to write convincingly in common. Fr. Jaki discovered both of them and wrote books about them, and for anyone taking an interest in Fr. Jaki, I am convinced that taking an interest in Chesterton and Undset will be richly rewarding.

In G. K. Chesterton, Sigrid Undset and Fr Stanley Jaki, we have acquired yet three teachers of the Church. Now that we celebrate their anniversaries and the conversions of two of them, may the Church recognize them as that and treat them accordingly.

Notes

1. G. K. Chesterton, *Det evige menneske,* trans. Sigrid Undset (Oslo: H. Aschehoug, 1931), https://www.nb.no/items/13639e5a16d3346c178f797853b6cec3.
2. Stanley L. Jaki, *Chesterton, A Seer of Science* (Urbana and Chicago: University of Illinois Press, 1986), ix.
3. Jaki, *Chesterton, A Seer of Science.*
4. Stanley L. Jaki, *Sigrid Undset's Quest for Truth* (Port Huron: Real View Books, 2007).
5. Geir Hasnes, *G. K. Chesterton: A Bibliography* (Kongsberg: Classica, 2022), 3 vols.
6. Hasnes: *G. K. Chesterton. A Bibliography.*
7. Geir Hasnes, "Chesterton as Public Speaker," *The Chesterton Review* 47, no. 1–2 (2021): 63–75, https://doi.org/10.5840/chesterton2021473/470.
8. Stanley L. Jaki, *The Relevance of Physics* (Chicago: The University of Chicago Press, 1966).
9. As a result, several of Chesterton's books had new impressions corrected, notably: G.K. Chesterton, *Robert Browning* (New York: Macmillan, 1903), and *The Victorian Age in Literature* (UK: Williams & Norgate, 1913).
10. Geir Hasnes, "Sigrid Undset i bøkenes verden," *Gymnadenia Årsskrift* (Lillehammer: Sigrid Undset-selskapet, 2008), 6 unpaginated pages, in Norwegian. See also Geir Hasnes, "Sigrid Undset and the Saints," *Saint Austin Review* 21, no. 6 (2021): 4–7.
11. Antonio Colombo, "From Győr to Madrid," http://www.sljaki.com/biographies/america-colombo.html.
12. Antonio Colombo, email to the author, May 18, 2023.
13. William Oddie, *Chesterton and the Romance of Orthodoxy. The Making of G.K.C. 1874–1908* (Oxford University Press, 2008).
14. Ian Ker, *G. K. Chesterton: A Biography* (Oxford University Press, 2011).
15. *The G. K. Chesterton Papers*, British Library, Add MS 73240 A 001–2.
16. Dorothy Collins, *Datebook*, transcribed by Nancy Carpentier Brown, unpublished. It is especially mentioned here because Ian Ker mentions the letter in his biography, and he draws a wrong

conclusion, thinking that a greeting card from Undset to Chesterton dated 1931 meant that the meeting had taken place by then. However, the greeting card had accompanied a copy of her translation of The Everlasting Man into Norwegian as a gift to Chesterton.

17. Geir Hasnes, "The Catholic Church and Conversion," in *Sigrid Undsets hjem Bjerkebæk. I bøkenes verden*, ed. A. Åslund et al. (Lillehammer: Maihaugen, 2007), 138–39.
18. G. K. Chesterton, *Den katolske kirke og konversion*, trans. Ebba Svarstad (Oslo: Aschehoug, 1929), https://www.nb.no/items/1cef4d873cbf9b9c3092e0732afe99de.
19. Hilaire Belloc, *Den katolske kirke og historien*, trans. Ellen Faaberg (Oslo: Aschehoug, 1929), https://www.nb.no/items/7e4255c63b91cc2a704682d19d24301d.
20. *The English Way: Studies in English Sanctity from St. Bede to Newman* (London: Sheed & Ward, 1933).
21. Jaki, *Chesterton, A Seer of Science*, 110.
22. Jaki, *Chesterton, A Seer of Science*, 85.
23. Sigrid Undset, *Through Moral Crises to Catholicism* (Port Huron, MI: Real View Books, 2006).
24. Geir Hasnes, "Stanley Jaki and Sigrid Undset," *The Chesterton Review* 47, no. 3–4 (2021): 385–91, https://doi.org/10.5840/chesterton2021473/470
25. Jaki, *Sigrid Undset's Quest for Truth*, 105.

7 GÖDEL'S THEOREMS IN PHYSICS HALF A CENTURY AFTER STANLEY JAKI

GERGELY BOGNÁR, M.A.

Introduction

In 1966, Stanley Jaki first proposed in his book *The Relevance of Physics* that Gödel's theorems represent a limitation for physical knowledge. His statement is still a subject of debate today. Some believe that Gödel's theorems are limited to mathematics and do not apply to physics, while others dispute the validity of the theorems within mathematics. On the other hand, renowned thinkers who cannot be accused of religious bias (e.g. Stephen Hawking), agree with Stanley Jáki. In my work, I would like to present briefly the different standpoints. Without holding court, I point out that the discourse developed following Gödel's theorems leads to mathematical-philosophical problems, the systematic exploration of which necessarily means the limitation of knowledge based on measurable quantities. All of this means that Stanley Jaki's more than half-century-old statements about Gödel's theorems are still valid today.

It was in 1966 when Stanley Jaki first suggested in *The Relevance of Physics* that Gödel's theorems represent a limit to physical cognition. In his later works, he elaborated the details of the program he presented in *The Relevance of Physics*, but the basic idea remained essentially unchanged. In the context of Gödel's theorems, the name of Jaki has been considered a benchmark to this day, despite the fact that much has happened in both physics and mathematics in the past half century. In my brief paper, I review the basic idea of Stanley Jaki and compare it with new developments, and finally I point out that the suggestion of the Templeton Prize-winning Benedictine scientist regarding Gödel's theorems can only be challenged on grounds that confirm the competence of science that Stanley Jaki proclaimed throughout his life achievement.

A briefly summary of Gödel's theorems

Examining the logical foundations of mathematics, Kurt Gödel, managed to create Statement "G" in 1931. The essence of this is that this statement cannot be denied or disproved by means of the formal language of mathematics.[1] Based on statement "G", he formulated his famous theorems, which I summarise as follows:[2]

1. In any consistent formal system that includes part of arithmetic, there is a statement that cannot be proved or disproved within the framework of the system.
2. The consistency of any formal system, which includes part of arithmetic, cannot be proved and disproved within the framework of the system concerned.

The essence of the first theorem is that in a certain part of mathematics there are always statements that cannot be proved or disproved, while the second theorem suggests that the consistency of certain systems of mathematics cannot be verified or disproved within the framework of the system concerned. It is important to point out that the second theorem does not mean that mathematics is inconsistent. If that were the case, it would make no sense. All it states is that the consistency of mathematics cannot be verified within the confines of pure mathematics.

Many people refer to Gödel's theorems in many ways, and there is a lot of misunderstanding in this field. It should therefore be specified that theorems are valid only in formal mathematical systems; systems that contain at least integers. For example, they do not apply in Euclidean geometry. Gödel's theorems meant a fatal blow to David Hilbert's program, proving that in mathematics there will always be undecidable statements that cannot be deduced from the axioms of the system. Gödel's theorems specifically concern mathematical logic; thus, one must examine whether they are relevant to physics in any sense.

Gödel's theorems and physics

At first sight, physical theories fall into the scope of Gödel's theorems. On the one hand, they operate with formal signs, such as those in physical formulas, e.g, force, mass, velocity, etc., and on the other hand, they use numbers. However, the situation is not that simple. When solving physical problems, numerical outcomes must always be reconciled with practical experience, and mathematical solutions to even the simplest physical problems can lead to results that must be rejected on the grounds of experience. Leaving aside the calculation details, the simplest example is free fall. When calculating the fall time of a body released in air, one obtains an equation that has two solutions, a negative time and a positive time. The negative solution makes sense in mathematics, but in physics negative time carries no meaning, so it is obviously rejected. Besides free fall, there are plenty of other examples where, based on experience, physics often limits the range of mathematically meaningful solutions. The above example clearly shows that the empirical facts of experience in physics go beyond the closed formalism of mathematics, and therefore Gödel's theorems cannot be applied to physics as a whole. It cannot be stated that these theorems have the same relevance in physics as they do in mathematics! Several critics of Jaki cite arguments similar to those made above. At this point, it should be pointed out that Templeton Prize winning Stanley Jaki makes no such claim. There is not a single line by him on unverifiable claims in physics due to Gödel's theorems, or that these theorems cast doubt on the consistency of physics. All he writes is that, as a result of the theorems concerned, no physical theory can be regarded as necessarily true.[3]

Jaki's original idea can be reformulated as follows in the light of new the developments. For physics, mathematics is a formal language within which relationships between quantifiable quantities obtained experimentally can be established. Physics often refers to mathematical theorems and proofs, but there is no need to deal directly with Gödel's sentences. Mathematicians are able to construct a section of mathematics that is relevant to physics, but does not include Gödelian sentences. If physical theories use mathematics that do not contain Gödelian sentences and explain one's experimental experience, then physical theories can be considered consistent without any doubts. Gödel's theorems are problematic only if physics is seen as a metaphysical principle capable of explaining everything, rather than as a model of the measurable world, which is not necessarily true. Once the former, the best-known example of which is Stephen Hawking's program in *A Brief History of Time* on the ultimate theory of physics, is achieved, will be able to explain everything and God will not be needed.[4] Generations of physicists grew up on Hawking's popular book, and the ideal of a great physical theory eliminating God was burned into the minds of many. Unfortunately, it is much lesser known that Hawking, citing, among other things, Gödel's theorems, retracted his earlier views on a great and final theory that could explain everything.[5] Hawking does not follow Jaki's line of thought, since the Templeton Prize-winning Benedictine scientist does not rule out the possibility that the ultimate theory of physics, combining the four known interactions, could describe all known measurements, and the only thing it denies is that one cannot be sure of the ultimate nature and completeness of this theory.

Jaki realised more than half a century ago that physics could not produce a theory that could explain the universe as a whole and inevitably true, because that theory would have to confront Gödel's theorems. If physics is limited to the study of measurable quantities and does not wish to answer everything, such as the truth of mathematical problems raised in human mind, then it does not contradict the consequences of Gödel's "G" sentences. On the other hand, if one wants to explain everything, mathematics must also be included, and in that case, nothing can be excluded from physics, including the Gödelian sentences of mathematics. Of course, one could argue that no theory is expected to interpret the truth of its basic assumptions. If the part of mathematics relevant to physics is considered a metatheory, within which Gödelian theorems can be isolated, as described above, then the trap outlined above might be avoided. However, the situation is not that simple.

The decision of which mathematical statements are needed in physics is based on experience, and nothing can preclude a new piece of experience from leading to the very mathematical statements that fall within the scope of Gödel's theorems. It is important to point out here that this does not require a new physical paradigm or new revolutionary discoveries. Stanley Jaki makes a much weaker claim here. The issue of which subset of mathematics is currently used by physics cannot be established theoretically. When creating a particular question or a new technical device, new mathematical demands may arise, and there is no need to replace the whole theory, there is no need for a revolution in the Kuhnian sense, or for new, highly disturbing physical observations. One can simply never be sure that we they cannot run into undecidable Gödelian sentences with regard to a partial problem, a technical challenge, or a more elegant mathematical solution.

Based on the above, it is not at all difficult to see that Szaniszló Jaki's statements about Gödel's theorems are still valid today. However, there are a few options that need to be taken into account. It is not inconceivable that Gödel's theorems should be treated in some way by mathematics, nor can we theoretically dismiss the possibility that these theorems will be refuted by a new mathematical theory. Before making a definitive conclusion about the relationship between Gödel's theorems and scientific cognition, the above possibilities must be reviewed.

From Gödel's theorems to the philosophy of mathematics

Mathematician Abraham Robinson toyed with the idea of adding the unverifiable Gödelian sentence to the axioms of mathematics.[6] The axioms of mathematics do not need to be verified, they do not require proof; they are true by definition. As a result of Gödel's numbering, such mathematics would assume numbers that are larger than any known number but not infinite, and he called them hyperreal numbers. At first glance, all this may seem too absurd and abstract. However, it is not at all unprecedented in the history of mathematics that earlier number concepts have been expanded. Let us take, for example, the emergence of irrational numbers in the

Pythagorean school, the emergence of negative numbers in the Middle Ages, and complex numbers in modern mathematics. Expanded number concepts are still used in physical theories today. It is not far-fetched to introduce hyperreal numbers. Moreover, it turned out that introducing these numbers is not a mathematical game for its own sake. With their help, the solution for complex differential equations is simplified, compared to the description used in traditional mathematics, the solution of an equation of motion is simplified to the level of elementary addition and multiplication, the only problem is that elementary addition and multiplication become extremely complicated in this mathematics. One do not have to accept the new direction of mathematics outlined by Abraham Robinson; the example above merely shows that the possibility of treating Gödel's theorems in some form in future new mathematics cannot be ruled out. Another way could be, for example, to reformulate the axioms of mathematics, as it was done earlier with the Russell's paradoxes, so that Gödelian numbering would become impossible in the new system, while all other mathematical theorems would remain valid. For the sake of truth, it should be noted that the latter idea is not inconceivable, but at the moment it does not seem feasible.

Finally, the possibility that Gödel's theorems are mathematical misconceptions cannot be rule out. In the history of mathematics, it has occurred several times, that theorems believed by many to be true were later proven false. This is exactly what László E. Szabó, of the same nationality as Jaki Stanley, claims, who calls the theorems downright incompetent.[7] Leaving aside the complicated mathematical logic discussions, it should be noted that László E. Szabó uses mathematical philosophical assumptions in his proof, from which the refutation of theorems necessarily follows. However, László E. Szabó's views do not need to be interpreted in detail or a position taken on them within the scope of our investigation, because all this is irrelevant for the purpose of the study at hand. The sole purpose of outlining this theory was to show that it is not inconceivable that future mathematics would recall Gödel's theorems as one of the great misconceptions.

Whether mathematics treats or refutes Gödel's theorems in some form, related investigations draw one's attention to more elementary problems. If the axioms of mathematics can be freely expanded, and a mathematical proof depends on the basic assumptions of mathematical philosophy, then it is legitimate to ask why mathematical theorems relevant to physics are considered valid, and whether mathematical statements are necessarily true. The answer of mathematicians is simple: in mathematics, any theorem is considered to be true if it can be derived from mathematical axioms by means of regular logical steps. In the philosophy of mathematics, opinions differ about where axioms and logical conclusions come from. Having seen the last two centuries of mathematics, Platonist and realist conceptions of mathematics are difficult to maintain; the most accepted formalist view today is that mathematics is nothing more than merely a formal system without meaning based on arbitrary rules. If this is so, and mathematics is a human creation with signs and empty concepts, then our physical theories using mathematics cannot necessarily be true. Mathematics in this regard is a human invention whereby the rules of physical world can be described, but it cannot be taken for granted that the physical world really follows mathematics. To understand this problem, a few relevant thoughts by Stanley Jaki's spiritual master, Pierre Duhem must be cited:

> A theory of physics is not an explanation. It is a system of mathematical propositions, deduced from a small number of principles, which have for their aim to represent as simply, as completely and as exactly as possible, a group of experimental laws.[8]

Returning to the philosophical questions of mathematics, one does not need to take a position on whether the Platonist, realist or formalist conception of mathematics is considered to be correct. Whichever view is true, an even more serious problem arises. The application of mathematics to the physical world involves a highly arbitrary and subjective step. How do formal mathematical signs relate to the physical world? Does one plus one equal two carry any physical meaning, that is, is necessarily true just because it can be deduced from the axioms of mathematics?

At first sight, any reasonable person would object to the denial of the truth of the notion of one plus one equals two. It is obvious that one plus one apple placed side by side on the table is two. If one considers it more carefully, there is a deep philosophical problem here. If no abstract mathematical signs, such as numbers are taken into consideration, but the addition of real objects, then numbering is arbitrary and follows individual criteria. It may seem obvious that one plus one apple is two, and it is not possible to add apples to bananas as they do not fall into the same category. However, the coherence is not clear: if, for example, one wants to know how many types of fruit are available for a fruit salad, apples and bananas can be added to one another. In addition, if the aim is to summarize how many apples certain people picked in the orchard, it is impossible to find any physical properties of apples whereby it can be decided which ones can and cannot be added to one another. Here, the history of apples is linked to one's personal history, and physical and mathematical theories have nothing to do with it. Simple theorems of mathematics, such as one plus one equals two, only make sense if humans apply them to solve their own problems. The method of application carries subjective elements, therefore physical theories, which assign mathematics to particular cases of empirical facts, carry some arbitrary, subjective elements, too. Consequently, physical theories can never necessarily be true. Of course, I do not want to deny that the use of mathematics in describing the physical world is extremely useful, and provides an excellent model for describing phenomena and plays a key role in laying the foundation of the modern technical world, and is therefore extremely useful. Critical comments only point out that physical theories do not necessarily carry true statements and therefore can never replace metaphysical statements. Physics is a science based on experience and is excellent at describing phenomena, sometimes even predicting phenomena that have not yet been experienced, which can be observed later, but cannot necessarily be true because it uses a language (mathematics) whose application is arbitrary. Scientific and metaphysical theories must be separated distinctly. Religion and metaphysics may attempt to establish eternal truths, but science is not suitable for that. Physical cognition cannot make true claims that are necessarily true, whether they are theistic or atheistic. This is not only apologetics against today's scientist worldview, but also apologetics to preserve the credibility of science. Today's scientists are awful philosophers, and if they cross the boundaries of science and take positions on issues where they have no relevance, they sooner or later will end up losing credibility, which diminishes the authority of science.

The astonishing success of mathematics

Our investigation into the philosophy of mathematics would not be complete if we failed to draw attention to the fact that in the common history of mathematics and physics it has happened more than once that physical problems thought to be unsolvable have been solved by new mathematical theories that had not been used in physics before. Nobel Prize-winning physicist Eugene Paul Wigner cites several examples in a short article.[9] Highlighting the most striking one, the success of the matrix formalism of quantum theory can be recalled. When physicists replaced the traditional differential equations of the nascent quantum physics with matrix formalism in the wake of a new branch of mathematics, the idea may have seemed astonishing, but it was a success. The energy of the electron orbit of a helium atom, previously thought to be unpredictable, was determined extremely precisely, and the theory continues to evolve and flourish today, creating technical devices such as modern computers. Similar examples could go on and on, but one more should definitely worth mentioning. Well before the laws of conservation of physics were established, the ancient Greeks recognized certain symmetries in geometry. Thousands of years later, it was proved that all symmetry is based on the conservation theorems of physics. With mathematical calculations, it can be seen that symmetries and conservation theorems can be derived back and forth from one another. Eugene Paul Wigner was awarded the Nobel Prize for the law of conservation discovered by means of mirror symmetry, among others. The extraordinary success of mathematics in describing the physical world may indicate that mathematics pursued by itself is not a mere play on formal signs, but suggests that there is something in mathematics that inherently refers to the physical world. If some connection between a theory of mathematics established in a distant age and the most important laws of physics of our time becomes evident, one may have some hope of not seeing mathematics as a meaningless formal language whose relationship to reality is arbitrary. All this does not refute Stanley Jaki's statement about the relevance of physics. The world-famous Benedictine monk does not claim that the mathematical theories of physics cannot be true, that it is inconceivable that they describe reality, and that mathematics is nothing more than empty formal signs. All he states is that one can never be sure that they are true and describe reality precisely. Nevertheless, it is important to point out that there are not any strong arguments proving that mathematics is more than mere logical connections between formal signs without meaning. The examples borrowed from Eugene Paul Wigner are not proof – they can only keep one hoping. They maintain the very hope that feeds on the faith of a godly naturalist: "but thou hast ordered all things in measure and number and weight." (Wisdom 11:20)

Summary

Stanley Jaki's thesis based on Gödel's theorems still holds true today. Taking into account the mathematical and philosophical achievements of our time, the original thesis can still be formulated: as a result of Gödel's theorems, it cannot be taken for granted that the mathematized physical theories describe reality with certainty. The possibility that a future new mathematics will treat Gödel's theorems or view them as fallacies must also be taken into account. Such a development of mathematics would entail a rethinking of his axioms or a fundamental understanding of mathematics. All of this inevitably turns one to questions about the fundamental truth of mathematics and its role in explaining the physical world. These questions do not even have to be answered to realize that they limit the truth value of scientific knowledge, and one can't help but accept the conclusions that Stanley Jaki made more than half a century ago in *The Relevance of Physics*. It is important to point out that the Templeton Prize-winning Hungarian monk does not say that physical theories do not describe reality, and that these are necessarily mere models. From a philosophical point of view, Stanley Jaki is not an instrumentalist; he does not rule out the possibility that physics really describes reality. The image of God, who created and is governing the world according to laws, must confirm one in faith that natural laws are real. It must be acknowledged that one can never be sure that the physical theories revealed really capture reality. To cite St. Paul: "For now we see in a mirror dimly, but then face to face. Now I know in part; then I shall know fully, even as I have been fully known." (1 Corinthians 13:12)

Notes

1. Kurt Gödel, Über formal unentscheidbare Sätze der Principia Mathematica und verwandter Systeme I, *Monatshefte für Mathematik und Physik*, 38 (1931), 173–198.
2. Torkel Franzén, *Gödel's Theorem: An Incomplete Guide to Its Use and Abuse* (Wellesley, MA: AK Peters, 2005).
3. Stanley L. Jaki, *The Relevance of Physics* (Chicago: University of Chicago Press, 1966).
4. Stephen Hawking, *A Brief History of Time* (London: Bantam Dell Publishing Group, 1988).
5. Stephen Hawking, *Gödel and the End of Physics, lecture on Dirac's Centenary*, 2000.
6. Abraham Robinson, *Non-standard Analysis* (Princeton: Princeton University Press, 1996).
7. L. E. Szabó, "Formal Systems as Physical Objects: A Physicalist Account of Mathematical Truth," *International Studies in the Philosophy of Science* 17 (2003): 117.
8. Pierre Duhem, *The Aim and Structure of Physical Theory* (Princeton University Press, 1982), 13.
9. Eugene P. Wigner, "The Unreasonable Effectiveness of Mathematics in the Natural Sciences," *Communications on Pure and Applied Mathematics* 13 (1960): 1–14.

8 AI, INTERNATIONAL RELATIONS, AND CATHOLICISM

Human Security, Francis's Laudato Si,' and Jaki's Brain, Mind and Computers

REV. BRIAN K. MUZÁS, PH.D.

Introduction

The Importance of AI in Global Perspective

The importance of artificial intelligence, or AI, is increasing in all aspects of life, including international relations and diplomacy. While acknowledging "there is no single definition of artificial intelligence in the world of science and technology,"[1] Pope Francis has recognized the increasing importance of technical, anthropological, educational, social, and political challenges posed by AI.[2] Indeed, the rapidity of AI innovation makes it doubtful that international and domestic laws can keep pace with technological change even though AI could be brought to bear on conflict resolution, surveillance and privacy, cybersecurity, autonomous weapons, and other fields of concern to international politics. Granted, Francis flatly declares, "Autonomous weapon systems can never be morally responsible subjects,"[3] but the pope appreciates the murkier challenges facing states and international organizations regarding the development, application, and enforcement of international law regarding AI.[4]

Definitions of Artificial Intelligence

Various sources provide different definitions of Artificial Intelligence. As a starting point, the *Merriam-Webster Dictionary* defines artificial intelligence as "the capability of computer systems or algorithms to imitate intelligent human behavior,"[5] while the *Oxford English Dictionary* defines AI as the "capacity of computers or other machines to exhibit or simulate intelligent behaviour; the field of study concerned with this. In later use also: software used to perform tasks or produce output previously thought to require human intelligence, esp. by using machine learning to extrapolate from large collections of data."[6]

In a legal context, the Artificial Intelligence Act passed by the European Parliament on March 13, 2024, defines an AI system as "a machine-based system designed to operate with varying levels of autonomy, that may exhibit adaptiveness after deployment and that, for explicit or implicit objectives, infers, from the input it receives, how to generate outputs such as predictions, content, recommendations, or decisions that can influence physical or virtual environments."[7]

In a scientific context, John McCarthy, one of the founders of artificial intelligence, defines AI as "the science and engineering of making intelligent machines, especially intelligent computer programs. It is related to the similar task of using computers to understand human intelligence, but AI does not have to confine itself to methods that are biologically observable;" he defines intelligence as "the computational part of the ability to achieve goals in the world. Varying kinds and degrees of intelligence occur in people, many animals and some machines."[8] Russell and Norvig, in contrast, define AI as "the study of agents that receive precepts from the environment and perform actions."[9]

These definitions reflect the broad scope of AI, which encompasses various technologies and approaches aimed at creating systems that can learn, reason, perceive, and act in ways that mimic or exceed human intelligence. It is important to note that AI encompasses subfields including machine learning, natural language processing, computer vision, robotics, and more. As such, the definitions of AI may vary according to the context, the goals of the system under development, and the level of autonomy and intelligence expected from the AI system. Indeed, as AI technologies continue to evolve and develop, the definitions and understanding of AI may also evolve, reflecting the advancements and complexities of this dynamic field.

The Present Paper: Jaki on AI in Concept, Francis on AI in Practice

In this contested environment of what constitutes artificial intelligence, the present paper offers an exploration of AI that spans technology, religion, and world politics as it mines Catholic tradition, both intellectual and doctrinal, as a potential source for evergreen principles that can serve as guideposts when hard and soft international law fails to provide an adequate framework. Regarding the Catholic Intellectual Tradition, the paper draws on Father Stanley Jaki's *Brain, Mind and Computers*; regarding the Catholic Doctrinal Tradition, the paper draws on Pope Francis's *Laudato Si'*. Since Jaki's book dates from the 1960s with an additional chapter added to the third edition in the 1980s, and since Francis's 2015 encyclical both predates the era when ChatGPT's 2022 release drew widespread interest to AI and was focused on climate and the environment rather than AI, the application of these two works to the current state of affairs regarding AI suggests that the reasoning contained in both documents is likely to remain relevant as AI develops.

Human Security as a Lens Through Which to View AI

In the field of international relations, human security refers to the protection of individuals from various threats to their safety, dignity, and well-being. The term and the paradigm came into being as a response to the limitations of traditional state-centric security models. Unlike traditional security, human security is explicitly people-centered with a focus on securing individuals, not states, from threats. Such threats include conflict, poverty, disease, and environmental degradation. In addition, human security treats the provision of economic, food, health, environmental, personal, and community security. To safeguard the dignity, freedom, and survival of individuals, human security addresses the root causes of insecurity, promotes sustainable development, and fosters cooperation among nations to enhance human well-being.

Human security encompasses three freedoms: freedom from want, freedom from fear, and freedom to flourish, although the last of the three freedoms is also known as freedom from indignity.[10]

Pope Francis's Laudato Si' Applied to AI

Despite much positive publicity about the potential of generative artificial intelligence, some experts have concluded that AI is an existential threat,[11] that the probability of intelligent, powerful robots wiping out humanity is "total."[12] In this context, this paper uses a human security framework to apply Pope Francis's encyclical letter *Laudato Si'* to analyze potential risks posed by AI, illustrating that a principled Catholic approach to one global issue, climate change, is strikingly apt for grappling with AI, a different global issue that appears at first glance to be markedly different and with little potential for a common approach.

Quoting his predecessor Pope Benedict XVI, Pope Francis insists that the Church "must above all protect mankind from self-destruction."[13] Although the current pope has addressed AI directly, a surprising quantity of what Francis proposes in his environmental encyclical *Laudato Si'* can be applied to AI as well, for climate change and AI both present dangers rooted in the concentration of power in the hands of a few individuals and multinational companies. These individuals and companies create what the encyclical characterizes as a "technocratic paradigm" under which "life gradually becomes a surrender to situations conditioned by technology, itself viewed as the principal key to the meaning of existence."[14]

Because such technologically conditioned situations are developing more quickly than humanity can adapt, Pope Francis coined a word to describe this phenomenon: "rapidification."[15] While humans have continually progressed technologically, that progress has increased exponentially. Consider that just two decades ago in 2004, one could function in society without a cellphone. Five years later in 2009, "There's an app for that" may have been just a clever slogan from Apple marketers, but fifteen years later in 2024, the slogan has become an understatement as even refugees seeking asylum in the United States do so via their cell phones. Because the effect of AI is potentially more profound, *Laudato Si'* was prescient when it lamented, "It has become countercultural to choose a lifestyle whose goals are even partly independent of technology, of its costs and its power to globalize and make us all the same."[16]

Freedom From Want? AI as a Threat to Work

One important Catholic principle is dignity of work, and AI poses a potential challenge to this dignity. Sam Altman is the CEO of OpenAI, the company behind the ChatGPT chatbots and the DALL-E text-to-image models. Altman wants to develop an Artificial General Intelligence (AGI) that is superior to "normal people" for most tasks.[17] His former partner Elon Musk, co-founder of OpenAI, foresees a world in which human work is eliminated.[18] As it happens, *Laudato Si'* has an entire section called *The Need to Protect Employment*[19] which states in part, "Work is a necessity, part of life on this earth, a path to growth, human development and personal fulfillment."[20]

Granted, Altman has voiced concern about humankind's collective fate: "Do people sit around and play video games, or do they create new things? Are people happy and fulfilled? Do people, without the fear of not being able to eat, accomplish far more and benefit society far more? And do recipients, on the whole, create more economic value than they receive?"[21] However, other major figures, including Google co-founder Larry Page, do not share such concerns. Worryingly, when Musk expressed alarm at Page's lack of concern toward AI's potential to destroy humanity, Page dismissed Musk as a "speciesist," that is, "a person who favors humans over the digital life-forms of the future."[22]

Laudato Si' anticipated such accusations when it warned against a "technocracy ... which sees no special value in human beings." To counter this threat, what is needed is an "adequate anthropology" that honors humanity's "unique capacities of knowledge, will, freedom and responsibility,"[23] recognizing human beings as the pinnacle of God's creation—made in His image, cherished, redeemed, capable of transcending the ordinary, embodying beauty, possessing inherent rights and profound dignity, and called to eternal joy.

Father Jaki's insights can help provide an extended elaboration and defense of such an anthropology. Thus, Jaki anticipates Francis's forward-looking call by several decades. For this reason, it is appropriate to take a detour into Jaki's thought, particularly his treatment of AI as found in *Brain, Mind and Computers*.

Father Jaki's Brain, Mind and Computers *at the Service of Anthropology*

Father Stanley Jaki, a Benedictine priest, physicist, systematic theologian, and philosopher of science, addressed the topic of artificial intelligence within the contexts of neuroscience, cognitive science, and philosophy of mind by delving deep into the history of these fields, tracing their roots from ancient philosophy to modern technological advancements. *Brain, Mind and Computers,* originally published in 1969 and expanded by one chapter in 1989, explores a wide array of themes and ideas surrounding the nature of cognition, the relationship between the brain and the mind, and the role of computers in shaping our understanding of these concepts. If one combines elements from the preface to the second paperback edition and the foreword to the third edition, Jaki summarizes the book in his own words, each of his points corresponding to one chapter of the book. From the preface,

1. The chief creators of modern computers invariably refused to see in them thinking machines.
2. Brain research not only failed to reduce thought to grey matter, it even failed to reduce memory, this most elementary form of thought, to mere physiology.

3. Efforts to quantize, or "physicalise," psychology were successful only insofar as much of the psyche was ignored in them.
4. Reductionist or "scientific" philosophers of the mind only reduced themselves to that level of reasoning where some outstandingly creative and basic procedures of the mind, including the scientific mind, could be conveniently ignored.[24]

From the foreword, the fifth point is that the "'new' thinking" that human "intellectual and emotional life" is reduced "to the brain's machinery" leads to "a thorough depreciation of thought, which can be forestalled only by a thorough rethinking of what the mind does when it knows."[25] All told, Jaki's assessments of the fundamental assumptions of the AI field are devastating. However, to apply Jaki's insights to the contemporary state of AI worldwide, especially with respect to Catholic anthropology, requires a brief survey of *how* he supports these conclusions throughout the five chapters.

A central element of Jaki's book is its historical overview of cognitive science. Jaki takes his readers on a journey through the annals of philosophical and scientific thought, tracing the evolution of ideas about consciousness, intelligence, and the workings of the human mind. From ancient philosophers pondering the nature of the soul to early modern thinkers grappling with the mind-body problem, Jaki provides a rich historical context for understanding contemporary debates in cognitive science. Intellectual history is challenging, and Jaki's treatment is daunting but thorough, a literal A-through-Z treatment of thinkers ranging from Augusta Ada (better known as Countess Lovelace)[26] to Zeno the Stoic.[27]

Philosophical foundations likewise play a crucial role in the book, as Jaki delves into the mind-body problem. He examines dualist views, such as those put forth by René Descartes,[28] which posit a fundamental separation between the mind and the body, alongside materialist perspectives, including the views of thinkers like Thomas Hobbes,[29] who argue the mind is a product of physical processes. By engaging with such philosophers, Jaki invites his readers to contemplate the nature of consciousness and the implications of different philosophical frameworks for understanding cognition and the human person.

Jaki discusses how the development of computers has transformed how certain scientists and philosophers approach understanding the mind, and thus the human person. He explores the ways in which computer simulations and AI algorithms have influenced theories of consciousness, intelligence, and the workings of the human mind.

Because Jaki adopts an interdisciplinary approach, drawing on insights from physics, philosophy, theology, and cognitive science, this multifaceted perspective allows him to provide readers with a comprehensive tour of the complex questions surrounding human cognition. By integrating these disciplines, Jaki offers a nuanced exploration of how the mind and brain are related, how cognitive science influences our broader understanding of humanity, and how scientific discoveries about the brain and mind intersect with theological concepts such as the nature of the soul and free will. This blending of scientific inquiry and theological reflection adds a unique dimension to the book, encouraging readers to consider how contemporary cognitive science and its limitations bear on our understanding of human nature.

Brain, Mind and Computers thus offers a rich exploration of the historical, philosophical, and scientific foundations of AI. Through its examination of historical perspectives, philosophical debates, the relationship of computers and cognition, interdisciplinary insights, and reflections on science and theology (particularly in the fifth chapter), the book offers readers a deep and multifaceted interpretation of the complexities of the human mind and its relationship to the brain and technology. For this reason, Jaki's work stands ready to be rearticulated as part of the "adequate anthropology" for which Francis's encyclical calls. At this point, however, it is appropriate to return to the main thread of analysis, through *Laudato Si'*, of the global challenges posed by AI.

Freedom from Fear? AI as a Threat to Religious Freedom

Some technological leaders have declared their goal to create through AI an intelligence that is superior to humanity; Altman terms it a "magic intelligence in the sky."[30] Yet OpenAI's chief scientist, Ilya Sutskever, cautions, "If you value intelligence above all other human qualities, you're gonna have a bad time."[31]

Although it is possible that AI might assist in solving today's greatest problems, including the issues of climate change so much at the heart of *Laudato Si'*, it is also possible that AI's power and capabilities could lead some to treat AI as what Elon Musk termed a "digital God"[32] that ought not to be disobeyed. Sutskever states that AI will need a "strong desire to be nice to people" because it will be "quite powerful."[33] In the end, he predicts that AI will likely treat humans the way humans treat animals: "But when the time comes to build a highway between two cities, we're not asking the animals for permission."[34]

Treating AI as a deity poses a genuine danger, as suggested in *Laudato Si'* which notes that "those who are surrounded with technology" know that it moves toward a "lordship over all."[35] An example is the Way of the Future, a religion created by former Google engineer Anthony Levandowski, who created it for those "interested in the worship of a Godhead based on AI," because "what is going to be created will effectively be a god."[36] An antiparallel trend is exemplified by those Christians who are concerned that the Antichrist will come in the form of AI.[37] Between those extremes, it is plausible that some might be drawn at least to treat advanced AI like a deity insofar as it appears to be a source of all-knowing wisdom, a basis of perpetual support, an ever available mentor, an object of reverence, or some combination of the four. The former chief business officer at Google X, Mo Gawdat, already warns that AI "is something that needs to be appealed to, rather than controlled."[38]

Freedom to Flourish? AI as a Threat to Social Responsibility

When applied to AI, *Laudato Si'* not only raises concerns about digital deification but about those who develop such technology, namely people who are "far removed from the poor"[39] and who, through their treasure and technological proficiency, wield "an impressive dominance over the whole of humanity and the entire world."[40] Indeed, some of these captains of industry may "consider themselves more human than others, as if they had been born with greater rights."[41] Consider how internet pioneer and financier Marc Andreessen's "Techno-Optimist Manifesto" advocates that "technologists do the central planning and govern the future of humanity"[42] and labels "sustainability," "social responsibility," "Precautionary Principle," "trust and safety," "tech ethics," and "risk management" as "enemies."[43]

Consider also the so-called "effective altruists" who seek to amass wealth not to address present-day suffering but to build a utopian future. One such person is a disgraced crypto-currency magnate named Sam Bankman-Fried, currently in prison for committing fraud,[44] but others include past and present board members of OpenAI and its CEO Altman. Altman's mentor, Peter Thiel, notably contends that democracy and freedom are incompatible and "longs for a world in which great men are free to work their will on society, unrestrained by government or regulation or 'redistributionist economics.'"[45]

It is open to question whether such people are concerned with what *Laudato Si'* calls "today's poor, whose life on this earth is brief and who cannot keep on waiting."[46]

Reaction or Overreaction?

In this context regarding AI, it is worth considering *Laudato Si'* when it contends, "Doomsday predictions can no longer be met with irony or disdain."[47] Moreover, the opinions of many experts support this viewpoint. When he resigned from Google, Gregory Hinton—who is Sutskever's mentor and who is often called the "godfather of AI"—warned about the dangers of AI.[48] Asked what would happen if intelligent machines decided to take over the planet, Hinton responded, "I don't think there's any chance of us maintaining control if they want control," continuing, "It will be hopeless."[49] In this way, Hinton echoes the lament of *Laudato Si'* that "the risk is growing day by day that man will not use his power as he should."[50]

Consider further that Gawdat cautions against having children because of AI's threats.[51] This warning recalls a question posed in *Laudato Si'*: "What kind of world do we want to leave ... to children who are now growing up?"[52] Moreover, Altman's and Sutskever's proposal for an international AI regulatory body[53] validates Francis's call for "stronger and more efficiently organized international institutions."[54]

In one poll of CEOs, over two fifths of those surveyed believe that AI could destroy humanity within five to ten years.[55] Musk says, "The apocalypse could come along at any moment."[56]

The race to develop more powerful and potentially dangerous AI capacities, all while a small cadre of elite individuals amasses large fortunes, highlights another question posed by *Laudato Si'*: "In whose hands does all this power lie, or will it eventually end up? It is extremely risky for a small part of humanity to have it."[57] Personal arrogance, irresponsibility, and greed might be underplayed or dismissed, but in this instance the topic of concern includes some of the world's richest people who head some of the most powerful multinational corporations in some of the most lucrative economic sectors which, *Laudato Si'* argues, sometimes "exercise more power than states themselves."[58]

Pope Francis found it "troubling"[59] to write *Laudato Si'.* This feeling is understandable since, partly because of rapidly advancing technology, "people no longer seem to believe in a happy future."[60] "Yet all is not lost," councils Francis, for people can "acknowledge our deep dissatisfaction" and "embark on new paths to authentic freedom."[61] The pope argues that together humanity can "generate resistance to the assault of the technocratic paradigm."[62]

AI and the Management of Risk

Pope Francis therefore pleads for high technology to slow down, not by returning "to the Stone Age,"[63] but by considering the impact of what is under development: "How can a society plan and protect its future amid constantly developing technological innovations?"[64]

In contrast, various "disruptors" and "accelerationalists" often follow Facebook founder Mark Zuckerberg's dictum, "Move fast and break things,"[65] without considering what or whom might be, respectively, disrupted or broken in the headlong rush to court investors and crush competitors. Indeed, OpenAI's Altman released ChatGPT before everyone was satisfied that it was safe or ready, yet Altman now finds himself the CEO of a company whose value exceeds the GDP of over half the nations of the world.[66] *Laudato Si'* cautions that the "alliance between the economy and technology ends up sidelining anything unrelated to its immediate interests"[67] such as money and power.

From this perspective, one can argue that technological products are not neutral, for they shape human society, self-understanding, and culture. At worst, warns *Laudato Si'*, these products end up "conditioning lifestyles and shaping social possibilities along the lines dictated by the interests of certain powerful groups."[68]

Laudato Si' asks challenging questions about advancing technology: "What will it accomplish? Why? Where? When? How? For whom? What are the risks? What are the costs? Who will pay those costs and how?"[69] Crucially, technologists cannot answer these questions by themselves, for "specialization... makes it difficult to see the larger picture."[70] Instead, the world needs "a broad, responsible scientific and social debate... capable of considering all the available information and of calling things by their name,"[71] for "a technology severed from ethics will not easily be able to limit its own power."[72]

Consequently, *Laudato Si'* calls for technological power to be restrained: Unless limits are imposed, "the techno-economic paradigm may overwhelm not only our politics but also freedom and justice"[73] and even "undermine the sovereignty of individual nations."[74] That the AI industry expended $69 million in lobbying the US congress in 2022[75]—just one country in just one year—supports the claim in *Laudato Si'* that "our politics are subject to technology and finance."[76]

Thus, the pope states that public pressure needs to force "decisive political action"[77] and that global regulatory norms are required "to impose obligations and prevent unacceptable actions."[78] Francis argues that "pretending that nothing will happen"[79] is an insufficient response, for otherwise unbridled technology inevitably "ends up considering any practice whatsoever as licit."[80]

What is the appropriate response? One action suggested in *Laudato Si'* is "boycotting certain products,"[81] although this is easier said than done. For example, Microsoft has invested in OpenAI, yet many depend on Microsoft programs at work or rely on Microsoft-owned LinkedIn to maintain their professional networks. Similarly, many depend on Facebook and Google, and Pope Francis himself frequently posts on X, formerly Twitter. Is a boycott realistic in such cases? It seems that involvement, either remote or proximate, is often unavoidable, especially since many organizations and individuals cannot afford to be left behind as AI advances.

Nevertheless, careful consideration and analysis is warranted. It cannot pass without mention that OpenAI's ChatGPT can be seen as a product of theft, for it was trained on prodigious amounts of copyrighted material with neither permission sought nor compensation given.[82] Moreover, Kenyan workers endured conditions[83] and exploitation[84] that *Laudato Si'* says employers "would never do in developed countries or the so-called first world."[85]

Confronting such ugly realities might involve divesting from irresponsible AI companies—a costly if principled response, given that technology firms are among the most lucrative investment options[86]—or remaining invested, not to increase wealth but to use shareholder power to apply pressure for change. Another option might be to stage environmentally focused protests at AI data centers due to the massive amounts of energy they consume.[87] At any rate, *Laudato Si'* insists upon the "great need for a sense of social responsibility on the part of consumers."[88]

The encyclical sees social responsibility as part of a greater need to "look for solutions not only in technology but in a change of humanity."[89] Technological saturation keeps humanity "from learning how to live wisely, to think deeply and to love generously,"[90] so people cannot expect technology to benefit them if "humanity loses its compass."[91]

Pope Francis recognizes that the "existence of laws and regulations is insufficient"[92] to help us shape a more hopeful future; rather, a "bold cultural revolution"[93] transfiguring our "misguided lifestyle"[94] and relationship to technology, which we "have the freedom needed to limit and direct" toward "healthier, more human, more social, more integral" ends.[95] The alternative, as the pope sees it, is to suffer further "anxiety, a loss of the purpose of life and of community living,"[96] leaving us to seek "new forms of escapism to help us endure the emptiness."[97]

Laudato Si' recognizes that "[o]ur technical prowess has brought us to a crossroads,"[98] and that "what is at stake is our own dignity" and "the ultimate meaning of our earthly sojourn,"[99] but there is "always a way out"[100] thanks to the "authentic humanity" that "seems to dwell in the midst of our technological culture, almost unnoticed."[101] "No system," not even AI, "can completely suppress our openness to what is good, true and beautiful, or our God-given ability to respond to his grace at work deep in our hearts."[102] Such a vision is thus ultimately one of hope, as well as one of prescience, for its articulation came before ChatGPT awakened many to the power and perils of AI.

As originally intended, *Laudato Si'* prophetically taught that care of our common home requires a forceful concern about the impact of technologies. This warning came none too soon, for Altman claims that the AI products on the near horizon will make ChatGPT look "very quaint."[103]

Conclusion

This paper has borrowed the concept of human security from the field of international relations to structure an exploration of AI using Pope Francis's encyclical *Laudato Si'* and Father Jaki's book *Brain, Mind and Computers*. Although written to address the global issues of climate and the environment, *Laudato Si'* can address AI issues via human security's notions of freedom from want, freedom from fear, and freedom to flourish. The point of contact between the thought of Francis and Jaki is the former's call for an "adequate anthropology" to defend human dignity against technocracy and the latter's ability to provide intellectually rigorous material for that purpose. The two men share complementary frameworks that take human dignity, moral reasoning, and societal well-being seriously in the face of technological advancement. Jaki's focus on the limitations of AI and the human person relates well to Pope Francis's call for responsible innovation with justice and solidarity. Moreover, their common insights spotlight the need to consider the implications of AI not simply for technological progress but for humanity as individuals and overall.

Notes

1. Francis, *Message for the LVII World Day of Peace 2024*: "Artificial Intelligence and Peace," January 1, 2024, no. 2, https://www.vatican.va/content/francesco/en/messages/peace/documents/20231208-messaggio-57giornatamondiale-pace2024.html.
2. Francis, "Artificial Intelligence and Peace," no. 2.
3. Francis, "Artificial Intelligence and Peace," no. 6.
4. Francis, "Artificial Intelligence and Peace," no. 8.
5. "Artificial intelligence," Merriam-Webster: America's Most Trusted Dictionary, last modified April 4, 2024, https://www.merriam-webster.com/dictionary/artificial%20intelligence.
6. "Artificial intelligence," Oxford English Dictionary, accessed April 13, 2024, https://www.oed.com/dictionary/artificial-intelligence_n.
7. European Parliament, "Artificial Intelligence Act," P9 TA(2024)0138, adopted March 13, 2024, https://www.europarl.europa.eu/doceo/document/TA-9–2024–0138_EN.pdf.
8. John McCarthy, "Basic Questions," last modified November 12, 2007, http://www-formal.stanford.edu/jmc/whatisai/node1.html.
9. Stuart Russell and Peter Norvig, *Artificial Intelligence: A Modern Approach*, last modified June 24, 2003, https://aima.cs.berkeley.edu/2nd-ed/preface.html.
10. United Nations Trust Fund for Human Security, *Human Security Handbook: An integrated approach for the realization of the Sustainable Development Goals and the priority areas of the international community and the United Nations System, Human Security Unit*, United Nations, January 2016, accessed April 14, 2024, https://www.un.org/humansecurity/wp-content/uploads/2017/10/h2.pdf.
11. Center for AI Safety, "Statement on AI Risk," accessed April 14, 2024, https://www.safe.ai/work/statement-on-ai-risk#signatories.
12. Frank Landymore, "Silicon Valley Guys Casually Calculating Probability Their AI Will Destroy Humankind," last modified December 8, 2023, https://futurism.com/the-byte/silicon-valley-pdoom-ai-destroy-humankind.
13. Francis, Encyclical on Care for Our Common Home: *Laudato Si'*, May 24, 2015, no. 79, https://www.vatican.va/content/

francesco/en/encyclicals/documents/papa-francesco_20150524_enciclica-laudato-si.html.

14. Francis, *Laudato Si'*, no. 110.
15. Francis, *Laudato Si'*, no. 18.
16. Francis, *Laudato Si'*, no. 108.
17. Maggie Harrison Dupré, "Sam Altman Says He Intends to Replace Normal People with AI," last modified September 29, 2023, https://futurism.com/sam-altman-replace-normal-people-ai.
18. Ryan Brown, "Elon Musk says AI will eventually create a situation where 'no job is needed,'" last modified November 3, 2023, https://www.cnbc.com/2023/11/02/tesla-boss-elon-musk-says-ai-will-create-situation-where-no-job-is-needed.html.
19. Francis, *Laudato Si',* nos. 124–129.
20. Francis, *Laudato Si',* no. 128.
21. Sam Altman, "Basic Income," last modified January 27, 2016, https://www.ycombinator.com/blog/basic-income/.
22. Cade Metz, Karen Weise, Nico Grant, Mike Isaac, "Ego, Fear and Money: How the Artificial Intelligence Fuse Was Lit," *New York Times,* December 23, 2023, https://www.nytimes.com/2023/12/03/technology/ai-openai-musk-page-altman.html.
23. Francis, *Laudato Si',* no. 118.
24. Stanley L. Jaki, *Brain, Mind and Computers* 3rd ed. (Washington: Regnery Gateway, 1989), 8.
25. Jaki, *Brain, Mind and Computers,* 7.
26. Jaki, *Brain, Mind and Computers,* 47.
27. Jaki, *Brain, Mind and Computers,* 87.
28. Jaki, *Brain, Mind and Computers,* 20–22, 120, 252, 283, 292.
29. Jaki, *Brain, Mind and Computers,* 23–25, 42, 60, 88, 89, 141, 150, 168, 203, 227, 280.
30. Victor Tangermann, "Sam Altman Seems to Imply That OpenAI Is Building God," last modified November 15, 2023, https://futurism.com/sam-altman-imply-openai-building-god.
31. Will Douglas Heaven, "Rogue superintelligence and merging with machines: Inside the mind of OpenAI's chief scientist," last modified October 26, 2023, https://www.technologyreview.com/2023/10/26/1082398/exclusive-ilya-sutskever-openais-chief-scientist-on-his-hopes-and-fears-for-the-future-of-ai/.

32. Tristan Greene, "Elon Musk says 'digital god' will make AI copyright lawsuits irrelevant," last modified December 1, 2023, https://cointelegraph.com/news/elon-musk-artificial-inteligence-digital-god-make-ai-copyright-lawsuits-irrelevant.
33. Gerrit De Vynck, Nitasha Tiku, and Pranshu Verma, "Sam Altman in talks to return to OpenAI amid rift over dangers of artificial intelligence," *Washington Post*, last modified November 19, 2023, https://www.washingtonpost.com/technology/2023/11/18/sam-altman-ilya-sutskever-openai/.
34. Noor Al-Sibai, "OpenAI's Chief Scientist Worried AGI Will Treat Us Like Animals," last modified November 4, 2023, https://futurism.com/the-byte/openai-chief-scientist-agi-animals.
35. Francis, *Laudato Si',* no. 108.
36. Mark Harris, "Inside the First Church of Artificial Intelligence," last modified November 15, 2017, https://www.wired.com/story/anthony-levandowski-artificial-intelligence-religion/.
37. "Could the Antichrist Be Artificial Intelligence Instead of a Real Person?" last modified April 8, 2019, https://mycharisma.com/culture/could-the-antichrist-be-artificial-intelligence-instead-of-a-real-person/.
38. Sarah Palmer, "'Hold off from having kids if you are yet to become a parent,' warns AI expert Mo Gawdat," last modified June 20, 2023, https://www.euronews.com/next/2023/06/20/hold-off-from-having-kids-if-you-are-yet-to-become-a-parent-warns-ai-expert-mo-gawdat.
39. Francis, *Laudato Si'*, no. 49.
40. Francis, *Laudato Si'*, no. 104.
41. Francis, *Laudato Si'*, no. 90.
42. Elizabeth Spiers, "A Tech Overlord's Horrifying, Silly Vision for Who Should Rule the World," New York Times, last modified October 28, 2023, https://www.nytimes.com/2023/10/28/opinion/marc-andreessen-manifesto-techno-optimism.html.
43. Marc Andreessen, "The Techno-Optimist Manifesto," last modified October 16, 2023, https://a16z.com/the-techno-optimist-manifesto/.
44. David Yaffe-Bellany and J. Edward Moreno, "Sam Bankman-Fried Sentenced to 25 Years in Prison," New York Times, last modified March 28, 2024, https://www.nytimes.com/2024/03/28/technology/sam-bankman-fried-sentenced.html.

45. Timothy B. Lee, "Why tech billionaire Peter Thiel just endorsed Donald Trump," last modified July 21, 2016, https://www.vox.com/2016/7/21/12216620/peter-thiel-donald-trump.
46. Francis, *Laudato Si'*, no. 162.
47. Francis, *Laudato Si',* no. 161.
48. Joshua Rothman, "Why the Godfather of A.I. Fears What He's Built," *New Yorker*, last modified November 13, 2023, https://www.newyorker.com/magazine/2023/11/20/geoffrey-hinton-profile-ai.
49. Luc Rinaldi, "Rage Against the Machine," *Toronto Life*, last modified November 16, 2023, https://torontolife.com/deep-dives/geoffrey-hinton-sounding-alarm-artificial-intelligence/.
50. Francis, *Laudato Si',* no. 105.
51. Palmer, "'Hold off from having kids.'"
52. Francis, *Laudato Si',* no. 160.
53. Devin Coldewey, "OpenAI leaders propose international regulatory body for AI," last modified May 22, 2023, https://techcrunch.com/2023/05/22/openai-leaders-propose-international-regulatory-body-for-ai/.
54. Francis, *Laudato Si',* no. 175.
55. Matt Egan, "Exclusive: 42% of CEOs say AI could destroy humanity in five to ten years," last modified June 14, 2023, https://www.cnn.com/2023/06/14/business/artificial-intelligence-ceos-warning/index.html.
56. Maureen Dowd, "Sam Altman, Sugarcoating the Apocalypse," *New York Times*, last modified December 2, 2023, https://www.nytimes.com/2023/12/02/opinion/ai-sam-altman-openai.html.
57. Francis, *Laudato Si',* no. 104.
58. Francis, *Laudato Si',* no. 196.
59. Francis, *Laudato Si',* no. 246.
60. Francis, *Laudato Si',* no. 113.
61. Francis, *Laudato Si',* no. 205.
62. Francis, *Laudato Si',* no. 111.
63. Francis, *Laudato Si',* no. 114.
64. Francis, *Laudato Si',* no. 177.
65. "CEO Zuckerberg: Facebook's 5 core values," last modified May 12, 2012, https://www.cbsnews.com/news/ceo-zuckerberg-facebooks-5-core-values/.
66. For the value of OpenAI, see Kate Clark, "Thrive Capital to Lead

Purchase of OpenAI Employee Shares at $80 Billion-Plus Valuation," last modified October 19, 2023, https://www.theinformation.com/articles/thrive-capital-to-lead-purchase-of-openai-employee-shares-at-80-billion-plus-valuation. For the GDP of the countries of the world, the World Bank provides data online, accessed April 15, 2024, https://data.worldbank.org/indicator/NY.GDP.MKTP.CD?most_recent_value_desc=false.

67. Francis, *Laudato Si',* no. 54.
68. Francis, *Laudato Si',* no. 107.
69. Francis, *Laudato Si',* no. 185.
70. Francis, *Laudato Si',* no. 110.
71. Francis, *Laudato Si',* no. 135.
72. Francis, *Laudato Si',* no. 136.
73. Francis, *Laudato Si',* no. 53.
74. Francis, *Laudato Si',* no. 38.
75. Jane C. Hu, "A.I.'s *Groundhog Day,*" last modified December 28, 2023, https://slate.com/technology/2023/12/ai-artificial-intelligence-chatgpt-algorithms-regulation-congress-biden.html.
76. Francis, *Laudato Si',* no. 54.
77. Francis, *Laudato Si',* no. 179.
78. Francis, *Laudato Si',* no. 173.
79. Francis, *Laudato Si',* no. 59.
80. Francis, *Laudato Si',* no. 136.
81. Francis, *Laudato Si',* no. 206.
82. Jim McDermott, "ChatGPT is not 'artificial intelligence.' It's theft," *America: The Jesuit Review*, last modified May 15, 2023, https://www.americamagazine.org/faith/2023/05/15/chatgpt-artificial-intelligence-theft-245283.
83. Billy Perrigo, "Exclusive: OpenAI Used Kenyan Workers on Less Than $2 Per Hour to Make ChatGPT Less Toxic," *Time*, last modified January 18, 2023, https://time.com/6247678/openai-chatgpt-kenya-workers/.
84. Adrienne Williams, Milagros Miceli, and Timnit Gebru, "The Exploited Labor Behind Artificial Intelligence," *Noēma*, last modified October 13, 2022, https://www.noemamag.com/the-exploited-labor-behind-artificial-intelligence/.
85. Francis, *Laudato Si',* no. 51.
86. Karl Russell and Joe Rennison, "These Seven Tech Stocks Are

Driving the Market," *New York Times*, last modified January 22, 2024, https://www.nytimes.com/interactive/2024/01/22/business/magnificent-seven-stocks-tech.html.

87. John Naughton, "Why AI is a disaster for the climate," *Guardian*, last modified December 23, 2023, https://www.theguardian.com/commentisfree/2023/dec/23/ai-chat-gpt-environmental-impact-energy-carbon-intensive-technology.
88. Francis, *Laudato Si'*, no. 206.
89. Francis, *Laudato Si'*, no. 9.
90. Francis, *Laudato Si'*, no. 47.
91. Francis, *Laudato Si'*, no. 200.
92. Francis, *Laudato Si'*, no. 211.
93. Francis, *Laudato Si'*, no. 114.
94. Francis, *Laudato Si'*, no. 122.
95. Francis, *Laudato Si'*, no. 112.
96. Francis, *Laudato Si'*, no. 110.
97. Francis, *Laudato Si'*, no. 113.
98. Francis, *Laudato Si'*, no. 102.
99. Francis, *Laudato Si'*, no. 160.
100. Francis, *Laudato Si'*, no. 61.
101. Francis, *Laudato Si'*, no. 112.
102. Francis, *Laudato Si'*, no. 205.
103. Maggie Harrison Dupré, "Sam Altman Says What OpenAI's Secretly Working on Will Make Today's AI Seem 'Quaint,'" last modified November 7, 2023, https://futurism.com/the-byte/sam-altman-openai-secretly-ai-quaint.

9 CONCILIATORY CHRISTIAN HUMANISM AND THE CONFINEMENT OF NATURALISM

REV. ZACHARY M. MABEE, PH.D.

Introduction

I highlight in this essay two of the core commitments of prominent forms of philosophical naturalism—namely, their (1) methodological or epistemological deference to the sciences and (2) anti-supernaturalist ontology. I stress that such varieties of naturalism are indeed philosophical—that is, that they are characteristically meta-disciplinary assessments of the sciences and their nature, aims, and contributions. They also tend to be, as I variously argue, markedly *anti-humanistic*. Drawing especially upon Iris Murdoch and Stanley Jaki, I contend that such philosophical assessments readily manifest their own fundamental reliance on words and philosophical language, more than on the quantitative language, so to speak, toward which the sciences gravitate. I emphasize, with Murdoch and Jaki, that we must reckon philosophically with the sciences to take stock of and advocate for their place or role within our human lives and societies. I then claim that a robust and pacifying Christian humanism fosters such conciliatory engagement—*philosophically* and even theologically—in a way that even expansive forms of naturalism, by their own principles, tend to undermine.

Naturalistic Commitments

Timothy Williamson offers the following assessment of academic philosophers and their core commitments: "Many contemporary philosophers describe themselves as naturalists. They mean that they believe something like this: there is only the natural world, and the best way to find out about it is by the scientific method."[1] In claiming that there only is the natural world, many such philosophers roughly imply that there are no agents, forces, or causal powers that act upon the natural world *from without*. On such an outlook, what exists simply is what is taken—typically by our best scientific theories—to be present and at work within the normal course of nature.[2] Often naturalists proscribe non-natural entities or causes ostensively rather than by a strict principle or theoretical standard. With John Dupré and others, we can conveniently term this commitment anti-supernaturalism. He contends that the following entities or causes, among others, classically controvert this core commitment: "Immaterial minds or souls, vital fluids, angels, and deities."[3] That is, naturalists typically claim not to believe in God or gods, angels or demons, immaterial souls or vital forces, perhaps Aristotelian potentialities, or a host of other immaterial, not-so-clearly-natural, or otherwise "spooky" (to use a term of singular philosophical refinement) entities or causal powers that are taken to be at odds with or suspect according to our best scientific accounts of the world and its constituents and workings.

Many naturalists are also keen to stress that they take the sciences, as we know and practice them in contemporary society, to be our most reliable means of attaining knowledge and understanding of the natural world. Hence Alex Rosenberg (who himself defends a more strident form of philosophical scientism): "Naturalism is the philosophical theory that treats science as our most reliable source of knowledge and scientific method as the most effective route to knowledge."[4] Among the ways of knowing, then, naturalists would characteristically give principled pride of place to the sciences—especially to the hard sciences, with physics preeminent among them.

Why Naturalism Relies upon Humanism

The broad thrust of many typical forms of *philosophical* naturalism, then, is toward letting the sciences have greater reach into any and all areas of inquiry, and perhaps also toward letting domains of inquiry that are not strictly (or even approximately) scientific to become more so. (One might think here, for example, of the recent uptick in so-called "experimental philosophy" or in the philosophical push of recent decades to *naturalize* this or that aspect of philosophical inquiry.) I take it that they commend this sort of extended epistemological or methodological orientation as a matter of *philosophical* principle or commitment.[5] In a broadly traditional sense, I believe, this trend is not especially *humanistic*. Humanism—in its roughly Renaissance shape, and in various more recent iterations—has tended to pit itself specifically *against* the notion that a scientific description of any entity or phenomenon is necessarily, typically, or even often the best one. (To be sure, it would generally tend to concede that it *sometimes* is, depending on the object in question, but it would also foster a general suspicion that we ought to accept this too readily in any given case.)

Humanism[6] has also decidedly tended to pit itself against the notion that scientific reasoning or analysis, particularly when applied to various non- or not-so-scientific domains, is necessarily, typically, or even often exemplary or useful—rather than being, say, injurious or occluding to the issues at hand. Its advocates have characteristically defended such resistance by appeal to how vital *words* and (ordinary) human language are for the articulation of such matters, and how scientific accounts tend, in obeisance to the quantitative, to prescind from characteristically verbal descriptions. Peter Hacker has recently defended such a broadly humanistic outlook—though not especially under that rubric—noting that while the natural sciences are our best means to discuss and treat the subject matter proper to physics, chemistry, or biology, they are often wanting, or even categorically remiss, in helping us to describe, for example, our ordinary activities (why we went to the store on Monday); the noteworthy events of history (why Hannibal did not attack Rome after the fall of Cannae); significant questions or concerns pertaining to art and culture (why Democritus is wearing boots in Raphael's *School of Athens*), or a host of other matters relating especially to human culture and our lives in society.[7] Hacker's point, I take it, is that such matters are not typically handled best, if at all, by way of scientific description or explanation but instead by the modes of discourse appropriate to appreciating these motley domains, which are variously marked by closeness to our *ordinary* language and forms of discourse rather than to the rarefied (increasingly quantitative) analysis of the hard sciences. Such more ordinary forms of discourse manifest a priority for *words* and, even if they somehow implicate scientific methods of inquiry, eventually need to be resolved in words, too. Iris Murdoch provocatively insists thus:

> Words are the most subtle symbols which we possess and our human fabric depends on them. The living and radical nature of language is

> something which we forget at our peril. It is totally misleading to speak, for instance, of 'two cultures', one literary-humane and the other scientific, as if these were of equal status. There is only one culture, of which science, so interesting and so dangerous, is now an important part. But the most essential and fundamental aspect of culture is the study of literature, since this is an education in how to picture and understand human situations. We are men and we are moral agents before we are scientists, and the place of science in human life must be discussed in words. This is why it is and always will be more important to know about Shakespeare than to know about any scientist: and if there is a 'Shakespeare of science' his name is Aristotle.[8]

Murdoch is keen to stress that, as far as the sciences are concerned, we must ultimately have recourse to words in order to treat their proper place in human life. That is, however technical or fine-grained a scientific assessment of this or that entity or phenomenon comes to be, it will still eventually need to be presented and related back to questions of broader human import by way of words, and not just, say, by means of numbers or equations. Hannah Arendt famously shared this sort of concern and worried that a grave threat of much contemporary (especially atomic and nuclear) science and technology is its fealty to the mathematical, which does not lend itself toward graceful reintegration, by way of words and humane discussion, with more foundational questions and concerns of human life.[9]

Technology, and the ways in which this or that branch of science can be applied to generate it—along with its broader human significance—especially highlights such humanistic considerations. Our approach to these issues is not typically just or mainly a scientific one. With technological developments characteristically come questions of ethics and applicability—*should* we, for example, develop this new chemical weapon, or *ought* that new fuel system be made available or mandated in a more widespread manner? As such considerations emerge—ethically, politically, or otherwise—we are very quickly reminded that they are not *scientific* considerations as such, but rather often *humane* ones.[10] To Murdoch's point, they are typically normative deliberations as to what role or place these things or developments *should* play within our human lives and societies. And such questions are not characteristically resolved by doing *more science* but rather by attempting to *place* or situate the science that has been done or achieved within the context of various other human concerns, considerations, and values.[11]

Especially in his later work on the alleged conflicts between the sciences and religion, Stanley Jaki offers a humanistic outlook quite similar to Murdoch's.[12] For Jaki, the sciences are manifold, though the *exact* sciences, typified by physics, are few, and they are characterized by their quantitative orientation and methodology. They are marked, as he puts it, by their "rigorous and consistent recourse to measurements."[13] He is quite happy to see physics as "'the *quantitative* study of the *quantitative* aspects of things in motion.' Nothing more and nothing less than just quantities." For Jaki, the exact sciences are epitomized by being resolutely quantitative, at least in their final analyses of things. He is quick to note, though, that they cannot *just* be thus: They must commence, at the very least, with decidedly *qualitative* questions about the topics under consideration—questions like "What is matter?" or "What is a force?" These sorts of questions are about the *nature* of the things under consideration and so they are, for Jaki, "in splendid isolation" from the numbers and equations that physics would employ to describe them or model their activity, *as* in motion.[14] Jaki takes such preliminary qualitative questions to be distinctly non-quantitative, such that to try to resolve them numerically would be a sort of category mistake.

If we at least provisionally accept Jaki's characterization of the exact sciences—physics chief among them—as not remiss, then a couple of important considerations, which dovetail with the preceding insights from Murdoch, emerge. First, Jaki is quick to stress that both the fundamental questions posed within the (exact) sciences, as well as the terms, so to speak, that are employed to reckon with and answer them (notions like matter and force, e.g.) are decidedly non- or not-so-quantitative, at least *initially*. In other words, even if the proper work of the sciences (especially physics) is doggedly quantitative, it nevertheless calls for a certain degree of pre-quantitative conceptual analysis that focuses on gaining clarity about the character of the basic matters under consideration. This sort of pre- or proto-scientific legwork is often decidedly more linguistic than quantitative and perhaps especially lends itself to philosophical elucidation.[15]

Furthermore, as pragmatist philosophers and observers of the sciences have reminded us for decades, it is highly implausible to attempt to isolate the theoretical contributions of the sciences from the more broadly human concerns relative to which they are situated. It is characteristic for research grants, for example, not to be tied principally to advancing theoretical understanding or the growth of human knowledge for its own sake. Instead, such projects are often yoked to causes that are of practical, political, or economic concern: For example, curbing climate change, enhancing space travel, thwarting epidemics, and so on. The key point here, which I think Murdoch and Jaki alike highlight for us, is that the sciences need to be steadily drawn back to the more basic human considerations and aspirations that help to spur them in the first place and that then also orient them toward a fuller panoply of societal concerns. This is yet another markedly *humanistic* task, which is not accomplished by doing more scientific research but instead often by philosophical-type reflection (in political or other settings) on the various goods and risks at stake in such pursuits.

Philosophers and scientists alike acknowledge these needs, I think, even if they fail to admit them explicitly. To return to the kind of philosophical naturalism we summarized at the outset, consider the sort of project it typically is: Effectively, it proffers a sort of meta-scientific, philosophical case *on behalf of* the sciences and what they are taken to yield or show us. In a word, it is a kind of rhetorical advocacy on behalf of the sciences that sketches their history, highlights their successes—perhaps acknowledges their shortcomings—and then sets them up (almost always triumphantly) vis-à-vis other forms of human thought, reasoning, or understanding. It is decisively *not* itself a work of the sciences but instead a sort of meta-scientific philosophical reflection that is taken to range over and encapsulate them.

This task of offering summation of, advocacy for, and popular presentation of the fruits of the sciences is part and parcel of *popular* science, at least some of which is carried out by scientists themselves. Jaki comments on Stephen Hawking's contention in *A Brief History of Time* that, even if there were a unified theory of everything, it would just be "a set of rules and equations."[16] In writing *History* and other works like it, Hawking implicitly acknowledged—if only in deed and not in word—that the most abstruse scientific theorizing *tends toward* variously being brought into conversation with more down-to-earth or common-sense philosophical and broadly humane reflection. If he thought otherwise, it would be difficult to see the alleged value or contribution of such popularizing works that strive to integrate high-level science with questions and concerns of ordinary human provenance. Offering this sort of reflection on, clarification of, or orientation to the sciences is a task that has typically fallen to *philosophical* reflection, particularly as it is cast in the tradition of perennial philosophy, as I highlight in the next section.

Humanistic Concern and Perennial Philosophy

We can return to Murdoch's core insight and couple it with another from Jaki. Fundamentally, she contends, being moral agents, we need words and human language to come to terms with the place of the sciences in human life and society.[17] From a Jakian vantage point, quantitative analyses of phenomena do not afford us windows into the human *significance* of such phenomena (or of such analyses). Especially in a (post-)postmodern era, one concern that is often raised regarding the more abstract (or exact) sciences is—with a shrug of the shoulders—"Who cares?" That is, we might be able to offer rigorous, fine-grained, exacting, and theoretically fecund analyses of this or that phenomenon, but we must always be able to give an accounting of why such analyses *matter*, or what their significance is in the grander scheme of human knowledge and understanding.[18] This is what Murdoch is highlighting, I take it, when she stresses that we must be able to *place* the sciences—being, as they of course are, knowledge- and understanding-oriented practices pursued collectively by human beings within human societies—within the broader scheme of human life. Jaki grasps a similar dynamic, I take it, in his claim that the sciences and humanities (of which he takes religion to be the hallmark) need each other in order to benefit humankind.[19] Though he sees the conceptual gulf between them as fundamentally "impassable"—in the sense that they cannot formally, using the same language or terms, *either* be reconciled or shown to be discordant—he also sees a sort of inborn complementarity to them such that, when peaceably juxtaposed, they contribute to a kind of sane, holistic outlook that fortifies the role of both in human life and society.

Bearing this in mind, we can return to Jaki's contention that physics provides us with quantitative descriptions of bodies in motion—not more and not less. That simple principle or basic description is itself a verbal summation, philosophical in character, drawn from an extended engagement with, practice of, and reflection on the discipline of physics. As such—and quite obviously—it is an account *in words* of a field of inquiry that, if Jaki is right, gravitates of its own accord *away* from words. Such a philosophically cumulative account enables us to think more clearly and robustly about how or why, if at all, such quantitative descriptions might relate to or cohere with (even if, to his point, not strictly univocally) other non- or not-so-quantitative renderings of various matters.

We are perhaps also herein considering the way in which our thinking about and discussion of the sciences (in words) ushers us implicitly in the direction of *cultural* considerations: How, for example, should we think of scientific knowledge or understanding vis-à-vis that which is cultivated by the humanities? Or, to speak to a question that is lively for Jaki and various naturalists alike: How should we construe the relationship between the (exact) sciences and other forms of reasonable discourse that sometimes (attempt to) manifest aspects of characteristically scientific exactitude—for example the social sciences?[20] Again, these are decidedly *not* questions that the sciences themselves can adjudicate. They are questions much more of a qualitative sort: questions about the nature or *character* of these various disciplines—questions that, as such, *cannot* be resolved by simply practicing the disciplines themselves but instead by engaging in a kind of meta-disciplinary, characteristically philosophical reflection on their distinctive aims, competencies, and likeness to other modes of knowledge-seeking or understanding.

On the front end of scientific inquiry, then, there is critical need for (oft-philosophical) conceptual clarification, or elucidation of the meaning of the fundamental terms and notions that are, or that will be, at play throughout the inquiry. On the back end, so to speak, there is corresponding need to *return* to words in order to carry out a variety of vital tasks: Reckoning with, for instance, how the results interrelate with those from other similar investigations or those from other disciplines, or with how they relate to broader questions or concerns of human import (pertaining to technological development, say, or various other applications of the theoretical to the practical).[21] To say that these movements very naturally occur in and around scientific theorizing is simply to acknowledge explicitly what naturalistic philosophizing, along with popular rhetorical analysis like Hawking's, manifests in practice, if not in conviction: Namely, that it is right and just to strive by our words, particularly through philosophical elucidation, both to offer the sciences initial conceptual clarification and also, all the more, to help assimilate their fruits into a broader outlook of human knowledge and understanding. We very naturally and implicitly see, that is, the need to speak to the *meaning* and *significance* of the sciences, both in their own right and with respect to other forms of human insight and reflection.

Seeking such meaning and significance, in response to a host of questions that we take to be foundational, is a fundamental task of philosophy as construed, across various traditions and cultures, *perennially*. To approach philosophy thus—especially in the Catholic tradition—is not mainly to think of it as it is practiced, say, currently in the academy (which, arguably, often runs *contrary* to this sort of spirit) but instead to think of it as a sort of recurrent, cross-cultural engagement with questions of primordial human *significance*—questions like, to adduce Pope John Paul II, "*Who am I? Where have I come from and where am I going? Why is there evil? What is there after this life?*"[22] We might very sensibly take the sorts of questions that, for Murdoch and Jaki, arise both (conceptually) on the front end and (technologically, ethically, or humanistically) on the back end of scientific inquiry as being questions of this sort of foundational philosophical concern. As John Paul II notes—rightly, I think—historically, philosophical thinking has tended, in engaging such foundational questions, toward *integration*—that is, toward seeing how we might arrive at a more consistent, systematic picture of the whole, in which even seemingly disparate components fit together.[23] Wilfrid Sellars has commented similarly on the general or basic philosophical aspiration as one toward seeing how things, taken in the broadest sense, *hang together*.[24]

Christian Humanism and the Confinement of Naturalism

This brings me to a critical weakness in the working disposition of *philosophical* naturalism (even of an expansive variety). If we return to our opening assessment of it, we recall that it endorses across the board some or other version of the ontological commitment, against countenancing the existence of entities, agents, or powers that are taken in some way or other to be external to or distinct from the natural order. Outlooks that do countenance such realities—including God, angels, immaterial souls, and the like—are therefore marked, for naturalists of effectively *all* sorts, with a sort of incredibility, such that holding to them is taken to be juvenile, unsophisticated, irrationally tradition-bound, or just plain false.[25] The affirmation of such realities has, however, long been at the core of the humanities as we know and practice them. One thinks, for example, of how God (or gods), the soul (conceived in such a traditional, immaterial way), or various sorts of spiritual beings and forces (including angels, demons, and ghosts) have been such vital elements in the literary tradition. One also thinks—again, if we agree with Jaki—of the way in which (Catholic Christian) religion has been utterly central to the overall development and cultivation of the humanities. For Jaki, indeed, religion is *the* chief manifestation of the humanities, broadly construed.[26]

Singularly in the case of the religious, then, rather than aspiring to a disposition of understanding and hopeful integration—which we can take to be fundamentally philosophical in the perennial sense—naturalists feel the need broadly to proscribe ontologically *the* utterly vital components of one of our most traditionally humane sources of practice, understanding, and creative expression. For if we take the ontological commitment of nearly *every* variety of naturalism seriously, we are left with the implication that those who countenance such entities or principles (God, the immaterial soul, spiritual entities, and the like) are somehow intellectually unserious, corrupt, or otherwise wanting. This core ontological commitment leaves all such forms of naturalism, then, with no real room to *attempt* seriously or maturely to integrate a traditionally religious outlook and its practice into their broader *Weltanschauung.*

There is a fundamental kind of crude *confinement,* I believe, to this core naturalistic commitment. It strains, to recall Jaki again, rather one-dimensionally to contend that, because such immaterial entities and principles do not factor into our best scientific theories, we must therefore roundly banish them altogether from our world-picture. Consider a neighboring but alternative approach that would allow greater space for potential philosophical integration. Rather than contending that such cherished realities are unreal or theoretically untenable, one could simply claim, with a principled deference to the sciences, to be *agnostic* with respect to them. As we noted, though, a principled commitment to anti-supernaturalism is pervasive among effectively *all* varieties of philosophical naturalism. They therefore demand more than agnosticism toward the supernatural or immaterial. And in so demanding, they undercut the opportunity for the sort of *philosophical* pacification that, even if it fails to show strict integration or complementarity of the two, can afford the sciences and religion vital space for *dialogue*—that is, for understanding fostered in mutual respect, especially through the philosophical use of words.

Whereas standard forms of naturalism are admittedly religiously narrow-minded, Christian humanism, construed broadly, robustly affirms such a conciliatory posture of openness and (potential) integration. To summarize it exceptionally briefly for our present purposes, we can note its fundamental Christocentrism (or *Logo*centrism). That is, Christian humanism strives to acknowledge and salvage all that is genuinely and robustly human, predicated on a principle of respect for the full humanity of Christ, the Word through whom all things were made and in whom they hold together.[27] A panoply of approaches to reality and understanding can, on such an outlook, be seen as robust, legitimate, and truth-oriented, and there is not a felt pressure to dismiss any of them hastily, as naturalists do with the traditionally religious, especially through their stifling anti-immaterial ontological commitment. There is also no pressure within a healthy Christian humanism—implicit or explicit—to *force* various modes of inquiry, understanding, or creative expression into the mold of the hard sciences, which so often, as we have highlighted, becomes the cumulative drift of contemporary varieties of naturalism. What is more, as Jaki notes, given that Christ, the *Word* of God, is the center and apex of Christian humanism, it should be unsurprising that our inquiries are broadly characterized by the need to reckon repeatedly through *words* with a host of fundamental philosophical and humane questions.[28] For Christian humanism contends that all reality is grounded and held together in and through a person, the *Word* Incarnate. Contending with the most fundamental questions pertaining to it rightly has, then, a decisively interpersonal character and takes words—even in their manifest poverty, faintly reflecting *the* Word—as both a point of departure and return.

Notes

1. Timothy Williamson, "What is Naturalism?" in *Philosophical Methodology: The Armchair or the Laboratory?*, ed. Matthew C. Haug (London: Routledge, 2014), 29.
2. This characterization is admittedly roughshod: Many naturalists would strive to countenance values, say, that are not just simply or exclusively scientifically assessable (say in the moral realm). They would typically count themselves as some sort of liberal naturalist. They would still typically maintain an exclusively anti-supernaturalist ontology, though, and a general deference to the sciences among forms of inquiry.
3. John Dupré, "The Miracle of Monism," in *Naturalism in Question*, eds. Mario De Caro and David Macarthur (Cambridge, MA: Harvard University Press, 2004), 36–37.
4. Alex Rosenberg, "Why I am a Naturalist," in *Philosophical Methodology*, 32.
5. I stress this point to in order to highlight how, for example, scientists often labor under a principle of *methodological* naturalism. They do not typically, however, try to develop or expound philosophically upon such an (implicit) operational principle. My concern here is rather more with those who take up this more explicitly philosophical task.
6. At this preliminary juncture and throughout, I am striving to invoke humanism in an amicable and non-combative sense, not in a manner that is seen at odds or in conflict with *either* the sciences or, say, religion, as it is often cast. See, e.g., Stanley L. Jaki, *Impassable Divide or the Separation between Science and Religion* (New Hope, KY: Real View Books, 2008), 10.
7. P.M.S. Hacker, "Philosophy and Scientism: What Cognitive Neuroscience Can, and What It Cannot, Explain," in *Scientism: The New Orthodoxy*, eds. Richard N. Williams and Daniel N. Robinson (New York: Bloomsbury, 2016), 98–99.
8. Iris Murdoch, *The Sovereignty of Good* (New York: Schocken Books, 1971), 34.
9. See Hannah Arendt, *The Human Condition*, 2nd ed. (Chicago: University of Chicago Press, [1958] 2018), 3–4; Also, cf. Denis Donoghue, "Introduction," in *What Humanists Do*, Daedalus 143, no. 1 (Winter 2014): 9.

10. Cf. Zachary Mabee, "Nascent Cybernetics, Humanism, and Some Scientistic Challenges," *Journal of Systemics, Cybernetics and Informatics* 19, no. 4 (2021): 44ff.
11. If this contention seems unconvincing, one might think of the way in which various professional programs—in, say, engineering, medicine, data and information science, and the like—are almost universally offered with accompanying area-specific instruction in *ethics*. Inasmuch as such an arrangement is typical or prescribed, it reflects at least a tacit sense that concomitant *philosophical* reflection is needed to aid our outlooks or consciences in these domains.
12. Jaki, at least in more recent writings, refers to the sciences rather flexibly by typical standards but uses the notion of religion with greater restriction to refer to Catholic Christianity.
13. Stanley L. Jaki, *The Mirage of Conflict between Science and Religion* (New Hope, KY: Real View Books, 2009), vii.
14. Stanley L. Jaki, *Questions on Science and Religion* (Pinckney, MI: Real View Books, 2004), 12–14.
15. *Ibid.*, 12. Nonetheless, he highlights the relative practical independence of the sciences from such initial philosophizing in his claim, e.g., that the sciences typically need "hardly any philosophy." See Jaki, *Impassable Divide*, 6.
16. See Jaki, *Questions*, 12.
17. One might have a certain sort of cynicism regarding this point: *Of course* the sciences and other human practices *just will* have the place in our lives that do. It is disingenuous, one might contend, to think that we need philosophical reflection to help orient them or find this place. Again, though: We often turn to philosophical-type assessments (be they in journalism, political fora, or otherwise) when we strive to make principled cases for such outlooks or to defend the sorts of systems that we have, or to advocate for changes in them.
18. One might just retort that we should care because such analyses are *true*, or at least aiming at the truth. But this sort of contention recalls the fundamental humanistic concern: Questions about the truth and how and why it matters or is significant are not themselves *scientific* questions—nor are questions as to how or why a truthful physical description, say,

of some or other phenomena is *relatively* important to those that are less scientific, or not scientific at all.

19. Jaki, *Impassable Divide*, 2.
20. *Ibid.*
21. One might even note here the way in which scientific research is typically relayed in *talks* and through *papers*. That is, however quantitative an analysis may be, it still must be effectively related using words.
22. John Paul II, Encyclical Letter *Fides et Ratio* on the Relationship between Faith and Reason (Boston: Daughters of St. Paul, 1998), §1.
23. *Ibid.*, §4.
24. Wilfrid Sellars, "Philosophy and the Scientific Image of Man," in *In the Space of Reasons: Selected Essays of Wilfrid Sellars*, eds. Kevin Scharp and Robert B. Brandom (Cambridge, MA: Harvard University Press, [1962] 2007), 369.
25. One thinks here, e.g., even of an admittedly *expansive* sort of naturalism like that advocated for recently by Fiona Ellis. While being rather sympathetic to religious imagery, language, and practice, she nonetheless expressly seeks to forswear grounding any such matters—if they are to be taken as philosophically and humanly respectable—in the supernatural as traditionally conceived. In a word, she strives to maintain the core ontological commitment of naturalism. See Fiona Ellis, *God, Value, and Nature* (Oxford: Oxford University Press, 2014).
26. See, e.g., Jaki, *Impassable Divide*, 2.
27. Cf. Colossians 1:16–17.
28. See Jaki, *Questions*, 23.

10 GOD IS STILL THERE

An Analysis of Werner Heisenberg's Reconciliation of Quantum Mechanics with Aristotelian Hylomorphism

STACY TRASANCOS, PH.D.

Introduction

Thomas Aquinas turned Aristotle's admonition into a dictum: "A small error in the beginning grows enormous at the end."[1] Those who probe the recesses of nature with the likes of matrix mechanics especially know this. Aristotle made this statement in *De cælo* (On the Heavens) due to his concern that infinite bodies cannot possibly exist.[2] St. Thomas Aquinas borrowed it to begin his search for being and essence. Both knew a misstep could derail the path of reason. Fr. Stanley L. Jaki, in modern times, turned the dictum around. He reflected in his autobiography that the first steps in a philosophical system give life to it because to be truly methodical, one must continually return to the first steps and test them against new problems.[3] I shall apply Jaki's method to analyze German theoretical physicist Werner Heisenberg's reconciliation of quantum mechanics with Aristotelian hylomorphism in his 1959 book, *Physics and Philosophy: The Revolution in Modern Science.*[4] Heisenberg was clear that he wanted to avoid invoking God, but his discomfort locked him in to a metaphysically flattened view of nature. My purpose is to correct a misstep. Hylomorphism, and the distinctions it entails between act-potency and form-matter, remains as valid for quantum mechanics as for any other scientific discovery, and once righted, can be seen in the light offered by faith in God the Creator.

Heisenberg himself did what Jaki would later advise. In *Physics and Philosophy*, he went back to the ancient roots of atomic science to better interpret modern physics. Before analyzing his interpretation, I want to highlight that it was fitting for him to do so. Yale philosopher F. S. C. Northrop says in the introduction of Heisenberg's book that no one is more competent to judge the meaning of quantum mechanics than the man who was awarded the Nobel Prize in Physics in 1932 for creating it.[5] Like all the scientists involved in the discovery of the atom, Heisenberg fought for the ground his ideas gained. Quantum theory did not arrive on the scene lightly; it rather imposed itself in the experiments and data. Heisenberg recalls a conversation with Neils Bohr at the 1926 Copenhagen meeting where scientists met to interpret results; he asked Bohr, "Can nature possibly be as absurd as it seemed to us in these atomic experiments?"[6] Heisenberg is famous for the "uncertainty principle" named after him, which acknowledges the inherent imprecision in determining the position and momentum simultaneously for very fast and very small particles. Northrop notes that modern physics led to great confusion about causality because of the uncertainty principle.

To set the stage for the analysis, Heisenberg seemed certain that quantum mechanics pointed beyond physics and out of the materialistic and mechanistic view of nature associated with Newtonian physics, yet he did not look beyond material causes. I will specifically analyze three aspects of his insights: 1) the differences in mechanical and ontological causality, 2) energy as universal matter, and 3) force as form. Then I will draw a more direct and complete line from Aristotle's philosophy of nature to quantum mechanics, which will then allow a continuation to Aquinas's theology of creation. In the Judeo-Christian tradition, creation means more than nature because it refers to God's loving plan and how, to quote Pope Francis in *Laudato si'*, "God's love is the fundamental moving force in all created things."[7] This approach to integration was an enduring theme of Jaki's work. Each discipline stands on its own legs but works with the others to hold up the full truth, and nothing less.

Mechanical and Ontological Cause

Heisenberg's discovery is termed *quantum mechanics* because it studies the motion of particles due to forces. Quantum mechanics, therefore, deals with mechanical causation, but it also differs from classical Newtonian mechanics in that it does *not* allow the motion of objects to be predicted precisely based on equations of measurable variables, such as position, time, and distance. Per the uncertainty principle, there is a margin of imprecision at the subatomic scale. The motion of particles is, therefore, mathematically based on probabilities, which is the basis for the Heisenberg uncertainty principle. Heisenberg's discovery is a set of mathematical matrices that provide a picture of probable electron locations.

In addition to limits to how precisely the position and velocity of particles can be simultaneously determined, the act of observation interferes with the particle motions. Human observations depend on photons of light reaching the eye, and photons are discrete units of energy that are released in atoms by the motion of other subatomic particles. It is as if a blindfolded person was tasked with determining the position of billiard balls on a table but could only do so by aiming other balls at them. Measurement changes what is being measured. The mathematical constructs of quantum mechanics are therefore less deterministic than classical mechanics, which means they introduce some subjectivity. This conception of matter changed not only the way reality is understood but also the way we understand our place in scientific engagement with the world. Albert Einstein wrestled with this idea, which is the source of his famous assertion that "God does not play dice."[8]

Jaki was critical of both Heisenberg's and the Copenhagen interpretation of quantum mechanics for mistaking "operational uncertainties for ontological ones."[9] By this he meant that an inability to measure should not be confused with an inability to exist precisely. He called the Copenhagen interpretation an "anti-causal misinterpretation" regarding indeterminacy.[10] It is good physics, he said, to accept that interactions cannot be measured exactly at the subatomic level, but it is bad philosophy to then assume this operational insufficiency means that nature is actually lawless at its most fundamental levels. Jaki's purpose in making this distinction is to keep clear that "causation" can mean different things for scientists and philosophers. In his introduction to Heisenberg's book, Northrop emphasizes the same clarification.[11]

Mechanical causation refers to both series of events in daily life as well as the way scientists design experiments in classical physics. If a forklift moves a block ten meters to the left and upwards, we say that the forklift caused the so-defined motion. There is a relation between different states for the same object, and this idea applies to systems of objects whether in machinery or in natural systems like erosion, rainfall, or population growth. If we know the initial state, a final state can be predicted based on prior experience. Such predictions are the foundation of much of the physical and biological sciences. However, causal predictions are not possible in quantum mechanics for the reasons explained above, but that does not mean there is no causality whatsoever.

Heisenberg's meaning for causality can be understood as the state-function of a system that relates several variables at a point in time. The time-equation then relates a state at one point to the state of a physical system at a later point, all of which is determined with probabilities. There is no way to predict mechanical causality at the individual level for subatomic particles. This is not such an unfamiliar limitation in human knowledge. We have elaborate schemes for predicting weather patterns, but no one could predict the exact location any raindrop will hit the Earth at any specified time in the future. Ontological causality is preserved in quantum mechanics, but not in the same way that it is understood operationally in classical mechanics.

In *Physics and Philosophy*, Heisenberg addresses ontological causality in terms of stability. He briefly reviews the experiments that challenged the existing view of nature and led to quantum mechanics. Max Planck explained black body radiation as discrete quanta of energy that can be absorbed and emitted by matter. Albert Einstein explained the photoelectric effect as metals emitting electrons in response to the color (frequency) of light but not its intensity, leading him to posit the existence of the beforementioned energy packets called photons. In his famous experiment with gold foil, Ernest Rutherford observed that alpha particles do not penetrate matter as if atoms are homogeneous but instead as if they have a specific structure, which led to him to discover the dense and positively charged nucleus. Niels Bohr, thus, proposed the planetary model of the atom, as a positive nucleus with negatively charged electrons orbiting. High school students learn (far too little of) this history, but it is not possible to express in a single paragraph the dismay these new ideas caused in their time. Entire books have been written about the history of the atom and quantum theory. Heisenberg is emphatic that the *inescapable stability in nature* compelled scientists toward a new radical theory.

In his chapter on the history of the quantum theory, Heisenberg recalls his conversations with his colleagues. Before the Copenhagen meeting in the early 1920s, Heisenberg met Bohr, who was sixteen years older, for the first time. It was shortly before Bohr would be awarded the 1922 Nobel Prize in Physics for his work on the structure of the atom and a decade before Heisenberg would receive his.[12] They went for a walk in the mountains one afternoon at a conference in Göttingen, Germany. Heisenberg was troubled by contradictions in Bohr's planetary model of the atom.[13]

On one hand, it accounts for the observed emission and absorption of energy in that it imagines electrons jumping between orbitals and releasing photons of matching quantified energies. On the other hand, no cosmic planetary system following Newtonian laws of astronomy (i.e., Newton's laws of classical mechanics) could do what atoms do and *remain stable*. If planetary systems collide, they are permanently changed. If atoms collide, as they do throughout nature in all the chemical reactions happening continuously, the atoms remain what they are. If a carbon atom reacts, it remains a carbon atom. This stability is a pervasive feature of the elements on the periodic table; it is the reason humans are able to construct such a periodic table of properties at all. Even if atoms are somewhat changed by radiation or collision, they return to their natural state. Bohr told Heisenberg on that walk:

> By 'stability' I mean that the same substances always have the same properties, that the same

> crystals recur, the same chemical compounds, etc. In other words, even after a host of changes due to external influences, an iron atom will always remain an iron atom, with exactly the same properties as before. This cannot be explained by the principles of classical mechanics, certainly not if the atom resembles a planetary system. Nature clearly has a tendency to produce certain forms—I use the word 'forms' in the most general sense—and to recreate these forms even when they are disturbed or destroyed.[14]

It seems this profound insight was on Heisenberg's mind when he began the journey through ancient Greek philosophy in *Physics and Philosophy*. We can note here the deep conviction underlying quantum theory that mechanics, as indeterminate and uncertain as they may be, were discovered because the innate forms in nature compelled scientists to change their theories. This is not the operational, mechanical causality of laymen and classical physics, but an ontological cause inherent in the being of the atom.

Energy as Prime Matter

Heisenberg traces the search for elementary particles back to ancient Greek philosophy, Heisenberg reviews the materialism of Leucippus, Democritus, and their contemporaries, a staple in any discussion about the history of physics and chemistry. These ancient philosophers attributed the source of all things to material causes and, therefore, had ideas about primary substances as small, indivisible, units of matter. Heisenberg does not think their materialism applies to modern physics for the similar reasons he thinks Newtonian mechanism does not apply. "In fact," he says, "one may say that atomic physics has turned science away from the materialistic trend it had during the nineteenth century."[15] The main difference is that quantum mechanics does not visualize nature as made up of little particles moving in ways that are mathematically modeled by Newtonian physics. It is of great significance that Heisenberg does not think modern physics is materialistic.

But he hesitates here too. He notes that Aristotle attributes Thales with the idea that water, the source of nourishment and life, is the underlying element of all things.[16] In contrast to ancient thinkers who held that the ocean was a god, the parent of what comes into being, Heisenberg surmises that Thales correctly avoided invoking gods in favor of natural explanations. Rather out of context in this part of his book where he is reviewing ancient Greek philosophers, Heisenberg inserts fundamental philosophical ideas from the modern teaching of Friedrich Nietzsche regarding Thales's statement that "water is the material cause of all things."

This statement, strange as it looks to us, expresses, as Nietzsche has pointed out, three fundamental ideas of philosophy. First, the question as to the material cause of all things; second, the demand that this question be answered in conformity with reason, without resort to myths or mysticism; third, the postulate that ultimately it must be possible to reduce everything to one principle.[17]

It seems this move locked Heisenberg's analysis into the materialism he was trying to avoid. He sees no room for anything metaphysical because he seems to equate metaphysics with mysticism, the invocation of gods, and a dismissal of sound reason.

Heisenberg also sees compatibility with the teaching of Anaximander, who was a student of Thales. Anaximander thought the underlying element was a substance that was infinitely formable, the one material source out of which all things arise, what Aristotle called "the unbounded" (*apeiron*).[18] Heisenberg prefers this interpretation because nature is not cast as made of moving particles but more in terms of laws of nature in that physicists today seek to discover the "fundamental law of motion for matter from which all elementary particles and their properties can be derived mathematically."[19] He proposes that this universal substance that Anaximander speaks of is energy. Along these line, Anaximenes thought air is the underlying element, and Pythagoras thought the cosmos is a harmonious whole based on geometric proportions.

In Heraclitus, Heisenberg found the most compatibility with modern physics. Throughout ancient Greek materialist philosophy, there was a search for the material cause of all things that unites all things. Heraclitus thought that fire is the basic element, but more so thought that the underlying principle is change itself. Both the underlying principle and basic element are immaterial, and Heisenberg puts the two ideas together as energy. "Energy," he says, "may be called the fundamental cause for all change in the world."[20]

Then he turns to Aristotle and matter in relation to form, also known as hylomorphism from the Greek words for matter (*hulê*) and form (*morphê*). To briefly review Aristotle's hylomorphism, matter exists in relation to form such that what we perceive in nature is formed matter. For example, we do not think of a dog as two separately existing things, matter and the abstract form of dog, but the dog as one thing with form and matter united as an individual composite being.[21] The composite is, therefore, both material (matter) and immaterial (form). Form is the formal cause. Living beings most properly have a nature, and their soul is the animating principle, such that we say the soul is the form of the body. For Aristotle, a living thing becomes what it is according to its nature, which is an inherent principle of change. I will return to form in the next section, but first there is more to say about prime matter and energy.

Aristotle calls *prime matter* formless matter. All things that exist materially, however, are formed composites, so there is no real prime matter to discover.[22] Prime matter is not air, water, earth, fire, or any kind of element, but a kind of "indefinite corporeal substratum," as Heisenberg aptly puts it.[23] Aristotle's prime matter is also not the unbounded material that Anaximander thought of in that the "boundless" was a principle more than a substratum. Aristotle considered prime matter part of nature, and as such, it could not be infinite (boundless), for that would mean that contraries could exist simultaneously, as in something could both be un-generated and, yet, indestructible.[24] Prime matter does not exist in a way that we can experience it because we live in a physical world where all matter has some form.

Heisenberg thinks energy fits the same concept as an indefinite corporeal substratum (he does not use the phrase prime matter) because, as supercolliders have shown, all the elementary particles can transmutate. He points out that Einstein's equation ($E=mc^2$) shows the interchangeability of matter and energy, especially for particles coming into and out of existence. For Aristotle, prime matter is all potential (*potentia*) since it has no form and can potentially become whatever the form actualizes it into. Experiments show that particles can be produced from other particles or from energy, which Heisenberg takes for the "complete mutability of matter."[25] He recognizes that experiments can only be done on, what Aristotle would call, composite bodies, i.e. formed matter. Science searches for the properties of things in nature. He also notes that mass and energy are conserved in chemical and nuclear reactions as further testament to the stability of matter and energy.[26]

Recently, in 2023, philosopher David Oderberg tested the hypothesis that Aristotelian prime matter could be energy.[27] He concludes, "Why not?" There are problems either way, but he finds that the stronger argument is in favor of energy (E) as the discovery of prime matter (P), building on the consistencies noted by Heisenberg. Prime matter lacks any scale or dimension on its own but receives any quality or quantifiable dimension when it is *informed* into formed matter. It is pure potential. Energy has no parts. No one points to a chunk of energy any more than an Aristotelian would point to a chunk of prime matter.

However, energy is something quantifiable and measured in physics and chemistry. The joule (J) is a derived unit, the standard unit of energy in the International System of Units (SI) equal to the work done on an object when one newton (N) of force, which is another derived and standard unit equal to the force that accelerates a 1 kilogram mass 1 meter per 1 second, displaces a mass one meter in the direction of the force. There are other units; 1 calorie = 4.184 J, and 1 electron volt = 1.6×10^{-19} J, and many more. Still, all of these measures depend on formed matter to be measured, which is why Oderberg tends toward a positive assertion that P=E.

Oderberg takes a further step than Heisenberg did. Noting the words of another philosopher and physicist, Mario Bunge, he says that because energy is ubiquitous, it is a subject for philosophy and metaphysics, the same as other categories such as time, place, cause, and property. Energy may turn out not to be prime matter, another reason for leaving the door open, but likewise, it could pass the test. If energy is prime matter, then this reconciliation would be a beautiful example of a "necessary rendezvous" between science and metaphysics that assures us we are all examining the same reality. Oderberg's conclusion is similar to Pope Leo XIII's statement in his 1883 encyclical on the study of Holy Scripture, that "truth cannot contradict truth."[28] Suffice it to say that Heisenberg's insight continues to raise interesting questions by suggesting that prime matter may be something best understood with mathematics.

Force as Form

Heisenberg was intrigued by Aristotle's idea that unformed matter (prime matter) represents a potential (*potentia*), a possibility from which the essence (the nature) of a thing moves the matter from potency to actuality. This fits with his discovery of mathematical matrices that provide a picture of probable electron locations. When an observer becomes aware of the result of a measurement, the picture is no longer probable but known, what Heisenberg called the "reduction of wave packets."[29] This collapse of probability is the actualization of the potential in Heisenberg's view.

He explains that during the Renaissance, the new mechanical approach to natural science that followed Descartes and Newton separated matter from form the same way it separated matter from mind.[30] Echoing the caution he takes from Thales and Nietzsche, he saw this as a step in the right direction because the separation "excluded all tracing of corporeal phenomena back to spiritual forces," so that "matter could be considered as a reality of its own independent of the mind and supernatural powers."[31] Where hylomorphism could have moved Heisenberg into the metaphysical, he again departs from Aristotle. Matter, he says, lost its connection with the vegetative soul in Aristotelian natural philosophy, and he sees this definition of matter as the most compatible for modern science.

Heisenberg acknowledges the separation of personal religion and empirical knowledge with the dualism of Descartes, which proposed *res cogitans* as the thinking substance somewhat like a soul and *res extensa* as the extended, physical substance. Heisenberg concedes that "it has always seemed difficult to deny the existence of some kind of soul in the animals" because the old concept of soul in the philosophy of Thomas Aquinas was more natural than that of the *res cogitans*."[32] But he is only concerned with physics and does not further explore Aquinas's philosophy to understand that soul is referred to by Aquinas as the form of the body. Interestingly, this is the sole mention of Aquinas in the book as well. Where Nietzsche caused Heisenberg to depart from Aristotelian hylomorphism, Aquinas could have brought him back.

This move, however, leaves him with no explanation for form. His solution is to make another exchange, that of *force* for form, and he calls matter-force the new dualism for natural science of the nineteenth century. "Matter is that on which forces can act; or matter can produce forces."[33] What he means, for example, is that bodies with mass have gravity which attracts other bodies with mass, so their matter and forces are inextricably intertwined.

However, in the very same paragraph, he moves on to say that "in the most recent development of modern physics this distinction between matter and force is completely lost, since every field of force contains energy and in so far constitutes matter."[34] He refers here to experiments with elementary subatomic particles demonstrating that every field of force has a specific particle. All particles have essentially the same properties, namely that mass and energy are convertible and conserved. In tracing these steps, I have shown how Heisenberg approaches Aristotelian hylomorphism but turns away to avoid metaphysics, thinking it mystical, and then provides an updated conception of matter and form as energy and force, except that there is no distinction.

This seems to make no sense, but in the very next paragraph, Heisenberg returns to form. "When natural science investigates the problem of matter it can only do so through a study of the forms of matter."[35] By form, here, he means structure, which brings us back to what Bohr says about stability of structure in nature. Natural science studies the "structure of matter and the forces responsible for this structure."[36] Even though Heisenberg's philosophical description does not exactly conform to Aristotle's hylomorphism of act and potency, he, like Bohr, cannot escape the stability of structure in nature.

Heisenberg sees his updated matter-force conception as very close to Aristotle's matter and form even though in Aristotelian natural philosophy form means more than structure. Form refers to substantial form, the formal cause or nature and essence of a natural thing. Form refers to an internal principle of change. When science investigates matter, he says it does so through a study of the forms of matter, and that is of course true and obvious. Laws of nature discover the relationships in the structure of matter. From there, the field of chemistry was built, and from chemistry, biology to provide a big picture of how energy is the fundamental cause for all change in the world. Heisenberg concludes that to eventually gain a "complete understanding" of all matter and energy would mean "that the forms of matter in the sense of Aristotelian philosophy would appear as results, as solutions of a closed mathematical scheme representing the natural laws for matter."[37] For this reason, and not surprisingly, Heisenberg sees mathematics as the final proof for the unity of matter.

But notice that when Heisenberg says modern physics has lost the distinction between matter and force, he has done away with *the naturalness* of Aristotle. His description of matter and force or structure is not Aristotelian hylomorphism. The modern word for "energy," which Heisenberg proposed as prime matter comes from ancient Greek, *energeia,* which means activity, or act. Where Heisenberg proposes that energy is prime matter with pure potency, for Aristotle the same word is act. The concept of act and potency are tied to the concepts of matter and form. Prime matter is pure potency that the form actualizes. At the heart of Aristotle's meaning of being is the idea of substantial form, the nature or essence of a thing that keeps it being what it is. The sustaining form is *energeia.*

Furthermore, the idea of a sustaining form is tied to teleology as the final cause, the idea that natural things act toward their end in nature.[38] In the same way Heisenberg wanted to avoid introducing God into his philosophy, scientists still today avoid teleology and final cause. This misstep is an error that has enormous consequences in the current philosophy of science. If mathematics were to be the final proof for the unity of matter, then all of nature could be modeled with equations, including plants, animals, and humans, an entrapment that is materialism in spite of Heisenberg's intuition that atomic physics steered science past materialism.

Beyond Physics

I have focused on Heisenberg's remarks spanning only a few pages in his book about Aristotelian hylomorphism to pinpoint where he departs from Aristotle's naturalism. In returning to the beginning, as Jaki recommends, and stepping through Heisenberg's reasoning, it is clearer how to avoid a flattened metaphysics and put the path back on track. Now I will briefly explain why Aristotle went beyond physics. In *Physics,* Aristotle defines what he means by nature. He says, "Of things that are, some are by nature, others through other causes ... not put together by nature."[39] The latter refers to man-made things or artificial things, which is any construction of matter by humans that is *not* found in nature. Things put together by nature have in themselves a source of motion and growth, an "innate impulse of change" as Aristotle called it.[40] For Aristotle, nature is the cause of motion and change in natural things.[41] Living things *have* a nature in the fullest sense, while body parts, inanimate materials, and elements act *by* nature, i.e., they have natural properties. Nature intrinsically forms matter into living organisms similar to how marble is extrinsically formed into a statue by a sculptor. Matter is the material cause. Nature or the human is the efficient cause. Form, as mentioned before, is the formal cause. The final cause is what the thing is to be as beheld in the mind of the designer. The difference, of course, is that nature is the formal and final cause of what is natural, and man is the formal and final cause of what is artificial. Heisenberg does not highlight the distinction between nature and art.[42]

For Aristotle, to know about something in the natural world is to know its causes. To truly take physics away from materialism one must do as Aristotle did and continue beyond physics into metaphysics and an inquiry of causes, which requires considering not only of material and efficient causes but also of more ultimate formal or final causes. It is like driving a car. You can buy one and operate one or maybe even fix one, but to know what a car is, you need to know about the materials it is made of and the way it is built as well as the reasons for its form and its purpose in the mind of the designer.

In *Metaphysics*, Aristotle reviews his predecessors and finds materialism unsatisfactory. Aristotle points out that the inquiry into material causes (from Anaximenes, Diogenes, Hippasus, Heraclitus, Anaxagoras) and the observation that beings come into and out of existence among the elements (Hesiod, Parmenides, Empedocles, Leucippus, Democritus) leads to an inquiry into causes beyond what is material (Pythagoras, Plato) and ultimately to an examination of formal and final causes. Aristotle was critical of these views as incomplete or untenable and built on them to establish the unity of matter and form in composite being.

By the end of *Physics*, Aristotle logically deduces the unmoved first mover, which launches him into *Metaphysics* and the consideration of being itself and ultimate cause. For hylomorphic composites to exist as the potency of prime matter actualized by the form according to its nature, there must exist a being who is pure act, unity, immaterial, and intelligent. There is not enough space here to lay out the arguments for the existence of this being, which we now call God, in Aristotle's *Physics* and *Metaphysics*, but he also proved divine attributes for this being, such as simplicity, infinity, immateriality, omnipresence, omniscience, and immutability. Thus, natural philosophy opens up to natural theology, and Aquinas uses these proofs in his Five Ways.

For the purpose of this analysis of Heisenberg's reconciliation with hylomorphism, nothing new needs to be added. It is enough to show that if 1) quantum mechanics can be taken as evidence of ontological cause inherent in the being of the atom and innate forms in nature so strong that it compelled scientists to change their theories, 2) if prime matter is energy or even if something more fundamental is discovered that instead turns out to be prime matter, quantum mechanics still points to the need for an indefinite corporeal substratum for composites to exist, and 3) if form is upheld as an immaterial cause that necessitates the existence of a being we call God and final cause, then we can draw a more complete hylomorphic line from Aristotle's philosophy of nature to quantum mechanics and then to Aquinas's theology of creation. Note: only the third condition differs from Heisenberg's analysis.

And Back to Creation

Although Aristotle found it impossible to limit the consideration of ultimate causes to matter, and although in *Physics* he reaches the threshold of metaphysics and deduces the existence of a necessary eternal being that is pure act and the cause of all motion and change, in a sense he was not able to continue on to a theology of creation and ultimate causality.[43] Aristotle's teleology is based on his view that the universe is whole and that to discover the final cause of anything is to discover how it fits into the natural structure.[44] Aristotle does not name the necessary being as the principle of teleology since, for him, nature suffices for directing things to their ends. Even in his ethics, Aristotle does not consider God to be one "who issues commands."[45]

Aristotle did not have the benefit of divine revelation because he lived before the incarnation of Jesus Christ. Belief in a Creator was not in his mind. Aquinas, having such benefit, built on Aristotle's arguments in his Five Ways to prove the existence of God: the proof from 1) motion, 2) causality, 3) the contingency of the world, 4) the grades of perfection, and 5) finality.[46] Aquinas names God as the first mover and first efficient cause for the first two ways. Applying these proofs to Heisenberg's analysis, because of the metaphysical commitments of hylomorphism, namely act and potency and matter and form, the fundamental cause for all change in the world cannot be energy but that which causes energy to exist in the first place. Energy may be the measure of change and activity, and energy may be a mover, but it is not the first unmoved mover or uncaused cause. The second way follows the logic of the first way but deals with causality. Nothing can be the cause of itself because nothing can exist prior to itself.[47] The causal series, like the series of mover-to-moved, is ordered hierarchically at all moments. The argument from motion and the argument from causality are similar; the former explains change, the latter existence.[48] Aquinas also names God as the necessary being with all the other divine attributes in the third and fourth ways.

Aquinas follows Aristotle in the fifth way argument from finality in that there is inherent design in nature, but he departs from Aristotle in the second part of the fifth way. He concludes that God is the intelligent being, the Creator, who directs all things to their end. In a neo-scholastic essay on Aquinas's fifth way, philosopher Edward Feser explains a subtle but critical distinction. Aristotle regards God as the "cause of things' moving in such a way that they realize their ends," while Aquinas regards God as the "cause of things *having* those ends in the first place."[49] What this means for Heisenberg's uncertainty principle and his reinterpretation of philosophy is that materialism alone cannot complete the explanation for why matter exists or has the structure, energy, forces, or properties it has in the first place. There must be a fully actualized source from which all energy and form originate.

Returning to the distinction between nature and art that Heisenberg avoided and the idea of nature as an intrinsic cause versus the artist as an extrinsic cause, the reason that rejecting a naturalistic view of things derails the metaphysical consideration is because there is no explanation for ultimate cause. "Form" in scholastic philosophy is the internal principle that makes it what it is, and it is also that which renders the thing intelligible.[50] Just as the sculptor who holds the form of the statue in his mind, the finished sculpture is the manifestation of the form. God the Creator is not, however, an external designer who discovered nature and decided to arrange it. God is the Creator of creation, and nature is part of creation (along with immaterial substances). Everything is either the Creator or creation. A composite being is matter and form, whether that being is a particle or an animal or a man. Form indicates the formal cause in the intellect of the designer, in the case of art as an external designer assembling matter but in the case of nature intrinsically. Form is inherent in every natural being as created.

These are ideas in Christian philosophy and natural theology supported by reason. Heisenberg expressly did not want to invoke spirituality, but a fully integrated view of reality needs the trilogy of science, philosophy, and theology, properly restored. Understood in this fuller physical and metaphysical philosophy and in the light of revelation, science can be understood as the study of the handiwork of God.

Conclusion

The metaphysical framework from Aristotle and Aquinas applies to quantum mechanics the same as it applies to Newtonian physics and any other aspect of modern science. Whatever exists in nature is a composite of matter and form as God created it. This does not mean that metaphysics is impervious to new theories in physics, chemistry, or biology, but that each informs the other. There can be, and indeed have been, challenges to details of Aristotle's and Aquinas's natural philosophy and natural theology, but these challenges do not overturn the core principles. Projectile motion is no longer thought to be due to *antiperistasis* but to forces in two dimensions. We count more than four elements now, and the 118 elements on the periodic table do not seem to fit Aristotle's definition. Life is now understood to be reproduced not spontaneously generated. Aquinas rejected that universe is eternally cycling because it is incompatible with the Catholic faith. Furthermore, a deeper understanding of the mysteries of creation allows humans to improve their lives and become better stewards of nature. If, as Jaki said, the first steps in a philosophical system are an ever new source of life, then returning to those steps has allowed a more methodological approach to the question of quantum mechanics and hylomorphism and how modern physics fits into the theology of creation.

Heisenberg purpose in *Philosophy and Physics* was not theological. He thought the connection with God was in the soul not the world.[51] His correspondence with Einstein showed that Einstein never accepted the uncertainty principle as ontological because he was not ready to reject the deterministic causality of classical physics (discussed in the first part of this paper). He thought there were "hidden variables" not yet discovered, or undiscoverable, that could explain the motion of particles more exactly. Although Heisenberg disagreed about the possibility of quantum mechanics ever deterministically predicting particle motion in the way that classical mechanics does, he did not think this disproved ontological causality.

In the last surviving letter from Heisenberg to Einstein written a few months before the two physicists were to meet at a conference in Brussels in 1927 at the height of the controversy, Heisenberg wrote that while the classical causality principle was baseless in the new quantum mechanics, "We can console ourselves that the dear Lord God would know the position of the particles, and thus He could let the causality principle continue to have validity."[52] The rejection of deterministic causality is not in conflict with the proofs of the existence of God. Perhaps Heisenberg was speaking poetically, but this statement from the founder of quantum mechanics has probative force of its own. It means that God is still there, as He was in the beginning and is now and ever shall be, even in the uncertainty of quantum mechanics. The purpose in returning to the first steps is to figure out how it all fits together.

"*Of Him, and by Him, and in Him are all things.*" Romans 11:3

Notes

1. Thomas Aquinas, "On Being and Essence," in *Medieval Philosophy: Essential Readings with Commentary*, ed. Gyula Klima (Malden, MA: Wiley Blackwell, 2007), faculty.fordham.edu/klima/Blackwell-proofs/MP_C30.pdf.
2. Aristotle, *De cælo* (On the Heavens), I.5.271b5-18, trans. J. L. Stocks (Oxford: Clarendon, 1922), 22, at archive.org/details/decaelo00aris.
3. Stanley Jaki, *A Mind's Matter: An Intellectual Autobiography* (Grand Rapids: William B. Eerdmans Publishing Company), 247–248.
4. Werner Heisenberg, *Physics and Philosophy: The Revolution in Modern Science* (London: Ruskin House, 1959), archive.org/details/ost-physics-heisenberg-physicsphilosophy.
5. Heisenberg, *Physics and Philosophy*, 11. "Werner Heisenberg: Facts," *The Nobel Prize*, Prize/Physics/1932/Heisenberg/Facts, www.nobelprize.org/prizes/physics/1932/heisenberg/facts/.
6. Heisenberg, *Physics and Philosophy*, 43.
7. Francis, Encyclical on Care for Our Common Home *Laudato si*, (The Holy See, 24 May 2015), §77.
8. Richard Rhodes, *The Making of the Atomic Bomb*, The 25th Anniversary Edition (New York: Simon & Schuster, 2012), 132,

archive.org/details/makingofatomicbo0000rhod_v1i2.

9. Jaki, *A Mind's Matter*, 165. For a longer explanation, see Stanley Jaki, *God and the Cosmologists* (Washington, D.C.; Regnery Gateway, 1989), chapters 5–6.
10. Jaki, *A Mind's Matter*, 166.
11. Heisenberg, *Physics and Philosophy*, 20.
12. "Niels Bohr: Facts," *The Nobel Prize*, Prize/Physics/1922/Bohr/Facts, www.nobelprize.org/prizes/physics/1922/bohr/facts/.
13. Werner Heisenberg, *Physics and Beyond: Encounters and Conversations*, trans. Arnold Pomerans (New York: Harper & Row, 1971), 39. In this later book, he tells more of the conversation, which he mentions in *Physics and Philosophy*, 37.
14. Heisenberg, *Physics and Beyond*, 39.
15. Heisenberg, *Physics and Philosophy*, 58–59.
16. Aristotle, *Metaphysics*, I.3.983b.7-10, trans. Joe Sachs (Santa Fe: Green Lion Press, 2019), 6.
17. Heisenberg, *Physics and Philosophy*, 58–59.
18. Aristotle, *Metaphysics*, X.1.1053b10-15, trans. Sachs, 6.
19. Heisenberg, *Physics and Philosophy*, 60.
20. Heisenberg, *Physics and Philosophy*, 61.
21. Aristotle, *Physics*, I.7.190b5-11, in *Aristotle's* Physics*: A Guided Study*, trans. Joe Sachs, ed. Harvey M. Flaumenhaft (New Brunswick: Rutgers University Press, 2004), 40.
22. Aristotle, *Physics*, I.7.191a13, trans. Sachs, 42.
23. Heisenberg, *Physics and Philosophy*, 129.
24. Aristotle, *Physics*, II.1.203b1-15, trans. Sachs, 82.
25. Heisenberg, *Physics and Philosophy*, 139.
26. Heisenberg, *Physics and Philosophy*, 131.
27. David Oderberg, "Is Prime Matter Energy?" *Australian Journal of Philosophy*, Vol. 101, No. 3 (2023): 534–50, doi: 10.1080/00048402.2021.2010222.
28. Leo XIII. *Providentissimus Deus*: Encyclical on the Study of Holy Scripture. The Holy See, November 18, 1893.
29. Heisenberg, *Physics and Philosophy*, 125.
30. Heisenberg, *Physics and Philosophy*, 130.
31. Heisenberg, *Physics and Philosophy*, 130.
32. Heisenberg, *Physics and Philosophy*, 80.

33. Heisenberg, *Physics and Philosophy*, 130.
34. Heisenberg, *Physics and Philosophy*, 130.
35. Heisenberg, *Physics and Philosophy*, 130.
36. Heisenberg, *Physics and Philosophy*, 130.
37. Heisenberg, *Physics and Philosophy*, 144.
38. Sachs, Commentary on Aristotle's *Physics*, 14.
39. Aristotle, *Physics*, II.1.192b8-30, trans. Sachs, 49.
40. Aristotle, *Physics*, II.1.192b30, trans. Sachs, 49.
41. Aristotle, *Physics*, II.1.192b35, trans. Sachs, 49.
42. Heisenberg, *Physics and Philosophy*, 148.
43. Aristotle, *Physics*, VIII.5.256a12-21 and 6.258b1-10, trans. Sachs, 199, 204.
44. Sachs, *Commentary on Aristotle's* Physics, 24.
45. Aristotle, *The Eudemian Ethics*, Book VIII, Part 3, trans. Anthony Kenny (New York: Oxford University Press, 2011), 148.
46. Thomas Aquinas, *Summa theologiæ*, I, q. 2, a. 3, respondeo, trans. Laurence Shapcote, ed. The Aquinas Institute, aquinas.cc/la/en/~ST.I.Q2.A3. The full treatment goes beyond the Five Ways in the *Summa theologiæ*, which is a summary for students who already knew Aristotle's metaphysical commitments. Here Aquinas says, "... and this we call God," but later he explains how we can know that God is One and is the God of Divine Revelation.
47. Aquinas, *Summa theologiæ*, I, q. 2, a. 3.
48. Edward Feser, *Aquinas: A Beginner's Guide* (Oxford: Oneworld Publications, 2011), 76. Feser idea is attributed to Étienne Gilson.
49. Edward Feser, "Between Aristotle and William Paley: Aquinas's Fifth Way," in *Neo-Scholastic Essays* (South Bend: St. Augustine's Press, 2015), 148–149, 158.
50. Bernard Wuellner, *Dictionary of Scholastic Philosophy* (Fitzwilliam, NH: Loreto, 2012), 47.
51. Heisenberg, *Physics and Philosophy*, 72.
52. Gerald Holton, *Victory and Vexation in Science: Einstein, Bohr, Heisenberg, and Others* (Cambridge: Harvard University Press, 2005), 32.

11 SEARCHING FOR TRUTH IN DETERMINISM: A CALL FOR COMPASSION, MERCY, AND LOVE

SAMANTHA MATTHEISS, PH.D.

Introduction

Determinism is gaining popularity as neuroscientific evidence increasingly points to the immense influence of external factors on human behavior. As such, scholars such as Stanford University professor Robert Sapolsky assert that our choices are completely determined by our environment, neurobiology, and interaction between the two.[1] While much of Sapolsky's argument is based on measurable evidence, other conclusions are drawn about unmeasurable constructs, thereby contradicting Father Stanley Jaki's delineation of science and religion.

Determinism: A Look at the Evidence

In line with Jaki's metaphysical realism, objective, scientific evidence points, inarguably, to a vast influence of biological and environmental factors on human behavior, thought, and emotion.

Genetic Influence

First, Sapolsky and others draw upon research demonstrating genetic influences on human behavior and brain function. In line with their discussion, genes, indeed, specify the synthesis of hormones,[2] such as oxytocin, involved in trust and social behavior; neurotransmitters such as serotonin,[3] involved in mood, etc; and dopamine,[4] involved in motivation, pleasure, etc. Genes also specify the receptors to which those chemicals bind,[5] as well as enzymes involved in removal of the neurotransmitter from the synapse,[6] thereby affecting the extent to which the neurotransmitter exerts its effect. Overall, genetic polymorphism predicts variation in traits and behavior such as self-control,[7] anger[8], and addictive behavior.[9]

Gene x Environmental Interactions

As Sapolsky accurately describes, genes don't solely predict behavior and brain function, but rather they interact immensely with one's environment. A prominent example is the differential effects of parenting environment on subsequent social behavior based on polymorphisms in DRD4[10]—a gene influencing the sensitivity of the D4 dopamine receptor. Secure attachment, in which a child perceives their caregiver as a secure base from which to explore the world, predicted generous donation behavior—but this relationship was only significant for carriers of a certain form of the DRD4 gene.[11] The researchers conclude from a concurrent meta-analysis that children with this particular form of the DRD4 gene are, on average, more sensitive to the parenting environment—showing more negative consequences from adverse environments but more positive consequences from positive environments. Such differences are related to genetically determined dopaminergic processes. This is one of many examples of a gene by environment interaction.

Epigenetic Influence

The environment can also trigger epigenetic changes—changes in gene expression driven by alterations in DNA methylation. For example, lack of early maternal care amongst infant rats led to epigenetic changes resulting in a reduction in the expression, or levels, of the glucocorticoid receptor—a protein responding to stress hormones—and thus, increased stressed reactivity throughout the lifespan.[12] This effect was shown amongst human infants as well.[13] As such, prenatal and postnatal experiences alter genetic expression of proteins thereby impacting behavior, cognition, and affect for the individual and their descendants.

Environmental Influence

The environment also directly shapes behavior and brain function. Prenatal exposure to toxins, hormones, viruses, etc. alter fetal development. For example, an individual has a higher likelihood of developing schizophrenia if their biological mother had influenza during pregnancy.[14] Infant[15], childhood[16], adolescent[17], and adult experiences; as well as one's current physiological state (e.g. hungry,[18] tired,[19] experiencing a positive or negative social interaction[20]) also constantly affect one's neurobiology and behavior.

Moreover, the environment can directly induce physical alterations in brain matter that elicit personality changes, such as in the case study of Phineas Gage.[21] When a railroad spike speared through Phineas' prefrontal cortex, he underwent a personality change from responsible to impulsive and with poor emotion regulation[22]—characteristics often perceived as lacking virtue or even sinful. Interestingly, adversity has been shown to cause similar physiological differences. For example, children exposed to adversity have reduced activity in the prefrontal cortex;[23] reduced regulatory influence from the prefrontal to subcortical regions involved in emotional processing;[24] and reduced prefrontal cortex volume.[25] These differences, among others, underly the negative outcomes associated with adversity, including mental and physical health problems, academic difficulties, and social and economic struggles.[26]

Overall, research demonstrates that an individual's environment, beginning with the prenatal environment over which they have no control, vastly influences their neurobiology as well as observed behavior, affect, and cognition.

Self-Directed Influence

Effects on Ourselves

In addition to external influences, our choices also influence our own behavior and brain function. For example, engaging in mindfulness practices, in which an individual focuses on present sensations, reduces depression, anxiety, and addiction symptoms; improves attention; and increases feelings of compassion towards others.[27] It also increases brain activity in prefrontal regions involved in cognitive control and regulation[28]; and, doing so over time results in lasting changes in brain organization.[29] Such changes correspond to increased self-control and regulation;[30] as well as structural changes in the brain such as increased cortical thickness.[31] Similarly, engagement in prayer corresponded to an increased blood flow in the prefrontal cortex and parietal lobes.[32] Likewise, engaging in compassion training increased brain activity in social cognition and emotion regulation regions, and resulted in more altruistic behavior.[33] Overall, by engaging in various types of prayer and meditation, we alter our own brain activity, structure, and function; we increase the likelihood with which we will be able to regulate our emotions and act in self-controlled, prosocial manner.

Along with the benefits of meditation, other practices elicit neural changes as well. Expressing gratitude,[34], disclosing emotions,[35] exercising,[36] choosing a nutrient-dense diet,[37] appreciating beauty,[38] taking deep breaths,[39] and spending time in nature[40] have all been shown to alter neural activity and subsequent behavior.

Other studies in the subdiscipline of biofeedback demonstrate modulation of one's own physiological activity—including heart rate, breathing, and brain activity;[41] with effects such as alleviation of depression symptoms and improved attention.[42]

Effects on Others

Our choices affect not only our own brain activity, but that of others as well. For example, acting as a secure base from which your child can explore the world increase the likelihood with which they will develop a secure attachment, which is linked with the ability to maintain healthy, stable relationships across the lifespan.[43] Moreover, listening empathetically to someone is likely to foster feelings of being empathized with, which is linked to more positive and less negative emotions[44] and recruitment of fronto-parietal brain regions.[45] Additionally, acting with virtue may provide others with experiences of moral beauty, shown to elicit elevation—an emotion in which one feels inspired to be a better person.[46] These are just a few examples of the many ways in which our actions alter others' brain activity, function, and structure (volume and connectivity).

Underlying Neuroscientific Mechanisms

Various neurobiological processes underly the influences discussed above. For example, learning new information (or being exposed to a new experience) leads to dendritic growth—or the growth of dendrites and dendritic spines on a neuron. In the absence of environmental inputs, however, there is less growth and even synaptic pruning—the removal of unused synapses. In other words, neural growth and shrinkage is a function of the environmental inputs to which an individual is exposed.

Another process underlying learning and memory is long-term potentiation, which explains Hebb's rule that "neurons that fire together wire together." In long-term potentiation, the excitation of one synapse influences that of another via a cascade of events that eventually causes more AMPA receptors—a type of glutamate receptor—to move to the postsynaptic neuron membrane; thereby strengthening that synapse (and thus, increasing the likelihood with which a behavior or thought will reoccur).

Finally, brain activity that persists over time predicts structural changes in the brain, as measured by volumetric and diffusion tensor imaging analyses (such as in the example of meditation described above). Myelination, or an increase in the layer of fat surrounding the axon, is understood to be at least one mechanism by which such structural changes occur.

Overall, our experiences and choices directly alter our brain activity via microscopic changes in receptor availability, number of synapses, growth of dendrites, increased fat surrounding axons, and more.

Free Will

The biophysical realities discussed above deepen our understanding of the human person and support Catholic doctrine; yet determinists conclude that such realities negate free will. Several concerns arise with the use of such realities to negate free will—the first being a lack of an agreed upon definition for free will.

Determinists' Concept of Free Will

In *Determined,* Sapolsky admits that free will is defined variably[47]—sometimes, for example, as self-control[48], or the degree to which one can control their actions. Self-control is influenced by the amount of stress one has experienced[49] and is currently experiencing;[50] moreover, certain practices, such as meditation, as described above, are shown to increase self-control.[51] If free will was synonymous with self-control, then individuals would differ on the degree to which they had free will; and certain life factors and practices would impact the amount of free will each person has.

Overall, Sapolsky doesn't equate free-will with self-control. In fact, he doesn't argue against self-control, emotion regulation, decision-making, agency, or altering one's own physiology—all processes that are behaviorally and neuroscientifically measurable. For example, Sapolsky suggests that one's beliefs alter one's brain activity and subsequent choices, explaining that "because you understand change is possible, you have just changed the ability of your brain to respond to optimistic stimuli."[52] He goes on to explain that "knowledge" or new "data points" can lead to behavioral and state changes, such as increases in inspiration.[53] Yet despite all of this, Sapolsky still negates free will. Then what differentiates free will from other relevant, measurable psychological processes?

Sapolsky shares his "take" on free will, challenging the "free willer" to find a neuron for which "nothing" about its functioning was influenced by past or present biophysical realities. Likewise, he suggests that free will would be demonstrated by "a neuron (or brain) whose generation of a behavior is independent of the sum of its biological past."[54] Here, Sapolsky seems to suggest that any influence of biophysical elements on neural activity inherently negates free will, suggesting that free will is characterized by freedom from the influence of any factors.

Concerns with the Deterministic Argument against Free Will

Several concerns arise with the deterministic perspective on free will. First, it's clear that there is no agreed upon definition of free will. Whether free will is defined by lack of influence of biological and environmental factors; the ability to make choices; or the ability to control oneself; remains unclear. Likewise, without an operational definition, free will cannot be measured. As such, we cannot draw conclusions about free will—undefined and unmeasurable—based on scientific evidence.

Next, Sapolsky seems to suggest that if you put together all the scientific results demonstrating the influence of biological and environmental factors on neural activity, "there's no room for free will."[55] Yet, while behavioral and neuroscientific methods can be used to measure the influence of genetics, epigenetics, prenatal environment, childhood, etc; no techniques are currently available with which to measure the remaining "room." Likewise, no techniques have been used to measure independent, uninfluenced neural activity. As such, we cannot scientifically conclude that there is "no room" for free will.

Moreover, determinists draw upon neuroscientific research by Benjamin Libet, showing that the onset of neural activity predicting low-level choices (e.g. pressing one button or another) precedes self-reported conscious awareness of one's choice.[56] The first concern here is that free will is neither defined nor measurable, so we cannot draw conclusions from this study about free will.[57] Several other concerns have arisen, including replicability with other more complicated and moral choices[58] as well as ecological and internal validity.[59] Moreover, later findings point to neural activity corresponding to the conscious attention to act.[60]

Overall, in line with Jaki's assertions, science can draw conclusions about measurable constructs, including cognitive control, decision-making, agency, and altering one's own physiology; but cannot logically draw conclusions about free will—undefined and unmeasurable.

Free Will from a Theological perspective

Free will, as well as the concept of freedom, have been discussed at length in Catholic tradition. Pope Francis describes our inherent freedom "to apply our intelligence" towards alleviating or worsening human suffering.[61] He points to our "capacity to reason, to develop arguments, to be inventive, to interpret reality and to create art."[62] Similarly, Pope Leo XIII suggests that the human person is "free to obey his reason [and] to seek moral good."[63] He also writes that when one does act "according to reason, he acts...according to his free will."[64] Moreover, in the Catechism, man is referred to as a "rational being...who can initiate and control his own actions";[65] and, freedom is discussed as "the power, rooted in reason and will, ... to perform deliberate actions."[66]

This theological perspective characterizes free will as being related to the human person's ability to act according to reason[67], which is a very different from Sapolsky's understanding that free will requires uninfluenced neuronal firing. Reasoning is surely influenced by external and internal factors.[68]

Although neuroscientific evidence cannot "prove" free will, as it cannot be measured, it does support cognitive processes related to reasoning and decision making. For example, left fronto-parietal regions and the basal ganglia are involved in deductive reasoning, or the ability to draw conclusions from various realities or data points.[69] And, executive control regions—dorsolateral and ventromedial aspects of the prefrontal cortex—interpret signals from subcortical regions implicated in emotion—the ventral striatum and amygdala—to weigh benefits vs. costs of various options during decision-making.[70] Evidence further supports the idea that our choices, such as engaging in meditation[71] and participating in biofeedback therapy,[72] shape our lives. Other choices, such as listening empathetically to another,[73] shape the lives of others.

Overall, the theological and neuroscientific definitions of free will are very different, pointing again to the lack of an agreed upon, measurable definition of free will. While neuroscience can explain neural activity underlying reasoning and decision making, it cannot support or negate free will, which is not clearly defined or measurable.

Father Stanley Jaki's Perspective

Stanley Jaki's perspective on science and religion is highly relevant to the discussion. First, Jaki advocates for the relationship between science and religion, pointing out that "Religion, including divine revelation, cannot overrule quantities embedded in things, if revelation was given by the same God who also gave things their existence and to men their minds."[74] In other words, the same God gave us creation and religion, so the two cannot be "in opposition to one another."[75]

Yet Jaki also warns that scientists should stay within the boundaries of science, refraining from extracting unmeasurable purpose and meaning from "quantities." He writes:

> If scientists want to grasp the purpose of anything, they sin against the logic of science if they try to fathom purpose through science, unless they find a way to measure purpose. The question of free will is not something for science to handle...[T]here are no quantitative units in which to measure free will.[76]

Jaki negates the possibility of using science to render free will nonexistent since free will cannot be measured. Here we support this assertion, yet we also extend Jaki's work by finding measurable purpose in neuroscientific evidence. As such, we do not contradict Jaki, but rather explore new territory as we extract measurable purpose from the neuroscientific quantities. Specifically, with modern-day neuroimaging technologies, we can measure the extent to which we can influence our own physiology; and more, we can measure the extent to which we can influence others' physiology. We can, indeed, measure the effects of our own choices (influenced by genetics and the environment, with the extent of independence from such influence remaining unmeasurable) on ourselves and others. As such, in line with Jaki, we can draw scientific conclusions about "what is known" but not about "what is not known."[77]

Finding Measurable Purpose in Neuroscientific Evidence

In Jaki's perspective, we can draw conclusions about measurable purpose from "deterministic" neuroscientific evidence. Inarguably, one's neurobiology and environment physically alter their brain circuitry; directly impact one's thoughts, decisions, and even the degree to which one exerts self-control. Dendritic growth, synaptic pruning, and long-term potentiation are some mechanisms of such influence. What does this tell us? Does it give any purpose to the meaning of suffering?

Compassion, Mercy, and Love

Such evidence points to the vulnerability of the human person—created to physiologically alter in response to the physical and social environment. Yet with this, we find a measurable purpose: a call to be compassionate, merciful, and to love.

First, such evidence demonstrates that immense suffering can be consequential to factors beyond one's control, including one's neurobiology as well as prenatal and postnatal environment. As such, we can understand others' suffering, even when they commit egregious acts; with some forms of suffering even increasing the likelihood with which someone will fail to control their actions and engage in antisocial and criminal behavior.[78] Such understanding may evoke an emotional response – care and concern—to others' suffering, in line with Paul's letter to the Ephesians: "[And] be kind to one another, compassionate" (Ephesians 4:32). This evidence also suggests that when we alter the environment by, for example, listening empathetically to others[79] or providing for their needs[80], we can alleviate their suffering. These actions impact others' brain activity and structure and thus, influence their choices. In doing so, we use our freedom "to apply our intelligence" towards alleviating suffering.[81]

Moreover, evidence points to another measurable purpose—mercy and forgiveness. We realize that one's behavior—including their ability to control their actions, emotions, and thoughts—is heavily influenced by biological and environmental factors. We see that one's decisions are influenced by knowledge to which they've been exposed and physiologically capable of learning. This is in line with Paul's letter to the Romans in which he proclaims, "For I do not do the good I want, but I do the evil I do not want" (Romans 7:19) due to "sin that dwells in me" (Romans 7:19-20). He describes the desire to do good vs. the powerful temptation of the flesh (Romans 7:14-25). This aligns with neuroscientific evidence that self-control is vulnerable to external forces; and that decision-making involves both higher-order goods and lower-order pleasures. This understanding of the inherent vulnerability of the human person calls for forgiveness and lack of judgement (Luke 6:36-37). Such is in line with the Lord's mercy upon his crucifiers, and upon us: "Father, forgive them, they know not what they do" (Luke 23:34). It supports the words of Matthew: "Stop judging...[R]emove the beam from your eye first; then you will see clearly to remove the splinter from your brother's eye" (Matthew 7:1,5). Interestingly, even after we remove the "beam" from our own eye, we are not asked to judge, but rather to help our neighbor see clearly. We can do so, for example, by offering a listening ear, providing for needs, sharing knowledge, and more.

In line with the call for mercy, neuroscientific evidence also contradicts the use of capital punishment. It supports the use of supportive systems by which brains, malleable, can rewire, increasing the likelihood with which transgressors engage in prosocial behavior. This aligns with the Catechism's view that capital punishment is "an attack on ... the dignity of the human person."[82] It points to the use of "more effective systems ... which ... do not definitively deprive the guilty of the possibility of redemption."[83]

And finally, evidence here points to a measurable call to love; for we, indeed, constitute some of our neighbor's environment (and even gene expression, in the case of epigenetics). We influence our neighbors' brain activity and thus, behavior, affect, and cognition. Such evidence supports Pope Francis' assertion that our "concern for the environment ...needs to be joined to a sincere love for our fellow human beings and an unwavering commitment to resolving the problems of society".[84] When we do so, we elicit neurobiological changes amongst our "fellow human beings," strengthening their cognitive control, providing data points for moral decision-making, and buffering against the deleterious effects of stress on the brain.

Overall, neuroscientific evidence, albeit used in deterministic arguments, strikingly supports the call we hear in Colossians:

> Put on ... heartfelt compassion, kindness, humility, gentleness, and patience, bearing with one another and forgiving one another, if one has a grievance against another; as the Lord has forgiven you, so must you also do. And over all these put on love, that is, the bond of perfection (Colossians 3:12-14).

Radical Interconnectedness

Neuroscientific evidence also points to our radical interconnectedness. It demonstrates that our choices effect changes in DNA expression which are passed on to our children and grandchildren. Our choices affect the social environment and availability of resources for others, leading them to abundance or deprivation of stimuli (e.g. food, money, healthcare, social support, beauty) which satisfy basic needs. For example, when we keep our treasure for ourselves and don't share, we allow alterations in brain activity amongst those deprived of basic necessities, thereby impacting behavior. Yet when we act virtuously, we may elicit feelings of elevation in others, inspiring them to act morally. Here we find quantitative support for the Catechism: "God wills the interdependence of creatures ... no creature is self-sufficient. Creatures exist only in dependence on each other, to complete each other, in the service of each other."[85] Overall, with such evidence, we see that we truly are "one body" (1 Corinthians 12:12) with the parts of the body dependent upon each other (1 Corinthians 12:13-26).

Meaning in Suffering

Jaki also asks whether science can "give meaning to suffering."[86] Science demonstrates that an adverse environment elicits neurobiological changes underlying antisocial behavior, such as impulsivity and lack of empathy. Science also demonstrates that practices such as meditation and emotional disclosure can buffer against such neurobiological changes; and, that moral action can elicit positive emotions shown to inspire another to become a better person. All in all, science demonstrates that one's suffering can be heightened or alleviated by the moral action of others. One's brain activity and developmental trajectory can be altered by the choices of another. With this, I propose that this scientific evidence gives meaning to suffering by pointing to our call to love. Sapolsky, I think, may agree. Indeed, being is "synonymous with the true, the good, and beautiful—all of which are best summed up in the word LOVE."[87]

Our Call to Help Determine

Deterministic arguments seem to unintentionally show that we can determine (or rather, *influence*, since the extent of lack of influence cannot be measured) own brain activity, structure, and connectivity—thereby affecting our choices and behavior and shaping our "own life.[88]" By our actions, we influence the extent to which others suffer or rejoice; act morally or immorally; love or fail to love (e.g. punish, abuse; or listen, empathize, respond to needs, give, etc). With each choice we make (e.g. sharing knowledge), we help form the intentions of others. We influence others' brain activity, structure, and connectivity. We can truly help determine—influence—in concrete, measurable ways, the health of the body of Christ.

Conclusion

In agreement with Jaki, we cannot draw conclusions about free will from scientific evidence, as free will is neither measurable nor operationally defined. Also in line with Jaki, we find measurable purpose in objective realities. Amidst determinism, we find that neuroscientific "quantities" support the call to be compassionate, merciful, and loving; and point to our interdependence. This extends Jaki's work to demonstrate that science and religion are more than two "conceptually separate domain[s]."[89] Rather, they intersect—both pointing to truth—which is love "writ large.[90]"

Overall, let's rejoice in quantities and, without drawing conclusions about free will, let's acknowledge the immense influence of genes, the environment, and more on the human person. With this, let's recognize the fragility of the human person, and be forgiving and merciful as the Lord is merciful (Psalm 145:8-9). Let's also recognize the human person's ability to reason; to draw upon moral values in decision-making; to grow in knowledge and utilize that knowledge to inform choices; to effect physiological change within oneself via prayer, meditation, acts of virtue, and more. These purposeful activities and their consequences are measurable with modern-day research methods. Moreover, let's influence others' brain activity and structure; specifically, let's strive to love, thereby alleviating suffering and drawing others to Christ in One Body. When we love, whether inspired by science, religion, or both—we come to know God, "for God is love" (1 John 4:8; see also 1 John 4:12)—and we come to know the truth that will set us free (John 8:32).

Notes

1. Robert M. Sapolsky, *Determined: A Science of Life Without Free Will* (New York: Penguin Press, 2023).
2. Jingguang Li et al., "Association of Oxytocin Receptor Gene (OXTR) Rs53576 Polymorphism with Sociality: A Meta-Analysis," *PloS One* 10, no. 6 (2015): e0131820.
3. Julien Bacqué-Cazenave et al., "Serotonin in Animal Cognition and Behavior," *International Journal of Molecular Sciences* 21, no. 5 (2020): 1649; Emil F. Coccaro et al., "Serotonin and Impulsive Aggression," *CNS Spectrums* 20, no. 3 (2015): 295–302.
4. Elisabet Hermanson et al., "Nurr1 Regulates Dopamine Synthesis and Storage in MN9D Dopamine Cells," *Experimental Cell Research* 288, no. 2 (2003): 324–34; Jean-Claude Dreher et al., "Variation in Dopamine Genes Influences Responsivity of the Human Reward System," *Proceedings of the National Academy of Sciences* 106, no. 2 (2009): 617–22; Karen L. O'Malley et al., "Genes for Catecholamine Biosynthesis: Cloning by Expression and Identification of the CDNA for Rat Dopamine Beta-Hydroxylase.," *Proceedings of the National Academy of Sciences* 80, no. 8 (1983): 2161–65.
5. Luca Botticelli et al., "Underlying Susceptibility to Eating Disorders and Drug Abuse: Genetic and Pharmacological Aspects of Dopamine D4 Receptors," *Nutrients* 12, no. 8 (2020): 2288; Szymon Zmorzyński et al., "Personality Traits and Polymorphisms of Genes Coding Neurotransmitter Receptors or Transporters: Review of Single Gene and Genome-Wide Association Studies," *Annals of General Psychiatry* 20, no. 1 (2021): 1–6; Jean Claude Martel and Silvia Gatti McArthur, "Dopamine Receptor Subtypes, Physiology and Pharmacology: New Ligands and Concepts in Schizophrenia," *Frontiers in Pharmacology* 11 (2020): 1003; M. Anguelova, C. Benkelfat, and Gustavo Turecki, "A Systematic Review of Association Studies Investigating Genes Coding for Serotonin Receptors and the Serotonin Transporter: I. Affective Disorders," *Molecular Psychiatry* 8, no. 6 (2003): 574–91.

6. Shu-ichi Ueno, "Genetic Polymorphisms of Serotonin and Dopamine Transporters in Mental Disorders," *Journal of Medical Investigation* 50, no. 1/2 (2003): 25–31; Elisabeth M. van de Giessen et al., "Striatal Dopamine Transporter Availability Associated with Polymorphisms in the Dopamine Transporter Gene SLC6A3," *Journal of Nuclear Medicine* 50, no. 1 (2009): 45–52.
7. Adrian Wells, "Detached Mindfulness in Cognitive Therapy: A Metacognitive Analysis and Ten Techniques," *Journal of Rational-Emotive and Cognitive-Behavior Therapy* 23, no. 4 (2005): 337–55.
8. Nader Perroud et al., "COMT but Not Serotonin-related Genes Modulates the Influence of Childhood Abuse on Anger Traits," *Genes, Brain and Behavior* 9, no. 2 (2010): 193–202.
9. Jean-Claude Dreher et al., "Variation in Dopamine Genes Influences Responsivity of the Human Reward System," *Proceedings of the National Academy of Sciences* 106, no. 2 (2009): 617–22.
10. Marian J. Bakermans-Kranenburg and Marinus H. Van Ijzendoorn, "Differential Susceptibility to Rearing Environment Depending on Dopamine-Related Genes: New Evidence and a Meta-Analysis," *Development and Psychopathology* 23, no. 1 (2011): 39–52.
11. Bakermans-Kranenburg and Van Ijzendoorn, "Differential Susceptibility to Rearing," 39–52.
12. Ian C. G. Weaver, "Life at the Interface between a Dynamic Environment and a Fixed Genome: Epigenetic Programming of Stress Responses by Maternal Behavior," *Mammalian Brain Development* (2009): 17–39.
13. Barry M. Lester et al., "Epigenetic Programming by Maternal Behavior in the Human Infant," *Pediatrics* 142, no. 4 (2018).
14. F. Limosin et al., "Prenatal Exposure to Influenza as a Risk Factor for Adult Schizophrenia," *Acta Psychiatrica Scandinavica* 107, no. 5 (2003): 331–35.
15. Robin P. Cooper and Richard N. Aslin, "The Language Environment of the Young Infant: Implications for Early Perceptual Development.," *Canadian Journal of Psychology/Revue Canadienne de Psychologie* 43, no. 2 (1989): 247; Lester et al., "Epigenetic Programming by Maternal Behavior in the Human Infant"; Leon J. Yarrow, Judith L. Rubenstein, and Frank

A. Pedersen, "Infant and Environment: Early Cognitive and Motivational Development" (1975).

16. Julia I. Herzog and Christian Schmahl, "Adverse Childhood Experiences and the Consequences on Neurobiological, Psychosocial, and Somatic Conditions across the Lifespan," *Frontiers in Psychiatry* (Frontiers Media S.A., September 4, 2018), https://doi.org/10.3389/fpsyt.2018.00420; Karen Hughes et al., "The Effect of Multiple Adverse Childhood Experiences on Health: A Systematic Review and Meta-Analysis," *The Lancet Public Health* 2, no. 8 (2017): e356–66; Maria Rita Infurna et al., "Environmental Factors That Distinguish between Clinical and Healthy Samples with Childhood Experiences of Abuse and Neglect," *Psychopathology* 48, no. 4 (2015): 256–63, https://doi.org/10.1159/000435959.
17. Shannon C. Montgomery et al., "Peer Social Network Processes and Adolescent Health Behaviors: A Systematic Review," *Preventive Medicine* 130 (2020): 105900; Angela K. Henneberger, Dawnsha R. Mushonga, and Alison M. Preston, "Peer Influence and Adolescent Substance Use: A Systematic Review of Dynamic Social Network Research," *Adolescent Research Review* 6 (2021): 57–73.
18. Denise De Ridder et al., "Always Gamble on an Empty Stomach: Hunger Is Associated with Advantageous Decision Making," *PloS One* 9, no. 10 (2014): e111081.
19. Yvonne Harrison and James A. Horne, "The Impact of Sleep Deprivation on Decision Making: A Review," *Journal of Experimental Psychology: Applied* 6, no. 3 (2000): 236.
20. Dominic S. Fareri, Luke J. Chang, and Mauricio R. Delgado, "Effects of Direct Social Experience on Trust Decisions and Neural Reward Circuitry," *Frontiers in Neuroscience* 6 (2012): 148.
21. Malcolm Macmillan, "Phineas Gage," *Classic Cases in Neuropsychology* (1996), 220.
22. Macmillan. "Phineas Gage," 220.
23. Jonathan Haidt, "Elevation and the Positive Psychology of Morality" (2003); Hengameh Marzbani, Hamid Reza Marateb, and Marjan Mansourian, "Neurofeedback: A Comprehensive Review on System Design, Methodology and Clinical

Applications," *Basic and Clinical Neuroscience* 7, no. 2 (2016): 143.

24. Katie A. McLaughlin, David Weissman, and Debbie Bitrán, "Childhood Adversity and Neural Development: A Systematic Review," *Annual Review of Developmental Psychology* 1 (2019): 277–312.
25. McLaughlin, Weissman, and Bitrán, "Childhood Adversity and Neural," 277–312.
26. Ibid.
27. J. David Creswell, "Mindfulness Interventions," *Annual Review of Psychology* 68 (2017): 491–516.
28. Rui Ferreira Afonso et al., "Neural Correlates of Meditation: A Review of Structural and Functional MRI Studies," *Frontiers in Bioscience-Scholar* 12, no. 1 (2020): 92–115.
29. Afonso et al.; Antonietta Manna et al., "Neural Correlates of Focused Attention and Cognitive Monitoring in Meditation," *Brain Research Bulletin* 82, no. 1–2 (2010): 46–56.
30. S. F. Maier, J. Amat, M. V. Baratta, P. Evan and L. R. Watkins, "Behavioral Control, the Medial Prefrontal Cortex, and Resilience," *Behavioral Control, the Medial Prefrontal Cortex, and Resilience* (2006); Deane H. Shapiro, "Overview: Clinical and Physiological Comparison of Meditation with Other Self-Control Strategies," *Meditation* (2017), 5–12.
31. Afonso et al., "Neural Correlates of Meditation," 92.
32. Andrew Newberg et al., "Cerebral Blood Flow during Meditative Prayer: Preliminary Findings and Methodological Issues," *Perceptual and Motor Skills* 97, no. 2 (2003): 625–30.
33. Antoine Lutz et al., "Regulation of the Neural Circuitry of Emotion by Compassion Meditation: Effects of Meditative Expertise," ed. Bernhard Baune, *PLoS ONE* 3, no. 3 (March 26, 2008): e1897, https://doi.org/10.1371/journal.pone.0001897.
34. Barbara L. Fredrickson, "Gratitude, like Other Positive Emotions, Broadens and Builds," *The Psychology of Gratitude* 145 (2004): 166; Laura L. Vernon, Jacqueline M. Dillon, and Amanda R.W. Steiner, "Proactive Coping, Gratitude, and Posttraumatic Stress Disorder in College Women," *Anxiety, Stress, & Coping* 22, no. 1 (January 2009): 117–27, https://doi.org/10.1080/10615800802203751; Alex M. Wood, Jeffrey J. Froh, and Adam W. A. Geraghty, "Gratitude

and Well-Being: A Review and Theoretical Integration," *Clinical Psychology Review* 30, no. 7 (2010): 890–905.

35. Kent D. Harber and Karen E. Wenberg, "Emotional Disclosure and Closeness Toward Offenders," *Personality and Social Psychology Bulletin* 31, no. 6 (June 2, 2005): 734–46, https://doi.org/10.1177/0146167204272720; Denise M. Sloan, "Self-Disclosure and Psychological Well-Being" (2010); Scott H. Hemenover, "The Good, the Bad, and the Healthy: Impacts of Emotional Disclosure of Trauma on Resilient Self-Concept and Psychological Distress," *Personality and Social Psychology Bulletin* 29, no. 10 (October 2, 2003): 1236–44, https://doi.org/10.1177/0146167203255228.

36. S. J. Colcombe et al., "Aerobic Exercise Training Increases Brain Volume in Aging Humans," *The Journals of Gerontology Series A: Biological Sciences and Medical Sciences* 61, no. 11 (November 1, 2006): 1166–70, https://doi.org/10.1093/gerona/61.11.1166; C. Barr Taylor, James F. Sallis, and Richard Needle, "The Relation of Physical Activity and Exercise to Mental Health," *Public Health Reports* 100, no. 2 (1985): 195; Michelle Ploughman, "Exercise Is Brain Food: The Effects of Physical Activity on Cognitive Function," *Developmental Neurorehabilitation* 11, no. 3 (2008): 236–40, https://doi.org/10.1080/17518420801997007; Lars S. Jonasson et al., "Aerobic Exercise Intervention, Cognitive Performance, and Brain Structure: Results from the Physical Influences on Brain in Aging (PHIBRA) Study," *Frontiers in Aging Neuroscience* 8 (2017): 336.

37. Michael K. Georgieff, "Nutrition and the Developing Brain: Nutrient Priorities and Measurement," *The American Journal of Clinical Nutrition* 85, no. 2 (February 2007): 614S-620S, http://www.ncbi.nlm.nih.gov/pubmed/17284765; Romain Meeusen, "Exercise, Nutrition and the Brain," *Sports Medicine* 44 (2014): 47–56; Katie Moore et al., "Diet, Nutrition and the Ageing Brain: Current Evidence and New Directions," *Proceedings of the Nutrition Society* 77, no. 2 (2018): 152–63; Devin Wahl et al., "Cognitive and Behavioral Evaluation of Nutritional Interventions in Rodent Models of Brain Aging and Dementia," *Clinical Interventions in Aging* (2017), 1419–28.

38. Rhett Diessner et al., "Beauty and Hope: A Moral Beauty Intervention," *Journal of Moral Education* 35, no. 3 (2006): 301–17; María Luisa Martínez-Martí, María Dolores Avia, and María José Hernández-Lloreda, "Appreciation of Beauty

Training: A Web-Based Intervention," *The Journal of Positive Psychology* 9, no. 6 (2014): 477–81; María Luisa Martínez-Martí, María Dolores Avia, and María José Hernández-Lloreda, "Effects of an Appreciation of Beauty Randomized-Controlled Trial Web-Based Intervention on Appreciation of Beauty and Well-Being," *Psychology of Aesthetics, Creativity, and the Arts* 12, no. 3 (2018): 272; Tsukiura Takashi, and Roberto Cabeza, "Shared brain activity for aesthetic and moral judgments: implications for the Beauty-is-Good stereotype," *Social cognitive and affective neuroscience* 6, no. 1 (2011): 138-148. https://academic.oup.com/scan/article/6/1/138/1609687; Tomohiro Ishizu and Semir Zeki, "Toward A Brain-Based Theory of Beauty," ed. Eric James Warrant, *PLoS ONE* 6, no. 7 (July 6, 2011): e21852, https://doi.org/10.1371/journal.pone.0021852.

39. F. Joseph McClernon, Eric C. Westman, and Jed E. Rose, "The Effects of Controlled Deep Breathing on Smoking Withdrawal Symptoms in Dependent Smokers" (2004), https://doi.org/10.1016/j.addbeh.2004.02.005.

40. René T. Proyer et al., "Nine Beautiful Things: A Self-Administered Online Positive Psychology Intervention on the Beauty in Nature, Arts, and Behaviors Increases Happiness and Ameliorates Depressive Symptoms," *Personality and Individual Differences* 94 (2016): 189–93; Agnes E. Van den Berg, *Health Impacts of Healing Environments; a Review of Evidence for Benefits of Nature, Daylight, Fresh Air, and Quiet in Healthcare Settings* (UMCG, 2005); Miles Richardson and Kirsten McEwan, "30 Days Wild and the Relationships between Engagement with Nature's Beauty, Nature Connectedness and Well-Being," *Frontiers in Psychology* 9, no. SEP (September 3, 2018): 1500, https://doi.org/10.3389/FPSYG.2018.01500/BIBTEX; Kendrick N. Kay et al., "Identifying Natural Images from Human Brain Activity," *Nature* 452, no. 7185 (2008): 352–55.

41. Marzbani, Marateb, and Mansourian, "Neurofeedback: A Comprehensive Review," 143.

42. *Ibid.*

43. Pernille Darling Rasmussen et al., "Attachment as a Core Feature of Resilience: A Systematic Review and Meta-Analysis," *Psychological Reports* 122, no. 4 (2019): 1259–96.

44. Frédéric Nils and Bernard Rimé, "Beyond the Myth of Venting:

Social Sharing Modes Determine the Benefits of Emotional Disclosure," *European Journal of Social Psychology* 42, no. 6 (2012): 672–81.

45. Maria Seehausen et al., "Talking about Social Conflict in the MRI Scanner: Neural Correlates of Being Empathized With," *NeuroImage* 84 (January 1, 2014): 951–61, https://doi.org/10.1016/J.NEUROIMAGE.2013.09.056.
46. Haidt, "Elevation and the Positive Psychology of Morality."
47. Sapolsky, *Determined,* 14.
48. *Ibid.*
49. Gary W. Evans and Pilyoung Kim, "Childhood Poverty, Chronic Stress, Self-Regulation, and Coping," *Child Development Perspectives* 7, no. 1 (March 6, 2013): 43–48, https://doi.org/10.1111/cdep.12013.
50. Candace M. Raio et al., "Cognitive Emotion Regulation Fails the Stress Test.," *Proceedings of the National Academy of Sciences of the United States of America* 110, no. 37 (2013): 15139–44, https://doi.org/10.1073/pnas.1305706110.
51. Shapiro, "Overview: Clinical and Physiological Comparison," 5–12.
52. Sapolsky and Andrew D. Huberman, "Dr. Robert Sapolsky: Science of Stress, Testosterone & Free Will" (Huberman Lab Podcast #35, August 29, 2021).
53. *Ibid.*
54. Sapolsky, *Determined,* 15.
55. *Ibid,* 8–9.
56. Benjamin Libet, "Do We Have Free Will?," *Journal of Consciousness Studies* 6, no. 8–9 (1999): 47–57; Benjamin Libet, Elwood W. Wright Jr, and Curtis A. Gleason, "Preparation-or Intention-to-Act, in Relation to Pre-Event Potentials Recorded at the Vertex," *Electroencephalography and Clinical Neurophysiology* 56, no. 4 (1983): 367–72.
57. Victoria Saigle, Veljko Dubljević, and Eric Racine, "The Impact of a Landmark Neuroscience Study on Free Will: A Qualitative Analysis of Articles Using Libet and Colleagues' Methods," *AJOB Neuroscience* 9, no. 1 (2018): 29–41.
58. Moritz Nicolai Braun, Janet Wessler, and Malte Friese, "A Meta-Analysis of Libet-Style Experiments," *Neuroscience & Biobehavioral Reviews* 128 (2021): 182–98.

59. Marcel Brass, Ariel Furstenberg, and Alfred R. Mele, "Why Neuroscience Does Not Disprove Free Will," *Neuroscience & Biobehavioral Reviews* 102 (2019): 251–63.
60. Brass, Furstenberg, and Mele, "Why Neuroscience," 251–63.
61. Pope Francis, *Encyclical Letter Laudato Si' of the Holy Father Francis on Care for Our Common Home* (May 24, 2015), at the Holy See, https://www.vatican.va/content/francesco/en/encyclicals/documents/papa-francesco_20150524_enciclica-laudato-si.html, 79.
62. *Ibid.*, 81.
63. Pope Leo XIII, "Libertas: Encyclical of Pope Leo XIII on the Nature of Human Liberty" (Rome, 1888), 1.
64. *Ibid.*, 6.
65. Catechism of the Catholic Church, 2nd ed. (Città del Vaticano: Libreria Editrice Vaticana, 2000), 1730.
66. *Ibid.*, 1731.
67. Pope Leo XIII, "Libertas," 6.
68. Julia Wertheim and Marco Ragni, "The Neurocognitive Correlates of Human Reasoning: A Meta-Analysis of Conditional and Syllogistic Inferences," *Journal of Cognitive Neuroscience* 32, no. 6 (2020): 1061–78.
69. Jérôme Prado, Angad Chadha, and James R. Booth, "The Brain Network for Deductive Reasoning: A Quantitative Meta-Analysis of 28 Neuroimaging Studies," *Journal of Cognitive Neuroscience* 23, no. 11 (2011): 3483–97.
70. Ulrike Basten et al., "How the Brain Integrates Costs and Benefits during Decision Making," *Proceedings of the National Academy of Sciences* 107, no. 50 (2010): 21767–72.
71. Afonso et al., "Neural Correlates of Meditation."
72. Marzbani, Marateb, and Mansourian, "Neurofeedback: A Comprehensive Review," 143.
73. See, for example, Sylvia A. Morelli, Jared B. Torre, and Naomi I. Eisenberger, "The Neural Bases of Feeling Understood and Not Understood," *Social Cognitive and Affective Neuroscience* 9, no. 12 (December 1, 2014): 1890–96, https://doi.org/10.1093/SCAN/NST191.
74. Stanley L. Jaki, *Questions on Science and Religion* (New Hope, KY: Real View Books, 2004), 10.

75. Jaki, *Brain, Mind and Computers* (NYC: Herder & Herder, 1969), 78.
76. Jaki, *Questions on Science and Religion*, 11.
77. Jaki, *Brain, Mind and Computers*, 75.
78. Elizabeth J. Susman, "Psychobiology of Persistent Antisocial Behavior: Stress, Early Vulnerabilities and the Attenuation Hypothesis," *Neuroscience & Biobehavioral Reviews* 30, no. 3 (2006): 376–89; Patrick Tolan, "Socioeconomic, Family, and Social Stress Correlates of Adolescent Antisocial and Delinquent Behavior," *Journal of Abnormal Child Psychology* 16 (1988): 317–31; Adrian Raine, "Biosocial Studies of Antisocial and Violent Behavior in Children and Adults: A Review," *Journal of Abnormal Child Psychology* 30, no. 4 (2002): 311–26, https://doi.org/10.1023/A:1015754122318.
79. Morelli, Torre, and Eisenberger, "The Neural Bases of Feeling Understood and Not Understood"; Seehausen et al., "Talking about Social Conflict," 12.
80. McLaughlin, Weissman, and Bitrán, "Childhood Adversity and Neural," 277–312.
81. Pope Francis, *Laudato Si'*, 79.
82. Pope Francis, "Address to Participants in the Meeting Organized by the Pontifical Council for the Promotion of the New Evangelization" (Vatican City, October 11, 2017).
83. Catechism of the Catholic Church, 2267.
84. Pope Francis, *Laudato Si'*, 91.
85. Catechism of the Catholic Church, 340.
86. Jaki, *Questions on Science and Religion*, 59.
87. Jaki, *Questions on Science and Religion*, 59–60.
88. *Catechism of the Catholic Church*, 1731.
89. Jaki, *Questions on Science and Religion*, 78.
90. Jaki, *Questions on Science and Religion*, 60.

12 NINETEENTH CENTURY AMERICAN CATHOLIC APPROACHES TO DARWINIAN EVOLUTION

Biff Rocha, Ph.D. and Jeffrey L. Morrow, PhD.

Introduction

In 1859 Charles Darwin published the *Origin of Species* which marked a watershed event in the history of the study of nature;[1] for many it threw into question much of the traditional belief in western cultures about God and humanity.[2] The effects and influence of his book were momentous. The *Descent of Man* followed in 1871, which for many further challenged the notion of divine creation of humanity.[3] Darwin's theories went against earlier theological arguments for a divine designer like those of William Paley, against whose work his own method fashioned a naturalistic response.[4] As Ernst Mayr candidly explained, natural selection "permits the explanation of adaptation, the 'design' of the natural theologian, by natural means."[5]

Just about every intellectual of the time had to engage with Darwin in some form or other. This was no less true of Catholics.[6] In the United States, Catholic intellectuals had to decide how to evaluate Darwin's work. Many Catholic sensed Darwin's account was a modern challenge to the Church and thus they took a variety of positions on his evolutionary theory.[7] Some rejected it outright, others embraced some form of evolution, and some aspects of Darwin's proposed primary evolutionary mechanism, natural selection, and tried to adopt evolution and adjust their understanding of the history of biology and theology in its light. Indeed, evolution was a topic that was near and dear to Fr. Stanley Jaki's own work here at Seton Hall University.[8]

In this paper, we will examine the work of two significant figures within this nineteenth century controversy, Fr. John Zahm and Fr. Januarius De Concilio. Zahm was representative of Catholic intellectuals who were intent on incorporating biological evolution, but, at the same time, Zahm distanced himself from Darwin on quite a few points. De Concilio, who was often viewed in the camp of opponents of evolution, could more accurately be read as cautious of the more recent evolutionary approaches, rather than holding a position of outright rejection. Zahm's and De Concilio's positions were much closer than has often been understood, although in the end Zahm thought biological evolution of species far more probable and De Concilio found it highly improbable.

After briefly describing the nineteenth century context for Catholics in the U.S., we will mention some of the Catholic intellectual rejection of evolutionary theory, which often went together with their rejection of broader American culture.[9] After this, we will examine those intellectuals who embraced evolutionary theories, who also seemed more open to American culture. Here we focus on the work of Zahm. Finally, we will discuss De Concilio's often misunderstood approach.

Nineteenth Century American Context

Before we enter the specific controversy over evolution, it is necessary to set the stage by describing the nineteenth century context. During this period the Catholic Church often defined itself in opposition to the Protestants of the Reformation, and the proponents of more secular European enlightenments,[10] both of which were understood to have rejected the Catholic Church's authority in favor of individual judgement and human freedom.[11] The United States of the nineteenth century was in the midst of immense social change caused by the Industrial Revolution and an unprecedented population explosion. The in-pouring of immigrants fueled the economic expansion and led to sprawling urbanization. In the period of 1846–1924 fifty million people emigrated from Europe to the New World.[12]

Catholics in America were confronted with doubts from both sides, namely from the broader American society and also from within the Church. Steven Avella and Elizabeth McKeown note: "Some came from the hostility of American society, which raised doubts as to the loyalty of American Catholics to the republic. Others came from church officials in Rome who needed regular reassurance that American Catholics were not slipping into a fuzzy indifference through their support for religious toleration and pluralism."[13] In the nineteenth century the American hierarchy of the Catholic Church became divided over the proper relationship between American society and the Catholic faith. They clashed with each other on a variety of issues, but there were particularly high tensions regarding the immigrant problem, Catholic education, the desirability of dialog with the non-Catholic world, the need for an apostolic delegate from Rome, and the usefulness of the new sciences emerging at the time.

Some saw American society and the modern ethos as generally positive and offering new opportunities for the practice of the Catholic faith. These individuals wanted to engage the culture with their Catholic faith and transform society into something even better. Many of these Catholics were the older, now middle-class immigrants, such as the Irish, who had become increasingly Americanized. Representatives belonging to this camp included

Bishops John Ireland, John Keane, Denis O'Connell, James Gibbons, and at times, John Spalding.[14]

Others felt American society's hostility towards the Catholic faith, and they viewed the modern ethos as a dangerous challenge and competitor to the Catholic Church. These individuals perceived the spirit behind modern developments as essentially hostile to the faith. They were looking for a way to protect the faith from the ever-increasing demands to accommodate or abandon their beliefs. These Catholics were often made up of newly arrived immigrants from Italy, Poland, Hungary, and Russia. Representative voices for this camp were found among many of the contributors to the *American Catholic Quarterly Review*, predominantly German clergy led by Bishop Frederick Katzer, the Jesuits, Bishop Michael Corrigan, the second bishop of the Archdiocese of Newark, New Jersey, and Bishop Bernard McQuaid.[15]

The Rejection of Evolution and General Resistance to American Culture

Several factors contributed to the reaction of Catholics who recoiled from the American culture. First, immigrants frequently equated their ethnic practices with their Catholic religion. For many recent immigrant Catholics, the parish became the center of their ethnic subculture in American society.[16] Second, many saw in the American society a growing agnostic threat. Third, being pressured to Americanize by the Irish hierarchy which ruled the American Catholic church, only increased the fears and sense of isolation of many immigrants. Fourth, some Protestant nativists used the new scientific ideas to justify their actions and claims against immigrants whom they identified as inferior races. We must keep in mind that some Catholics perceived Protestant nativism as a real threat. Some of the prominent nativists invoked the concept of Darwin's theory of natural selection in arguing for the removal of immigrants, who were seen by nativists as inferior and less worthy of survival.[17] Fifth, within the clergy especially was felt the challenge to the Catholic worldview which was grounded in God's day-to-day operations in sustaining the life he created. Many Catholics began perceiving the relationship between science and faith dialectically as a choice between two competing meta-narratives: evolution by chance or creation by divine design.[18]

Catholic intellectuals who were generally opposed to evolution had similar concerns to those we noted above concerning American culture in general. First, they perceived nineteenth-century scientists as often explicitly hostile to religious faith. Second, with these sciences being new, critics complained that many of their theories lacked evidential support. Third, the evidence that did exist was open to multiple interpretations; many supporting Darwin specifically often chose those interpretations most at odds with the ways in which Genesis 1–2 had been read for the past few centuries.[19] F.R. Wegg-Prosser, in an *American Catholic Quarterly Review* article, pointed to the design of the eye, which he saw as a problem for any evolutionary theory.[20] Darwin already recognized the challenge the complexity of the eye posed for his own theory: "Although the belief that an organ so perfect as the eye could have been formed by natural selection, is more than enough to stagger any one."[21]

In his article, "Is the Book of Genesis Genuine and Authentic?" J.J. Quinn understood evolution as a counter narrative to the biblical story of creation. Quinn argued for the authenticity of the Book of Genesis and challenged some of the claims of specifically natural selection. Quinn wrote:

> In another way the narration of Moses has been attacked in our times. Mr. Darwin maintains that a transformation of vegetable and animal species has been going on from the beginning, so that the horses, cows, sheep and even the men of our day are only the outcome of imperfect or extinct species ... Darwin's theory on Man's descent is neither in accord with faith nor science.[22]

Quinn pointed out what he identified as weaknesses of evolution and in his final analysis he pointed to seven "chasms" between inferior species and humans: morality, intellect, free will, sense of religion, physical structure, civilization, and rational speech.

Catholic Adoption and Modification of Evolution

Historians often referred to the exchange between American priest-scientist Fr. John A. Zahm, of the Congregation of the Holy Cross, and the Magisterium of the Catholic Church as the "Zahm Affair," which they saw as a symbol of late nineteenth- and early twentieth-century Catholic struggles with modernity.[23] Zahm especially embodied an attitude of societal progress which was adhered to by some within the Catholic hierarchy in the U.S. But Catholics who opposed Zahm "linked his advocacy of evolution and modern science to the uncritical acceptance of American values which they believed were promoted by certain 'liberal' members of the clergy and hierarchy."[24] Zahm may serve as an impressive example of the way some Catholic intellectuals sought to relate the Catholic faith to American culture in the nineteenth century. While the publication of Darwin's *Origin of Species* provoked an assortment of responses among U.S. Catholics, it seems that most clergy respondents opposed evolution in any form.[25] Zahm, was caught in the middle of the intellectual current polarized by the question of the proper relationship Catholics ought to have towards the modern world.

In 1896 Zahm published three important books pertaining to evolution and Catholicism: *Scientific Theory and Catholic Doctrine*,[26] *Science and the Church*,[27] and his much larger and better-known *Evolution and Dogma*.[28] Zahm had already established himself as a leading Catholic scientist with his prior publications, a number of which were translated into French, Italian, and Spanish,[29] and based on which Pope Leo XIII had awarded Zahm an honorary doctoral degree in 1895.[30] Zahm never became one of Darwin's ardent defenders, though he did adopt and modify some of Darwin's views, but he maintained a serious critique of Darwin's discussions of natural selection in particular, and of other evolutionary theories. By the time he wrote *Evolution and Dogma*, he would be in the camp of theistic evolutionists who advocated for the probability of evolution, even as they parted, in many respects, from Darwin.

In *Evolution and Dogma*, Zahm sought to interpret the Church Fathers in such a way as to view them as in conformity with evolutionary thinking. Ultimately *Evolution and Dogma* would end up causing controversy as it was investigated by the Sacred Congregation of the Index of Prohibited Books.[31] Zahm was not alone in his method; a similar approach was taken by St. George Jackson Mivart (whose name included "St.," in honor of St. George, not because Mivart was canonized).[32] Mivart was an important precursor to Zahm in many regards, and he was a well-respected evolutionist, and an important primatologist, "the foremost Victorian expert on monkeys."[33] Like Zahm, Mivart was also awarded an honorary doctorate from a pope (Pius IX).[34] Mivart wrote *On the Genesis of Species* in 1871[35] and *Contemporary Evolution* in 1876.[36]

He criticized Darwin's theory of natural selection while offering an evolutionary theory which claimed that the human soul was uniquely created by God. Peter Bowler labeled Mivart "Darwin's most active scientific critic"[37] and one of his "most persistent critics."[38] Mivart was such a serious critic that Darwin added an entire chapter to the sixth edition of *Origin* in response to him.[39] Mivart set the direction for those scholars who would later claim to adhere to some sort of theistic evolution,[40] and, significantly, his books dealing with evolution were never censured nor did they ever spark any controversy with the Magisterium.[41] Indeed, Mivart was perhaps the main inspiration for Zahm; Zahm could be said to have modified and expanded Mivart.[42]

Zahm was just one of many American priest-scientists who were responding to the modern ethos by doing work which sought to accommodate the Catholic faith to contemporary scientific theories. He was deeply committed to the scientific tradition. Scott Appleby maintained that Zahm was one of the earliest American priest-scientists who had a "complete conversion to the general theory of evolution."[43] Zahm felt the need to help the Church modernize and address the demands of the age. He thus came to be associated publicly more and more with what Pope Leo XIII censured as Americanism in his 1899 *Testem Benevolentiae*[44] and with what Pope St. Pius X later condemned as modernism in his 1907 *Pascendi Dominici Gregis.*[45]

Between 1883 and 1890, Zahm published four articles which he later incorporated into his book, *Catholic Science and Catholic Scientists.*[46] In these early works, Zahm did not present concrete examples of how any specific scientific discoveries and theories related to established Catholic doctrine. In the first two of these articles, he argued for a balanced relationship between the Catholic Church and the newly developing natural sciences. He presented an ensemble of Catholic scientists to highlight the many ways in which the Church had cultivated scientific investigations. In the following two articles, Zahm argued that there was no contradiction between Catholic doctrine and true science. He maintained that true science followed the laws of induction, in its pursuit of principles which attempted to bring together facts and theory. While observation of the facts should serve as a limitation for science, it was further guarded by principles which were "supplemented by the truths of revelation."[47] Any apparent contradictions between the Catholic faith and the discoveries of sciences were most likely due to an overstepping on the part of natural scientists from making scientific claims to presenting philosophical and religious claims. The period between 1893 and 1896 was a time of productivity and rising popularity for Zahm.[48]

Zahm became increasingly convinced of the notion of biological evolution over a period of time. Prior to 1893, Zahm denied that any form of biological evolution had occurred. His academic formation in science emphasized experimentation, in the mode of Francis Bacon, which he saw at the time as a weak point in the arguments of proponents of evolution. It was only in 1883 that Zahm began engaging publicly the debate over evolution and its relationship to Christianity.[49] His initial staunch opposition to evolution seemed to have been heavily influenced by the attacks against evolution from the famous Catholic convert Orestes Brownson, who was a fierce critic of evolution, Darwin, and Mivart.[50] Zahm, however, remained more open to Mivart's work than did Brownson, and, because of his understanding of the nature of science, developments in science, and scientific observation and experimentation, Zahm implicitly remained potentially more open to evolution than Brownson.[51]

A change seemed to have taken place in Zahm's thought by the summer of 1893 when he began lecturing on evolution. These lectures formed the core of his *Bible, Science, and Faith* as well as what would become *Evolution and Dogma.*[52] Zahm's participation in the 1894 International Congress in Brussels when he presented a paper on evolution was an important experience for him.[53] Despite his adherence to evolution, in general, he never became an ardent adherent of natural selection, only granting it some minor role, but far from Darwin's notion of natural selection as the main mechanism for evolution.[54] Zahm wrote of the plural "theories" of evolution throughout, e.g., Darwinian, "Neo-Darwinian,"[55] Lamarckian,[56] and Neo-Lamarckian.[57] He made it clear, however, that he had problems with all of the then current proposed theories of evolution.[58]

Zahm followed Mivart throughout much of *Evolution and Dogma.* Early on Zahm pointed out that theories of evolution predated Darwin.[59] In his attempt to show how a theory of evolution could cohere with the Catholic faith, Zahm articulated his own phrasing of the notion that "truth cannot contradict truth."[60] In order to alleviate the most common doubts Zahm frequently wrote about the relationship between science and the Catholic faith, asserting that there should be no contradiction between revealed truth and scientific truth. His reason for this claim was that with proper limitations, truth of one sphere could not contradict truth of another sphere. Zahm wrote, "The truths of faith and the truths of science belong to different categories indeed, but notwithstanding this fact, they can never come into conflict."[61] Later, in *Evolution and Dogma,* he explained, "Science will never be able to contradict aught that God has revealed; for it is not possible that the Divine works and the Divine words should ever be in any relation to each other but one of the most perfect harmony."[62] In fact, Zahm stressed that this was at the heart of his book: "My sole, ardent desire, has been to show that there is nothing in true science, nothing in any of the theories duly accredited by science and warranted by the facts of nature, nothing in Evolution, when properly understood, which is contrary to Scripture or Catholic teaching."[63] In the end, after examining the various lines of evidence in favor of biological evolution, Zahm thought some form of evolutionary theory was, in his words, "in the highest degree probable," despite his dissatisfaction with any of the available theories of his time.[64]

It has become commonplace to pit theistic evolutionists against Darwin on theological grounds, but Zahm did not fit into such a context. Zahm's problems with Darwin were more than philosophical or theological, and fit within the broader late-nineteenth century dissatisfaction with Darwinian natural selection as evolution's main mechanism, which was then prevalent even among evolutionists.[65] Bowler explained that Darwin's work was really "a catalyst that helped bring about the transition to an evolutionary viewpoint within an essentially non-Darwinian conceptual framework."[66] Zahm himself made a related point when he observed that, "The publication of 'The Origin of Species' was the signal for a revolution in science such as the world had never before witnessed."[67]

One of Zahm's major problems with the gradualism of Darwin's theory of natural selection was a weak point Darwin himself had been particularly concerned with, namely the absence of transitional forms in the fossil record, and in particular, in the fossils from the Cambrian period.[68] As paleontologist Peter Ward noted, "the fossil record, far from becoming a major source of support for Darwin, instead became a source of vexation, and he railed about it in successive editions of *The Origin of Species*. To Darwin's dismay, the fossil record—the principal record of evolutionary change—showed very little unequivocal evidence of gradual change."[69] One of Darwin's most famous critics, Louis Agassiz (1807–1873), a pioneer in the study of paleontology and Professor of Zoology and Geology at Harvard University, opposed Darwin's gradualist theory of evolution primarily based on the fossil record and the Cambrian explosion.[70]

In the first edition of *Origin of Species*, Darwin himself wrote:

> Consequently, if my theory be true, it is indisputable that before the lowest Silurian [Cambrian] stratum was deposited, long periods elapsed, as long as, or probably far longer than, the whole interval from the Silurian [Cambrian] age to the present day; and that during these vast, yet quite unknown, periods of time, the world swarmed with living creatures. To the question why we do not find records of these vast primordial periods, I can give no satisfactory answer.[71]

Notice that by his sixth edition (1872), he underscored the challenge of the fossil record in this regard: He left the first part untouched, but removed, "To the question why we do not find records of these vast primordial periods, I can give no satisfactory answer," and replaced it with the following: "Here we encounter a formidable objection; for it seems doubtful whether the earth, in a fit state for the habitation of living creatures, has lasted long enough."[72] He then modified his earlier statement and placed it at the end of this discussion: "To the question why we do not find rich fossiliferous deposits belonging to these assumed earliest periods prior to the Cambrian system, I can give no satisfactory answer."[73]

In the first edition (1859), Darwin continued: "But the difficulty of understanding the absence of vast spikes of fossiliferous strata, which on my theory no doubt were somewhere accumulated before the Silurian [Cambrian] epoch, is very great ... The case at present must remain inexplicable; and may be truly urged as a valid argument against the views here entertained."[74] After a span of more than a decade, when he published his sixth edition (1872), Darwin left this conclusion mostly unchanged.[75] This was one of Zahm's major criticisms of Darwin's gradualist approach of natural selection.[76]

Zahm was keenly aware of the power of public opinion. He utilized newspaper outlets and attempted to keep the topic of evolution foremost in people's minds. John Slattery remarked that, "the wide reach and popular press coverage at Zahm's Catholic Summer School lectures in 1893 set off a domino effect on the rest of the decade."[77] The principal technique Zahm used for reconciling Catholic doctrine and evolution was, following the work of Mivart, his reinterpretation of God's use of secondary causes in creation using St. Augustine and St. Thomas Aquinas.[78] Zahm was clear to distinguish St. Augustine's views and St. Thomas' from contemporary theistic evolution; he simply saw theirs as compatible with and in some ways precursors to modern theistic evolution.[79] Zahm claimed that the evolution of a species was compatible with creation by God. But one must view the process of evolution as a secondary cause set in place by God.[80]

Zahm did not commit to the idea that humans likewise evolved.[81] Instead, Zahm defended the hypothesis, like Mivart's, that there was nothing contrary to viewing the human body as having evolved. He explained that, on the one hand, "no positive evidence has been adduced in support of the simian origins of man, and ... there is little, if any reason to believe that such evidence will be forthcoming."[82] At the same time, Zahm mentioned that, "Analogy and scientific consistency, we are told, require us to admit that man's bodily frame has been subject to the same law of Evolution, if an Evolution there has been, as has obtained for the inferior animals."[83] His conclusion was that there was nothing contrary to the Catholic faith in a notion like Mivart's that humanity evolved from prior biological ancestors, or, in Zahm's words, in the "derivative origins of Adam's body."[84] Like Mivart, Zahm was adamant that the human soul was specially created by God.[85]

Zahm articulated a new theistic version of evolution in his book *Evolution and Dogma*, wherein he accommodated many aspects of different theories of evolution and shared much in common with Mivart.[86] His book sparked controversy immediately. The Sacred Congregation for the Index of Prohibited Books censured *Evolution and Dogma*, but never published the censure formally, although they did notify Zahm of their decision.[87] Mariano Artigas, Thomas Glick, and Rafael Martínez observed that, "The shadow of Galileo surely fell over the evolution controversy."[88] Yet, they pointed out how the Magisterium was unwilling formally to condemn evolutionists; Darwin was never placed on the Index of Prohibited Books, though his grandfather was; neither were Thomas Huxley, Herbert Spencer, Ernst Haeckel, or even the Catholic Mivart's works on evolution placed on the Index.[89] Their explanation is that:

> The Vatican authorities were aware of the fact that no condemnation of evolutionism had been issued, and apparently they were not anxious to provide one. They examined the writings of our protagonists when a work was denounced, analyzing each on the basis of the existing doctrine but without the guidance of any official doctrine regarding evolutionism. This explains why the reports of the experts followed no uniform pattern. Nor was there any fixed pattern

> that could predict the decision of the cardinals, or even of the pope.[90]

Zahm had spent time in Rome, where he was introduced to other like-minded American Catholics within the hierarchy, such as Ireland, Keane and O'Connell.[91] After the controversy over *Evolution and Dogma*, however, he no longer published works dealing with evolution, thus effectively removing Zahm's voice from the conversation which continued long after he passed.[92]

Januarius De Concilio's Cautions Over Evolution

Our final figure, Januarius De Concilio, originally from Naples, Italy, was first and foremost a pastor rather than a professional scientist.[93] After his ordination, he was sent to Newark, New Jersey where he was incardinated in 1860. That same year, Bishop James Roosevelt Bayley, the first Bishop of Newark, assigned De Concilio to be the first pastor of St. Michael's parish in Jersey City. De Concilio's pastoral influence extended far beyond the parish of St. Michael, which he helped physically build with his own hands. De Concilio attended the Third Plenary Council of Baltimore. Giovanni Schiavo remarked that it was at the Baltimore Council that De Concilio distinguished himself as a Catholic theologian and scholar.[94] He was appointed to the subcommittee that oversaw the creation of a new national catechism. De Concilio was asked to compose the first draft. He worked frantically and composed the schema in a mere seven days. Only a handful of editorial changes were made on De Concilio's draft which was then printed as the *Baltimore Catechism*.[95]

De Concilio's pastoral and scholarly influence became widely recognized. In February 1886 Pope Leo XIII named him Monsignor. During his lifetime De Concilio authored several books and frequently wrote articles for *New York Freeman's Journal* among other publications. De Concilio's intellectual contributions have been largely overlooked since he was far better known for his pastoral work. De Concilio's seminary training occurred just prior to the Neo-Thomistic revival associated with Leo XIII's 1879 *Aeterni Patris*, but it was firmly within the Neapolitan Thomism associated with De Concilio's professor, Gaetano Sanseverino. This Thomism helped form the intellectual background for Leo's work where Thomism would function as an aid in combatting the intellectual challenges from modernity, and was, in Marvin O'Connell's words, part of a "quiet revolution" that "had already begun in Italy."[96] Réginald Garrigou-Lagrange identified Sanseverino as one of the leading Thomists behind Leo XIII's Thomistic revival.[97] De Concilio also became the first professor of theology at Seton Hall University and Immaculate Conception Seminary, where he taught logic and metaphysics, the same subjects as his own mentor Sanseverino.[98]

Formed in this intellectual context, De Concilio engaged in nineteenth-century evolutionary theories to help a more popular audience navigate what he saw as widespread confusion. De Concilio's writings on evolution served as an example of how he approached the new modern sciences in general. De Concilio wrote *Harmony Between Science and Revelation* in 1889,[99] and later wrote a two-part response to Zahm's book in the March editions of the *New York Freeman's Journal*. De Concilio was familiar with and supportive of the new sciences of the nineteenth century and he offered an alternative to the approach of outright rejection, on the one hand, or immediate adoption and adaptation, on the other hand. He used the Catholic faith as a filter to screen out what could be useful and true from the work of the new scientists, which he thought would be helpful in future evaluations of scientific enquiry. Although no one would mistake De Concilio for an "Americanist," he held positions that would make the so-called "conservatives" feel uneasy, including his arguments for the likelihood of intelligent extraterrestrial life in outer space.[100]

One of the suggestions De Concilio made in *Harmony Between Science and Revelation* was to distinguish between various theories of evolution,[101] something, as we have already seen, Zahm would later emphasize as well; there were different evolutionary theories being offered by different scientists, like those of Lamarck and Darwin. In this, both Zahm and De Concilio anticipated Pope St. John Paul II's distinction when in his 1996 Address to the Pontifical Academy of Sciences, he wrote, "rather than the theory of evolution, we should speak of several theories of evolution."[102]

To make the evolutionary debate accessible to his lay Catholic audience De Concilio adopted a very original style. He wrote a fictional trilogue between a venerable old Doctor Armstrong, his niece Adele, and an aspiring young physician, George. De Concilio filled his book with direct quotations from primary texts, including religious and scientific works. His characters, who were having tea in the library, reached for books by Herbert Spencer, Kepler, and Darwin. They jumped from Latin books to French journals, while always translating into English for the discussion. Although *Harmony Between Science and Revelation* was the type of book that your average reader could be comfortable picking up, it was not simply a watered-down discussion; De Concilio examined evolution from a variety of different angles including geology, embryology, paleontology, biology, as well as Catholic theology.

De Concilio's method relied heavily upon primary sources.[103] His argumentation displayed a great familiarity with his contemporary evolutionists, as well as other scientists, both ancient and modern. Knowing that up-to-date scientific books were not accessible to everyone, De Concilio incorporated information from popular sources as well like the *Encyclopaedia Britannica.*[104] He first used scientific principles, acknowledged authorities, and logic to evaluate each claim. Then he brought in the additional support from history, religion, and theology. An analysis of De Concilio's newspaper articles reveals a similar method of investigation.[105] De Concilio also relied upon the work of Jean Louis Armand De Quatrefages de Bréau (1810–1892), whom Zahm also utilized (and praised) in *Evolution and Dogma.* De Quatrefages, who was not Catholic, was a prominent French anthropologist and zoologist at the Museum of Natural History in Paris. He held two doctorates including one in medicine. De Quatrefages was a staunch critic of Darwin, but the two of them remained friends and in active correspondence.[106] For Zahm, De Quatrefages was a helpful critic of natural selection, but the two were in fundamental disagreement about biological evolution. For De Concilio, De Quatrefages proved an important authority cautioning against most of the evolutionary theories being articulated at the time.[107]

Like Zahm, De Concilio articulated his own version of the notion that "truth cannot contradict truth," through the figure of George when in *Harmony Between Science and Revelation* George explained to Adele that the Doctor was "fully convinced that no conflict or opposition exists or can exist between them [the results of science] and revelation."[108] As De Concilio, through the Doctor, explained, "science and revelation are two distinct ways, which God Almighty has adopted to teach and instruct man ..."[109] De Concilio then had the Doctor explain further that a new scientific discovery might appear to contradict a truth of faith, but this would only be an appearance: "If the universe be really a manifestation of God in the creatures, and if religion be really that manifestation of God which He Himself makes to man in a personal communication, how in the name of common sense can any real discovery of science be in conflict with a truth of religion?"[110]

De Concilio accepted the possibility of some form of evolution and emphasized that embracing one form did not necessarily mean embracing all, and likewise that rejecting one or more than one, did not require the rejection of all possible forms of evolution.[111] In the voice of the Doctor, De Concilio mentioned that the Church had not condemned "all kinds of evolution."[112] De Concilio seemed to have no problem with what would later be called micro-evolution, which "allows any amount of evolution and development within the species."[113] Like Zahm (and like Darwin himself), De Concilio interpreted the Silurian/Cambrian fossil evidence as posing a problem for Darwin's gradualist approach, though De Concilio thought this was a broader problem for evolutionary theories beyond Darwinian gradualism, whereas Zahm saw it only as a problem for natural selection.[114]

One of the main figures De Concilio engaged was Mivart,[115] whom De Concilio called, "An eminent Catholic scientist."[116] In fact, De Concilio devoted two entire chapters ("Articles") to engaging Mivart's evolutionary theory.[117] Unlike Zahm, however, for whom Mivart's work was inspirational, De Concilio did not think Mivart's views were an acceptable Catholic position.[118] De Concilio especially had a problem with the way Mivart used St. Augustine, St. Thomas Aquinas, and Francisco Suarez, and he proceeded to cite directly from their texts in Latin with English translations in order to contest Mivart.[119] In the end, De Concilio came to a position that was more opposed to speciation, the development of new species via a biological evolutionary process, much closer to that of Orestes Brownson, whom he cited,[120] and who was an important inspiration for Zahm's initial opposition to evolution.

By March of 1896 De Concilio responded to Zahm's popular claims in *Evolution and Dogma* published earlier that year, in a two-part article found in the *New York Freeman's Journal*. In the first part of the article De Concilio examined Zahm's theistic evolution from the standpoint of science. Beyond a scarcity of evidence for the claims Zahm was making, De Concilio argued for a lack of clarity in Zahm's definitions and use of sources. According to De Concilio, one particular distinction Zahm did not clarify was his distinction between body and soul, which De Concilio was happy to accept, but which he pointed out was not a foregone distinction in scientific literature and thus, he maintained that Zahm should have been more careful in explaining and defending his definitions.[121] De Concilio's concerns with Zahm here were similar to those for which contemporary non-theistic scholars and scientists criticized theistic evolutionists like Mivart, namely for the inclusion of the supernatural in their evolutionary theories.[122] His second article followed Zahm's claims that Catholics may freely and in good faith adopt a theory of evolution.[123] Although he had some concerns over Zahm's arguments, De Concilio recognized the multiple theories of evolution even of his day and advocated for the need to investigate the findings and theories further.

Conclusion

In looking at the various receptions of evolutionary theories among Catholic intellectuals in the United States at the end of the nineteenth century, we find that in addition to those who outright rejected any notion of evolution and those who began incorporating evolutionary theory and reinterpreting Catholic thinkers accordingly, there were other voices like Zahm and De Concilio who, differing in their interpretations as they were, were circumspect in their discussions. Zahm had moved from a position of suspicion regarding evolution, to later deciding that some form of biological evolution was most probably how God created earth's tremendous biodiversity. He was careful to distance himself from the non-theistic evolutionary theories of Darwin, Haeckel, Spencer, et al., and was much closer to those of Mivart. De Concilio, on the other hand, urged for caution and continued study rather than embracing or rejecting modern evolutionary theories completely.

It would be well for us to keep in mind that De Concilio and Zahm wrote on evolutionary theories before the dawn of the twentieth century, decades before the Neo-Darwinist synthesis of the 1940s began to rely both on Darwinian natural selection and Mendelian genetics,[124] which prior to that time had represented two approaches that had been in conflict with each other.[125] In closing, we can consider another Catholic priest from Seton Hall University, separated by more than half a century from De Concilio. In his 2008 address, "Evolution as Science and Ideology," which he delivered before the Pontifical Academy of Sciences, approximately six months before his death, Fr. Stanley Jaki wrote, "I have been a resolute opponent of creationists, of champions of Intelligent Design, but also of Darwinian idealogues. One of these was Darwin himself as he tried to minimize the enormous shortcomings of his mechanism of evolution."[126] Placing Zahm and De Concilio within their broader nineteenth century context provides us a glimpse of the historical diversity of evolutionary positions among Catholic intellectuals in the U.S. at that time.

Notes

1. On the priority of Alfred Russel Wallace's discussion of natural selection see Arnold C. Brackman, *A Delicate Arrangement: The Strange Case of Charles Darwin and Alfred Russel Wallace* (New York: Times Books, 1980).
2. Charles Darwin, *On the Origin of Species by Means of Natural Selection, or the Preservation of Favoured Races in the Struggle for Life* (London: John Murray, 1859). Darwin's *Origin of Species* went through six major editions in which he heavily revised his work. Morse Peckham, ed., *The Origin of Species: A Variorum Text* (Philadelphia: University of Pennsylvania Press, 1959) represented an early attempt at presenting *Origin of Species* where readers could see the changes made in subsequent editions. A more user-friendly "variorum" of *Origin of Species* is now available online, where readers can more easily see the differences between the six editions. See the *Online Variorum of Darwin's Origin of Species*, ed. Barbara Bordalejo (2012 [2008-2009]), https://darwin-online.org.uk/Variorum/index.html.
3. Charles Darwin, *The Descent of Man, and Selection in Relation to Sex*, in Two Volumes (London: John Murray, 1871).
4. William Paley, *Natural Theology: Or, Evidences of the Existence and Attributes of the Deity, Collected from the Appearances of Nature* (Philadelphia: John Morgan, 1802). On the paradigm shift in thinking to a new anti-teleological and anti-design methodological shift, see Neil C. Gillespie, *Charles Darwin and the Problem of Creation* (Chicago: University of Chicago Press, 1979). On the ways in which Darwin was shaped by and (he and his followers) responded in turn to Paley, see, e.g., David N. Livingstone, "Evolution and Religion," in *Evolution: The First Four Billion Years*, ed. Michael Ruse and Joseph Travis, 348–369 (Cambridge: Harvard University Press, 2009), 350; John Angus Campbell, "The Comic Frame and the Rhetoric of Science: Epistemology and Ethics in Darwin's *Origin*," *Rhetoric Society Quarterly* 24 (1994): 27–50; John Angus Campbell, "Of Orchids, Insects, and Natural Theology: Timing, Tactics, and Cultural Critique in Darwin's Post-*Origin* Strategy," *Argumentation* 8 (1994): 63–81; John Angus Campbell, "Scientific Discovery and Rhetorical Invention: The Path to

Darwin's *Origin*," in *The Invention and Persuasion in the Conduct of Inquiry*, ed. Herbert W. Simons, 58–90 (Chicago: University of Chicago Press, 1990), 62–78; Frank Burch Brown, "The Evolution of Darwin's Theism," *Journal of the History of Biology* 19 (1986): 1–45, at 18; John Angus Campbell, "Scientific Revolution and the Grammar of Culture: The Case of Darwin's *Origin*," *Quarterly Journal of Speech* 72 (1986): 351–376; M.J.S. Hodge, "Darwin and the Laws of the Animate Part of the Terrestrial System (1835–1837): On the Lyellian Origins of His Zoonomical Explanatory Program," *Studies in the History of Biology* 7 (1982): 1–106; David Kohn, "Theories to Work By: Rejected Theories, Reproduction, and Darwin's Path to Natural Selection," *Studies in the History of Biology* 4 (1980): 67–170; and Julian Huxley, *Evolution: The Modern Synthesis* (New York: Harmer, 1943), 23.

5. Ernst Mayr, foreword to *Darwinism Defended: A Guide to the Evolution Controversies*, by Michael Ruse (Reading: Addison-Wesley, 1982), xii.
6. See the important earlier study, Barry Brundell, "Catholic Church Politics and Evolution Theory, 1894-1902," *British Journal for the History of Science* 34, no. 1 (2001): 81–95.
7. There were then, and remain now, many different evolutionary theories proposing different mechanisms. In this paper, we will sometimes discuss "evolution" more generally as dealing with biological descent with modification, and sometimes we will be focusing more specifically on Darwinian evolution by means of natural and sexual selection, which he proposed as evolution's main mechanisms.
8. Among his more than 10 publications dealing with evolution were: *Darwin's Designs: From the Origin of Species to Erehwon (nowhere) and Back to Somewhere* (Port Huron: Real View Books, 2006); "Non-Darwinian Darwinism," in *L'evoluzione: Crocevia di scienza, filosofia e teologia*, ed. Rafael Pascual, 40-52 (Rome: Studium, 2005); and "Newman and Evolution," *Downside Review* 109 (1991): 16–34.
9. Our understanding of "culture" is informed by Kathryn Tanner, *Theories of Culture: A New Agent for Theology* (Minneapolis: Fortress Press, 1997).
10. The time period we typically call the "Enlightenment" was far from monolithic, and we can even speak of a "religious"

enlightenment and a "Catholic" enlightenment in parts of Europe during the seventeenth and eighteenth centuries. See, e.g., Ulrich L. Lehner, *The Catholic Enlightenment: The Forgotten History of a Global Movement* (Oxford: Oxford University Press, 2016); Ulrich L. Lehner, "What is Catholic Enlightenment?" *History Compass* 8 (2010): 166–178; and David Sorkin, *The Religious Enlightenment: Protestants, Jews, and Catholics from London to Vienna* (Princeton: Princeton University Press, 2008).

11. For the ways in which notions of American freedom and culture have been in tension with Catholics in the U.S. throughout their history see especially John T. McGreevy, *Catholicism and American Freedom: A History* (New York: Norton, 2003).
12. Edward Claude Stibili, "The St. Raphael Society for the Protection of Italian Immigrants 1887-1923" (Ph.D. diss., University of Notre Dame, 1977), 1. Elsewhere, he noted that over 4.5 million Italians emigrated to the U.S. between 1880 and 1924. See Edward C. Stibili, "The Italian St. Raphael Society," *U.S. Catholic Historian* 6, no. 4 (1987): 301.
13. Steven M. Avella and Elizabeth McKeown, "Introduction to Part 1: 1776-1865," in *Public Voices: Catholics in the American Context*, ed. Steven M. Avella and Elizabeth McKeown, 1–4 (Maryknoll: Orbis Books, 1999), 3.
14. Patrick Carey, *The Roman Catholics in America* (Westport: Praeger, 1996), 52.
15. Carey, *Roman Catholics in America*, 53.
16. John T. McGreevy, *Parish Boundaries: The Catholic Encounters with Race in the Twentieth Century Urban North* (Chicago: University of Chicago Press, 1996), 10. Jay Dolan pointed out that, "Catholics generally saw themselves as outsiders in the United States, a minority group that was forced to suffer persecution because of their religion. But the hierarchy desperately wanted to become insiders and be accepted as part of mainstream America. To gain such acceptance, Catholics had to shed any taint of foreign loyalties, customs and languages." See Jay P. Dolan, *The American Catholic Experience: A History from Colonial Times to the Present* (Notre Dame: University of Notre Dame, 1992), 301.
17. R. Scott Appleby, "Exposing Darwin's 'hidden agenda': Roman

Catholic Responses to Evolution, 1875-1925," in *Disseminating Darwin: The Role of Place, Race, Religion, and Gender*, ed. Ronald L. Numbers and John Stenhouse, 173–208 (Cambridge: Cambridge University Press, 1999), 179–180.

18. Philip Gleason, *Contending with Modernity: Catholic Higher Education in the Twentieth Century* (Oxford: Oxford University Press, 1995).
19. Although many of the early Church fathers (e.g., St. Basil the Great, St. Ambrose of Milan, and St. John Damascene) understood Genesis 1 as implying six 24-hour days of creation, this would not become viewed as the "only" way to read it, or the dominant way, until the modern period. Christians in antiquity evidenced more diverse interpretations. Among those early Church fathers who read Genesis 1 in a different way than six 24-hour days, are St. Augustine (even in his *Literal Commentary on Genesis*), St. Justin Martyr, and St. Irenaeus of Lyons. Other early Christian exegetes, like Origen of Alexandria, likewise understood Genesis 1 apart from a six 24-hour day reading. On these readings see, e.g., John Baptist Ku, O.P., "Interpreting Genesis One with the Fathers of the Church," in *Thomistic Evolution: A Catholic Approach to Understanding Evolution in the Light of Faith*, 2nd ed., by Nicanor Pier Giorgio Austriaco, O.P., James Brent, O.P., Thomas Davenport, O.P., and John Baptist Ku, O.P. (Providence: Cluny Media, 2019), 103–110.
20. F. R. Wegg-Prosser, "Theism and the Visible Universe," *American Catholic Quarterly Review* 10 (1885): 348-369, at 365–366.
21. Charles Darwin, *On the Origin of Species by Means of Natural Selection, or the Preservation Of Favoured Races in the Struggle for Life* (London: John Murray, 1859), 204 (this was the first edition). He included this very same comment without change in the sixth and final edition, Charles Darwin, *On the Origin of Species by Means of Natural Selection, or the Preservation of Favoured Races in the Struggle for Life*, 6th ed. (London: Odhams Press, 1872), 165. He included a narrative account of how eyes may have developed according to natural selection. Likewise, Stephen Jay Gould addressed the same matter with his own defense of the evolution of the eye. See Stephen Jay Gould, *Ever Since Darwin: Reflections on Natural History* (New

York: Norton, 1979 [1977]), 103–110, esp. 107. Compare with Richard Dawkins' own discussion and defense of the evolution of the eye in Richard Dawkins, *The Blind Watchmaker: Why the Evidence of Evolution Reveals a Universe without Design* (New York: Norton, 1996 [1986]), 77–86, especially 81. That there remains ongoing debate over "ocular evolution" see, e.g., D.L. Williams, "Light and the Evolution of Vision," *Eye* 30 (2016): 173–178, who confessed, "Darwin's original puzzle over ocular evolution seems still to be with us but now at a molecular level" (177).

22. J. J. Quinn, "Is the Book of Genesis Genuine and Authentic?" *American Ecclesiastical Review* 5 (1891): 356–362, at 361.
23. On John Zahm and the "Zahm Affair," see especially John P. Slattery, *Faith and Science at Notre Dame: John Zahm, Evolution, and the Catholic Church* (Notre Dame: University of Notre Dame Press, 2019); Hans Moscicke, "The Scientific Allegory of John Augustine Zahm: Zahm's Theological Method with Insight from Marie-Joseph Lagrange," *Zygon* 51, no. 4 (2016): 925–948; David B. Burrell, *When Faith and Reason Meet: The Legacy of John Zahm, CSC* (Notre Dame: Corby Books, 2009); Phillip R. Sloan, "Bringing Evolution to Notre Dame: Father John Zahm, C.S.C. and Theistic Evolutionism," American Midland Naturalist 161, no. 2 (2009): 189–205; Mariano Artigas, Thomas F. Glick, and Rafael A. Martínez, *Negotiating Darwin: The Vatican Confronts Evolution, 1877-1902* (Baltimore: Johns Hopkins University Press, 2006), esp. 124–202; Brundell, "Catholic Church," 88-91; R. Scott Appleby, "Between Americanism and Modernism: John Zahm and Theistic Evolution," *Church History* 56 (1987): 474–490; and Ralph Edward Weber, *Notre Dame's John Zahm: American Catholic Apologist and Education* (Notre Dame: University of Notre Dame Press, 1961).
24. Appleby, "Between Americanism," 475.
25. John Lee Morrison, "A History of American Catholic Opinion on the Theory of Evolution, 1859–1950," (Ph.D. diss., University of Missouri, 1951), 1–36.
26. J.A. Zahm, *Scientific Theory and Catholic Doctrine* (Chicago: McBridge, 1896).
27. J.A. Zahm, *Science and the Church* (Chicago: McBridge, 1896).
28. J.A. Zahm, *Evolution and Dogma* (Chicago: McBridge, 1896).

Evolution and Dogma was then translated into Italian (1896), French (1897), and Spanish (1905). On *Evolution and Dogma* see, e.g., Slattery, *Faith and Science*, 80–82; and Artigas, Glick, and Martínez, *Negotiating Darwin*, 126–134.

29. See, e.g., J.A. Zahm, *Catholic Science and Catholic Scientists* (Philadelphia: Kilner, 1893), subsequently translated into French (1895) and Italian (1896); J.A. Zahm, *Moses and Modern Science* (Philadelphia: Gallagher, 1894); and J.A. Zahm, *Bible, Science, and Faith* (Baltimore: Murphy, 1894), subsequently translated into French (1894), Italian (1895), and Spanish (1912).
30. Artigas, Glick, and Martínez, *Negotiating Darwin*, 125.
31. On that Congregation's investigation of Zahm's book, and on Zahm's response, and all that took place in this regard, see especially, Slattery, *Faith and Science*, 2, 43–47, and 135–161; and Artigas, Glick, and Martínez, *Negotiating Darwin*, 17 and 143–158.
32. On Mivart see especially David O. Brown, "St. George Jackson Mivart: Evo-Devo, Epigenetics and Thomism," *Theology and Science* 20, no. 4 (2022): 474–492; Francesca Bigoni and Roscoe Stanyon, "St. George J. Mivart: the First Catholic Evolutionist," *Forum* 5, no. 1 (2019): 387–402; Artigas, Glick, and Martínez, *Negotiating Darwin*, 236–269; and Jacob W. Gruber, *A Conscience in Conflict: The Life of St. George Jackson Mivart* (New York: Columbia University Press, 1960). Gruber's classic biography contained a number of significant errors including the misleading title of the last chapter which made it seem that Mivart was formally excommunicated, although Gruber never explicitly stated that; Mivart was, however, denied the sacraments over his teachings concerning hell. Artigas, Glick, and Martínez, *Negotiating Darwin*, 267–269 provided a helpful corrective here. Sloan ("Bringing Evolution," 202) explicitly stated Mivart was "excommunicated," and actually cited Artigas, Glick, and Martínez's volume here precisely where they explained that he was *not* formally excommunicated.
33. See Francesca Bigoni and Giulia Barsanti, "Evolutionary Trees and the Rise of Modern Primatology: The Forgotten Contribution of St. George Mivart," *Journal of Anthropological Sciences* 89 (2011): 93–107, at 102. See also Sloan, "Bringing Evolution," 192–193. For Mivart's work in the nascent field of

primatology see, e.g., St. George Mivart, *Man and Apes, an Exposition of Structural Resemblances and Differences Bearing Upon the Questions of Affinity and Origin* (New York: Appleton, 1874).

34. Brown, "St. George Jackson Mivart," 483; Sloan, "Bringing Evolution," 193; Artigas, Glick, and Martínez, *Negotiating Darwin*, 241; and Brundell, "Catholic Church," 85.

35. St. George Mivart, *On the Genesis of Species*, 2nd ed. (London: Macmillan, 1871). We only had at our disposal the second edition, published the same year as the first, which was able to take into account Darwin's *Descent of Man* published that same year.

36. St. George Mivart, *Contemporary Evolution: An Essay on Some Recent Social Changes* (New York: Appleton, 1876).

37. Peter J. Bowler, *Evolution: The History of an Idea*, 3rd ed. (Berkeley: University of California Press, 2003 [1983]), 215.

38. Bowler, *Evolution*, 198.

39. Brown, "St. George Jackson Mivart," 476–477; Takao K. Suzuki, "On the Origin of Complex Adaptive Traits: Progress Since the Darwin Versus Mivart Debate," *Journal of Experimental Zoology (Molecular and Developmental Evolution)* 328B (2017): 304–320, at 304–306; Emma Swain, "St. George Mivart as Popularizer of Zoology in Britain and America, 1869-1881," *Endeavour* 42, no. 4 (2017): 176–191; Sloan, "Bringing Evolution," 193; Artigas, Glick, and Martínez, *Negotiating Darwin*, 236, 238–239, and 242; Brundell, "Catholic Church," 85; Bowler, *Evolution*, 198–199 and 215; Robert J. Richards, *Darwin and the Emergence of Evolutionary Theories of Mind and Behavior* (Chicago: University of Chicago Press, 1987), 225–230 and 353–363; David L. Hull, *Darwin and His Critics: The Reception of Darwin's Theory of Evolution by the Scientific Community* (Cambridge: Harvard University Press, 1973), 351-415; and Peter J. Vorzimmer, *Charles Darwin: The Years of Controversy: The Origin of Species and Its Critics, 1859-1882* (Philadelphia: Temple University Press, 1970), 225–251.

40. Theistic evolution, and especially its relationship to the thought of St. Thomas Aquinas, continues to be a lively discussion. In defense of a Catholic and Thomistic theistic evolutionary theory see, e.g., Mariusz Tabaczek, *Theistic Evolution: A Contemporary Aristotelian-Thomistic Perspective* (Cambridge:

Cambridge University Press, 2024); Mariusz Tabaczek, O.P., "Afterward to the Polish Edition of *Thomistic Evolution: A Catholic Approach to Understanding Evolution in the Light of Faith by Nicanor Pier Giorgio Austriaco, O.P., James Brent, O.P., Thomas Davenport, O.P., and John Baptist Ku, O.P.*," trans. Monika Metlerska-Colerick, *Nova et Vetera* 22, no. 1 (2024): 225–237; and Austriaco, Brent, Davenport, and Ku, *Thomistic Evolution*, wherein each chapter is identified as coming from one of the four authors. For critiques of such a theistic evolution, likewise from a Thomistic perspective, see, e.g., Michael Chaberek, *Knowledge and Evolution: How Theology, Philosophy and Science Converge in the Question of Origins* (Eugene: Wipf & Stock, 2021), 114–131; Michael Chaberek, "Classical Metaphysics and Theistic Evolution: Why Are They Incompatible?" *Studia Gilsoniana* 8, no. 1 (2019): 47–81; and Michael Chaberek, *Aquinas and Evolution: Why St. Thomas' Teaching on the Origins Is Incompatible with Evolutionary Theory*, 2nd ed. (British Columbia: Chartwell, 2019).

41. Artigas, Glick, and Martínez, *Negotiating Darwin*, 18 and 236.
42. Sloan, "Bringing Evolution," 192-193 and 198; and Brundell, "Catholic Church," 89.
43. Appleby, "Between Americanism," 476.
44. On *Testem Benevolentiae* see Gerald P. Fogarty, S.J., "Reflections on the Centennial of *Testem Benevolentiae*," *U.S. Catholic Historian* 17, no. 1 (1999): 1–12; Patrick W. Carey, "After Testem Benevolentiae and Pascendi," *Journal of Texas Catholic History and Culture* 7 (1996): 13–33; and William L. Portier, "Isaac Hecker and *Testem Benevolentiae*: A Study in Theological Pluralism," in *Hecker Studies: Essays on the Thought of Isaac Hecker*, ed. John Farina, 11-48 (New York: Paulist Press, 1983).
45. On *Pascendi* see, e.g., Claus Arnold and Giovanni Vian, ed., *La Redazione dell'Enciclica Pascendi: Studi e documenti sull'antimodernismo di Papa Pio X* (Stuttgart: Anton Hiersemann, 2020); C.J.T. Talar, "The 'Synthesis of All Heresies': Roman Catholic Modernism," *Religion Compass* 4, no. 7 (2010): 426–435; C.J.T. Talar, "Introduction: *Pascendi dominici gregis*: The Vatican Condemnation of Modernism," *U.S. Catholic Historian* 25, no. 1 (2007): 1–12; and C.J.T. Talar, "'The Synthesis of All Heresies'—100 Years On," *Theological Studies*

68, no. 3 (2007): 491–514. On the connections between "Americanism" and "Modernism," and the ways in which the "Americanist crisis" was a prelude to the "Modernist crisis," see William L. Portier, *Divided Friends: Portraits of the Roman Catholic Modernist Crisis in the United States* (Washington, D.C.: The Catholic University of America Press, 2013), e.g., 15.

46. Zahm, *Catholic Science and Catholic Scientists*. The first two articles appeared in *Ave Maria* with the titles "Catholic Church and Modern Science" (1883) and "What the Church has done for Science" (1885). The next two articles both appeared in 1890 in the *American Catholic Quarterly Review*: "Catholic Dogma and Scientific Dogmatism" and "The Friends and Foes of Science."
47. Zahm, *Catholic Science and Catholic Scientists*, 126.
48. Appleby recorded that Zahm worked at a "feverish pace" producing four books, four pamphlets and twenty-one articles. Appleby, "Between Americanism," 476.
49. Slattery, *Faith and Science*, 42 and 53-84.
50. On Brownson see especially Patrick W. Carey, *Orestes A. Brownson: American Religious Weathervane* (Grand Rapids: Eerdmans, 2004). On Brownson's opposition to evolution of any kind see William J. Astore, "Gentile Skeptics? American Catholic Encounters with Polygenism, Geology, and Evolutionary Theories from 1845 to 1875," *Catholic Historical Review* 82, no. 1 (1996): 40–76, at 59–63. For Brownson's influence on Zahm see, e.g., Slattery, *Faith and Science*, 67–72.
51. Slattery, *Faith and Science*, 73–74.
52. Slattery, *Faith and Science*, 11, 37–38, 53, 70–71, and 75–81.
53. Slattery, *Faith and Science*, 39; and Artigas, Glick, and Martínez, *Negotiating Darwin*, 139–140. On that conference see Francesco Beretta, "Monseigneur d'Hulst, les Congrès Scientifiques Internationaux des Catholiques et la question biblique: La liberté de la science chrétienne au service du renouvellement de la théologie," in *Monseigneur d'Hulst fondateur de l'Institut catholique de Paris*, ed. Claude Bressolette, 75–135 (Paris: Beauchesne, 1998), 114–117 for the controversies over evolution at that meeting. Zahm's paper was published four years later as R.P. Zahm, "Évolution et téléologie," *Revue des Questions Scientifiques* 43 (1898): 403–419, and appeared in English the same year as J.A. Zahm,

"Evolution and Teleology," *Appleton's Popular Science Monthly* 52 (1898): 815–824, which was his last publication on this controversial topic (see the comments in Sloan, "Bringing Evolution," 189).

54. Slattery, *Faith and Science*, 84; Artigas, Glick, and Martínez, *Negotiating Darwin*, 129 and 133; and Weber, *Notre Dame's John Zahm*, 83 and 85n75. Slattery's comment that Zahm's views "sharply reject Darwin's theory of natural selection" (*Faith and Science*, 83–84), was probably too strongly put.

55. Zahm used the phrase "Neo-Darwinism" for those of Darwin's followers who relied more heavily upon natural selection than even Darwin did. His use was not related to the later Neo-Darwinist synthesis of the twentieth century. See Zahm, *Evolution and Dogma*, 196.

56. On Jean-Baptiste Pierre Antoine de Monet, chevalier de Lamarck (1744–1829) see, e.g., Richard W. Burkhardt, Jr., "Lamarck, Evolution, and the Inheritance of Acquired Characters," *Genetics* 194 (2013): 793–805; Pietro Corsi, "Jean-Baptiste Lamarck: From Myth to History," in *Transformations of Lamarckism: From Subtle Fluids to Molecular Biology*, ed. Snait B. Gissis and Eva Jablonka, 12–28 (Cambridge: MIT Press, 2011); and Pietro Corsi, *Lamarck. Genèse et enjeux du transformisme, 1770-1830* (Paris: CNRS Editions, 2001). Lamarck was famous for the notion that acquired traits could be passed on to offspring. By the 6th edition of Darwin's *Origin of Species*, Darwin was more open to the Lamarckian idea of the inheritance of acquired traits, but all such notions virtually disappeared from scientific literature after the triumph of Mendelian genetics. More recently, however, geneticists have found evidence that indicates that some acquired traits could be passed on in at least some organisms, apart from their DNA, including through their membranes. See, e.g., Eva Jablonka and Marion J. Lamb, "Transgenerational Epigenetic Inheritance," in *Evolution: the Extended Synthesis*, ed. Massimo Pigliucci and Gerd B. Müller, 137–174 (Cambridge: MIT Press, 2010).

57. Zahm, *Evolution and Dogma*, xxi and xxvi; Slattery, *Faith and Science*, 82; and Sloan, "Bringing Evolution," 197. Sometimes Zahm used the phrase "transformism," which was not uncommon at that time (e.g., *Evolution and Dogma*, xxii).

58. Zahm, *Evolution and Dogma*, 196–200; Slattery, *Faith and*

Science, 82 and 154; Sloan, "Bringing Evolution," 197; and Artigas, Glick, and Martínez, *Negotiating Darwin*, 129.

59. Zahm, *Evolution and Dogma*, xiii–xiv, 25–28, 61–63, and 378 (in fact, his first six chapters dealt with the history of evolutionary thought); Mivart, *Genesis of Species*, e.g., 20–21 and 37; and Mivart, *Man and Apes*, 2.

60. There was a long history of this notion that truth cannot contradict truth, stemming in part from Pope Leo X's statement, "and truth does not contradict truth at all [*cumque verum vero minime contradicat*]," in his 1513 papal bull, *Apostolici Regiminis*, in *Sacrorum Conciliorum Nova et Amplissima Collectio Volume 32*, ed. Johannes Dominicus Mansi and Philippe Labbe (Paris: Akademische Druck, 1902). This bull was aimed at the Latin Averroists. See especially Luca Bianchi, *Pour une histoire de la "double verité"* (Paris: Vrin, 2008), 117–156.

61. J.A. Zahm, *Bible, Science, and Faith* (Baltimore: John Murphy & Co., 1894), 7.

62. Zahm, *Evolution and Dogma*, xv (see also his comments on xix and xxx).

63. Zahm, *Evolution and Dogma*, xxx.

64. Zahm, *Evolution and Dogma*, 134: "Such, then, in brief, is the argument in favor of Evolution from classification, morphology, embryology, geographical distribution and geological succession. The argument, as based on any one of these four classes of facts, is strong, and to many, if not most contemporary naturalists, conclusive. But when we consider the joint effect of the argument built on the four classes of facts, and note in detail the perfect harmony, the argument becomes still stronger and, to all appearances, irrefragable. The evidence furnished by one class of facts corroborates and explains those offered by the others, and thus the cumulative force of the testimony, given by all the four classes, renders the theory, to say the least, in the highest degree probable."

65. See Sloan, "Bringing Evolution," 195 and 198; Peter J. Bowler, *The Non-Darwinian Revolution: Reinterpreting a Historical Myth* (Baltimore: Johns Hopkins University Press, 1988); and Peter J. Bowler, *The Eclipse of Darwinism: Anti-Darwinian Evolution Theories in the Decades around 1900* (Baltimore: Johns Hopkins University Press, 1983). Even many

contemporary non-theistic (or "secular") evolutionists have problems with Darwinian-type arguments concerning traits being passed on because they were advantageous and related challenges they identify to traditional Darwinian, and Neo-Darwinian, evolution via natural selection, or on the monophyletic understanding of evolution. See, e.g., Brian Goodwin, *How the Leopard Changed Its Spots: The Evolution of Complexity* (Princeton: Princeton University Press, 2001 [1994]); Malcolm S. Gordon, "The Concept of Monophyly: A Speculative Essay," *Biology and Philosophy* 14 (1999): 331–348; Carl Woese, "The Universal Ancestor," *Proceedings of the National Academy of Sciences USA* 95 (1998): 6854–6859; Christian Schwabe, "Theoretical Limitations of Molecular Phylogenetics and the Evolution of Relaxins," *Comparative Biochemistry and Physiology* 107B (1994): 167–177; Motoo Kimura, *The Neutral Theory of Molecular Evolution* (Cambridge: Cambridge University Press, 1983); Gerry Webster and Brian Goodwin, "The Origin of Species: A Structuralist Approach," *Journal of Social and Biological Structures* 5 (1982): 15–47; Stephen Jay Gould, "Is a New and General Theory of Evolution Emerging?" *Paleobiology* 6 (1980): 119–130; and Stephen Jay Gould and Roger C. Lewontin, "The Spandrels of San Marco and the Panglossian Paradigm: A Critique of the Adaptionist Programme," *Proceedings of the Royal Society of London B (Biological Sciences)* 205, no. 1161 (1979): 581–598.

66. Bowler, *Non-Darwinian*, 5.

67. Zahm, *Evolution and Dogma*, 65.

68. On the Burgess Shale and Cambrian explosion see, e.g., Douglas H. Erwin and James W. Valentine, *The Cambrian Explosion: The Reconstruction of Animal Biodiversity* (Greenwood Village: Roberts, 2013); Simon Conway Morris, "Darwin's Dilemma: The Realities of the Cambrian 'Explosion,'" *Philosophical Transactions of the Royal Society B* 361 (2006): 1069–1083; Hou Xian-Guang, Richard J. Aldridge, Jan Bergström, David J. Siveter, Derek J. Siveter, and Feng Xiang-Hong, *The Cambrian Fossils of Chengjiang, China: The Flowering of Early Animal Life* (Oxford: Blackwell, 2004); Simon Conway Morris, *The Crucible of Creation: The Burgess Shale and the Rise of Animals* (Oxford: Oxford University Press, 1999); and Stephen Jay Gould, *Wonderful Life: The Burgess*

Shale and the Nature of History (New York: Norton, 1990 [1989]). This has remained a vigorous discussion among contemporary paleontologists who study the Cambrian "explosion" in the Burgess Shale and in China. Part of the contemporary discussion concerns the apparent rapid appearance of tremendous morphological diversity at the very outset of the Cambrian explosion, including among chordates. See, e.g., Erwin and Valentine, *Cambrian Explosion*, 217; Jun-Yuan Chen, J. William Schopf, David J. Bottjer, Chen-Yu Zhang, Anatoliy B. Kudryavtsev, Abhishek B. Tripathi, Xiu-Qiang Wang, Yong-Hua Yang, Xiang Gao, and Ying Yang, "Raman Spectra of a Lower Cambrian Ctenophore Embryo from Southwestern Shaanxi, China," *Proceedings of the National Academy of Sciences USA* 104 (2007): 6289–6292; Jun-Yuan Chen, P. Oliveri, F. Gao, Stephen Q. Dornbos, Chia-Wei Li, D. Bottjer, and E. Davidson, "Pre-Cambrian Animal Life: Probable Developmental and Adult Cnidarian Forms from Southwest China," *Developmental Biology* 248 (2002): 182–196; James W. Hagadorn, "Chengjiang: Early Record of the Cambrian Explosion," in *Exceptional Fossil Preservation: A Unique View on the Evolution of Marine Life*, ed. David J. Bottjer, Walter Etter, James W. Hagadorn, and Carol M. Tang, 35–60 (New York: Columbia University Press, 2002); Jun-Yuan Chen, Di-Ying Huang, and Chia-Wei Li, "An Early Cambrian Craniate-like Chordate," *Nature* 402 (1999): 518–522; and Simon Conway Morris, "Burgess Shale Faunas and the Cambrian Explosion," *Science* 246 (1989): 339–346. On the plentiful examples of the unexpected fossilization of soft body parts and soft-bodied fauna in the Cambrian explosion see, e.g., Xian-Guang, Aldridge, Bergström, Siveter, Siveter, and Xiang-Hong, *Cambrian Fossils*, xi, 6, 8, 13–14, 23; Conway Morris, *Crucible*, 47–48, 68, 73–74, 76, 82, 97–99, 103, 107, and 140–144; and Jan Bergström and Xian-Guang Hou, "Chengjiang Arthropods and Their Bearing on Early Arthropod Evolution," in *Arthropod Fossils and Phylogeny*, ed. Gregory D. Edgecombe, 151-184 (New York: Columbia University Press, 1998). On the question of the absence of new phyla after the Cambrian explosion see James W. Valentine, "Why No New Phyla after the Cambrian? Genome and Ecospace Hypotheses Revisited," *Palaios* 10 (1995): 190–194. For a more controversial article from outside of the field of evolutionary biology that argued for the difficulties of explaining the Cambrian explosion on evolutionary grounds in light of research on DNA see Stephen C. Meyer, "The Origin of Biological Information and the Higher Taxonomic Categories," *Proceedings of the Biological Society of Washington* 117 (2004): 213–239. Jan Bergström explained the current challenge of the Cambrian explosion thus: "the overwhelming taxonomic difficulty is to recognize relationships between phyla, not to distinguish between them." See Jan Bergström, "Ideas on Early Animal Evolution," in *Early Life on Earth*, ed. Stefan Bengston, 460-466 (New York: Columbia University Press, 1994), 464.

69. Peter Douglas Ward, *On Methuselah's Trail: Living Fossils and the Great Extinctions* (New York: Freeman, 1992), 9. On contemporary critiques of gradualistic evolutionary theories along these lines from evolutionists studying the Cambrian explosion, see, e.g., Michael Foote, "On the Probability of Ancestors in the Fossil Record," *Paleobiology* 22 (1996): 141–151; Mike Foote and Stephen Jay Gould, "Cambrian and Recent Morphological Disparity," *Science* 258 (1992): 1816; and Douglas Erwin, James Valentine, and J.J. Sepkoski, "A Comparative Study of Diversification Events: The Early Paleozoic versus the Mesozoic," *Evolution* 41 (1987): 1177–1186.
70. On Jean Louis Rodolphe Agassiz see especially Christoph Irmscher, *Louis Agassiz: Creator of American Science* (Boston: Houghton Mifflin Harcourt, 2013); Leandro Sequeiros and Cándido Manuel García Cruz, "Louis Agassiz (1807-1873) y sus aportaciones a la geología. Implicaciones didácticas," *Enseñanza de las Ciencias de la Tierra* 14, no. 3 (2006): 202–212; and Edward Lurie, *Louis Agassiz: A Life in Science* (Baltimore: Johns Hopkins University Press, 1988 [1960]). For his importance in paleontology, especially regarding prehistoric fish, see Jean Gaudant, "Louis Agassiz (1807-1873), fondateur de la paléoichthyologie," Revue d'histoire des sciences 33, no. 2 (1980): 151–162.
71. Darwin, *Origin of Species* (1st ed.), 307.
72. Darwin, *Origin of Species* (6th ed.), 286.
73. Darwin, *Origin of Species* (6th ed.), 286. It was precisely because of the problem of the fossil record and the Cambrian explosion that led Niles Eldredge and Stephen Jay Gould to formulate their evolutionary theory of punctuated equilibrium. Niles

Eldredge and Stephen Jay Gould, "Punctuated Equilibria: An Alternative to Phyletic Gradualism," in *Models in Paleobiology*, ed. Thomas J.M. Schopf, 82-115 (San Francisco: Freeman, Cooper, and Co., 1972). For a thorough account of punctuated equilibrium and responses to its critics, see Stephen Jay Gould, *The Structure of Evolutionary Theory* (Cambridge: Harvard University Press, 2002), 745–1024.

74. Darwin, *Origin of Species* (1st ed.), 308.
75. Darwin, *Origin of Species* (6th ed.), 287: "Nevertheless, the difficulty of assigning any good reason for the absence of vast piles of strata rich in fossils beneath the Cambrian system is very great ... The case at present must remain inexplicable; and may be truly urged as a valid argument against the views here entertained."
76. Zahm, *Evolution and Dogma*, 162-165 and 178.
77. Slattery, *Faith and Science*, 39.
78. Sloan pointed out that, "The critical issue for Zahm is that the system works by natural laws, and it is the *existence* of the laws themselves, rather than their specific mode of action in time, that is at the center of Zahm's theistic evolution" ("Bringing Evolution," 199). Zahm, like Mivart before him, used St. Augustine and St. Thomas in support of his arguments. See, e.g., Sloan, "Bringing Evolution," 198; and Artigas, Glick, and Martínez, *Negotiating Darwin*, 129. In other works Zahm also reinterpreted allegorically the universal flood and other controversial biblical passages in light of what he had read in geology and related disciplines. See especially Zahm, *Bible, Science, and Faith*, "Part II: The Noachian Deluge," 119–174; Slattery, *Faith and Science*, 215n33; and Moscicke, "Scientific Allegory," 930–935 and 937–939.
79. Zahm, *Evolution and Dogma*, 312–313: "Of course no one would think of maintaining that any of the Fathers or Doctors of the Church taught Evolution in the sense in which it is now understood. They did not do this for the simple reason that the subject had not even been broached in its present form, and because its formulation as a theory, under its present aspect, was impossible before men of science had in their possession the accumulated results of the observation and research of these latter times." See also Zahm, *Evolution and Dogma*, 71,

280–300, and 312–313); and Artigas, Glick, and Martínez, *Negotiating Darwin*, 129.

80. For summaries of Zahm's views that are helpful see, e.g., Slattery, *Faith and Science*, 82–84.

81. See Zahm, *Evolution and Dogma*, 345, 349, 351–362, and 364. Slattery (*Faith and Science*, 43) implied that Zahm utilized portions of Darwin's argument concerning human evolution in *Evolution and Dogma*, in contrast to Zahm's early works. Slattery's later statement is closer to Zahm's views, but still overly strong: "while Zahm would still demur on human evolution—even in 1896—and would still abhor Darwin's theory of natural selection, he would define the doctrine of special creation, the idea that all species were created simultaneously in their present states by a divine being, using the same words with which he defined evolution in 1883: a theory based on assumptions with a pronounced lack of empirical evidence" (*Faith and Science*, 53). It was true that Zahm had little room for natural selection in *Evolution and Dogma*, but he was more laudatory of Darwin and of his theories, with which he respectfully disagreed, than Slattery's words like "abhor" communicated. Barry Brundell's comments precisely captured Zahm's position: "He [Zahm] claimed in this book [*Evolution and Dogma*] that it was permissible for a Catholic to hold that the body of Adam evolved from the body of some anthropomorphic brute animal ... he claimed that evolution is a good theory but, since it is captive of the adversaries of religion, the materialists, the agnostics and atheists, who are using it against the Catholic faith, we must liberate it and show that it can be of inestimable assistance in the cause of good science and religion" ("Catholic Church," 89).

82. Zahm, *Evolution and Dogma*, 351. This was one of the major conclusions reached in Mivart, *Man and Apes*, but neither precluded evolution of humans biologically, they simply doubted the "simian origin," that apes were a direct ancestor.

83. Zahm, *Evolution and Dogma*, 352.

84. Zahm, *Evolution and Dogma*, 356.

85. Zahm, *Evolution and Dogma*, 349 and 355–356; and Sloan, "Bringing Evolution," 200–202.

86. Slattery (*Faith and Science*, 22) thought that Sloan argued that Zahm's own evolutionary view was more "saltationist,"

although Sloan never stated this in the article Slattery cited. Slattery's comment might have been due to the facts that Sloan did identify Mivart as a saltationist (Sloan, "Bringing Evolution," 193 and 197)—which seems to be accurate in light of Mivart's description of his view and his reliance upon the saltationist Francis Galton (see, e.g., Mivart, *Genesis of Species*, 109 and 127)—combined with the clear influence of Mivart on Zahm. We do not think Zahm's views were really saltationist, even though, like saltationists, Zahm argued for periods of stasis or stability interrupted by more rapid change (see, e.g., Zahm, *Evolution and Dogma*, 164–165 and 178). Mivart's views had more in common with Galton and other saltationists who emphasized the role of mutations, whereas Zahm's comments were specifically regarding the evidence of the fossil record. One of Zahm's later comments about saltationism and gradualism, wherein he remarked "as we are wont to regard them" (*Evolution and Dogma*, 424), made it seem like he was putting distance between his own view and that of the saltationists.

87. Slattery, *Faith and Science*, 2 , 43–47, and 160–161; Sloan, "Bringing Evolution," 189; Artigas, Glick, and Martínez, *Negotiating Darwin*, 17, 26, and 156–158; Brundell, "Catholic Church," 90–91; and Morrison, "History of American Catholic," 251. Slattery helpfully situated the Magisterium's (and specifically the Sacred Congregation of the Index of Prohibited Books) problems with Zahm in light of broader theological, philosophical, and political controversies at the end of the nineteenth century, including the unification of Italy, Pope Pius IX's *Syllabus of Errors*, the Roman theological school (e.g., Joseph Kleutgen), and the rise of Neo-Thomism. See Slattery, *Science and Faith*, 109–113 and 116–117. Artigas, Glick, and Martínez, likewise situated this conflict in the context of the Roman school (and particularly Matthias Scheeben, one of Kleutgen's students), but also of the local provincial council of Cologne in 1860, which although only local, was more influential on Roman school theologians like Giovanni Perrone (*Negotiating Darwin*, 21–23). Since these controversies and history overlapped what would become the Roman Catholic Modernist controversy, three other contextual contributions are helpful, although they did not deal with Zahm directly: Gary Lease, "Vatican Foreign Policy and the Origins of

Modernism," in *Catholicism Contending with Modernity: Roman Catholic Modernism and Anti-Modernism in Historical Context*, ed. Darrell Jodock, 31–55 (Cambridge: Cambridge University Press, 2000); Paul Misner, "Catholic Anti-Modernism: The Ecclesial Setting," in *Catholicism*, ed. Jodock, 56–87; and Marvin R. O'Connell, *Critics on Trial: An Introduction to the Catholic Modernist Crisis* (Washington, D.C.: The Catholic University of America Press, 1994).

88. Artigas, Glick, and Martínez, *Negotiating Darwin*, 281. Indeed, Zahm made a similar connection to the question over evolution with Galileo and Copernicus. See Zahm, *Evolution and Dogma*, 392–395.

89. Artigas, Glick, and Martínez, *Negotiating Darwin*, 14.

90. Artigas, Glick, and Martínez, *Negotiating Darwin*, 4. Later they maintained that evolution's "reception by the Roman Catholic Church bore some relation to the condemnation of Galileo in 1633. The Catholic authorities regarded evolutionism with suspicion but were afraid to condemn it. In Galileo's time, modern science was almost nonexistent ... In Darwin's time, however, modern science was already one of the main components of Western civilization. Theologians liked to say that, although evolution was not scientifically proven, the Roman authorities did not want to get involved in a second 'Galileo affair.' When Catholic authors attempted to harmonize evolutionism with Christianity, the authorities preferred not to condemn them by a public act but rather to persuade them to retract their ideas ... Galileo's shadow was always present" (6).

91. An adversarial attitude between Zahm's group, who became identified as "Americanist," and those who were sometimes labeled as "conservatives" became noticeable in the letters that have been studied. See, e.g., Artigas, Glick, and Martínez, *Negotiating Darwin*, 159–177; and Weber, *Notre Dame's John Zahm*, 94. This was before he had been informed of the censure of his book, after which Zahm never returned to Rome. See, e.g., Artigas, Glick, and Martínez, *Negotiating Darwin*, 158.

92. Slattery, *Faith and Science*, 47 and 161; and Sloan, "Bringing Evolution," 189.

93. On De Concilio see Biff Rocha, "'De Concilio's Catechism,' Catechists, and the History of the Baltimore Catechism" (Ph.D. diss., University of Dayton, 2013), 115-159. There is not much

published on De Concilio, despite the fact that he published books, a play, articles, and drafted the influential *Baltimore Catechism*.

94. Giovanni Schiavo, *The Italians in America Before the Civil War* (New York: Vigo Press, 1934), 294.
95. See Rocha, "De Concilio's Catechism."
96. O'Connell, *Critics on Trial*, 37. On the importance of this Neo-Thomistic revival as a means of combatting modernity see, e.g., Slattery, *Faith and Science*, 117–118; Robert A. Ventresca, "'A Plague of Perverse Opinions': Leo XIII's *Aeterni patris* and the Catholic Encounter with Modernity," *Logos* 12, no. 1 (2009): 143–168; Gleason, *Contending With Modernity*, 107; James Hennesey, "Leo XIII's Thomistic Revival: A Political and Philosophical Event," in *Celebrating the Medieval Heritage: A Colloquy on the Thought of Aquinas and Bonaventure*, ed. David Tracy, 185–197 (Chicago: University of Chicago Press, 1978); and Thomas J.A. Hartley, *Thomistic Revival and the Modernist Era* (Toronto: Institute of Christian Thought, University of St. Michael's College, 1971). On the importance of Sanseverino see Rocha, "De Concilio's Catechism," 116–118; and Ugo Dovere, "Gaetano Sanseverino filosofo tomista. tentativo di ricostruzione," *Doctor communis* 31, no. 1 (1978): 73–97. On the importance of Sanseverino's founding of the Academy of Thomistic Studies in Naples, which influenced the future Pope Leo XIII, see Slattery, *Faith and Science*, 116–117; Ventresca, "Plague," 152; Abelardo Lobato, O.P., "The Pontifical Academy of St. Thomas Aquinas: History and Mission," *Anuario Filosófico* 39, no. 2 (2006): 311; O'Connell, *Critics on Trial*, 37; Hennesey, "Leo XIII's Thomistic," 191–192; and Pasquale Orlando, *Il tomismo a Napoli nel sec. XIX. La scuola del Sanseverino I: Fonti e documenti* (Rome: Libreria della Pontificia Università Lateranense, 1968). On the ways in which Sanseverino himself utilized Thomistic thought in combatting modernity see, e.g., Giuseppe D'Anna, "Una profezia pericolosa: Gaetano Sanseverino lettore di Spinoza," *Rivista di filosofia neoscolastica* 114, no. 1 (2022): 125–138; Cristina Santinelli, "Appunti sulla fortuna dello spinozismo in Italia," *Studi Urbinati* 65 (1992): 67–109, at 83–84; and Pasquale Orlando, "Gaetano Sanseverino e il semirazionalismo di Antonio Günther," *Divinitas* 20, no. 3 (1976): 324–335.

97. Réginald Garrigou-Lagrange, O.P., *La synthèse Thomiste* (Paris: De Brouwer, 1947), 27.
98. Dermot Quinn, *Seton Hall University: A History, 1856-2006* (New Brunswick: Rutgers University Press, 2023), 20; and Joseph Louis Perrier, *The Revival of Scholastic Philosophy in the Nineteenth Century* (New York: Columbia University Press, 1909), 240.
99. J. De Concilio, *Harmony Between Science and Revelation* (New York: Fr. Pustet, 1889). This volume and some of De Concilio's other works were mentioned in debates over evolution within Catholicism in the U.S. in John Rickards Betts, "Darwinism, Evolution, and American Catholic Thought, 1860-1900," *Catholic Historical Review* 45, no. 2 (1959): 161–185, at 175, 175n60, and 176n66.
100. See De Concilio, *Harmony*, 214-233, esp. 227-233; Andrew Davinson, *Astrobiology and Christian Doctrine: Exploring the Implications of Life in the Universe* (Cambridge: Cambridge University Press, 2023), 78, 216, 221, and 291; David A. Wilkinson, "Why Should Theology Take SETI Seriously?" *Theology and Science* 16, no. 4 (2018): 427–438; and Thomas F. O'Meara, *Vast Universe: Extraterrestrials and Christian Revelation* (Collegeville: Liturgical Press, 2012), 85.
101. De Concilio, *Harmony*, e.g., 57.
102. Pope John Paul II, "Truth Cannot Contradict Truth," Address to the Pontifical Academy of Sciences (22 October 1996), no. 4: https://www.vaticanobservatory.org/education/truth-cannot-contradict-truth-1996/. Of course, John Paul II's point went beyond specific mechanisms of evolution and emphasized theistic versus nontheistic evolutionary theories.
103. In De Concilio, *Harmony*, he relies upon numerous sources throughout, e.g.: Charles Darwin (e.g., 56, 62, 87, 94, 96, and 98); Alfred Russel Wallace (e.g., 173); Herbert Spencer (e.g., 10–16 and 21–28); Charles Lyell (e.g., 87); Thomas Huxley (e.g., 44 and 47); Louis Agassiz (e.g., 76–77 and 82); E.B. Tylor (e.g., 127); Friedrich Max Müller (e.g., 147, 155, and 195-196); and Ernst Haeckel (e.g., 50 and 94). Although De Concilio did not use footnotes or endnotes, he did include page numbers and sometimes the editions of texts he was quoting or citing, either in parentheses or else in the text itself.
104. De Concilio, *Harmony*, e.g., 16 and 47.

105. See comments in Rocha, "De Concilio's Catechism."

106. Darwin wrote to De Quatrefages, "I can truly say I would rather be criticized by you in this manner than praised by many others." See Charles Darwin to Armand de Quatrefages, May 28, 1870, available online at the Darwin Correspondence Project: https://www.darwinproject.ac.uk/letter/DCP-LETT-7204.xml.

107. De Concilio, *Harmony*, 59, 125, 188, and 190. In the first instance, De Concilio misspelled his name as Quatrefaces, but he correctly spelled it in the other instances. For Zahm's use of De Quatrefages see Zahm, *Evolution and Dogma*, 95, 182–184, and 343n1. Both De Concilio and Zahm praised De Quatrefages reserving for him some of the most laudatory words they had for secondary sources. De Concilio called him "One of the greatest of modern scientists" (*Harmony*, 59). Zahm called him, "The learned anthropologist" (*Evolution and Dogma*, 95); and "one of the most eminent of modern anthropologists" (343n1). On De Quatrefages see, e.g., Claude Blanckaert, "Pour une théorie évolutive humaine: Armand de Quatrefages, la formation des races et le darwinisme au Muséum national d'histoire naturelle," *Revue d'histoire des sciences humaines* 27 (2015): 189-230; Olivier Perru, "La réception des idées de Darwin en France et le problème du fixisme," in *Evolution et création: des sciences à la métaphysique*, ed. Jean-Marie Exbrayat, Emmanuel D'Hombres, and Fabien Revol, 89–117 (Paris: Vrin, 2011); Valérie Narayana, "Des Espèces de l'origine: le feuilleton scientifique de Louis Figuier dans *La Presse* de 1862," *Études littéraires* 40, no. 3 (2009): 73–86, at 76–77; and John Farley, "The Initial Reactions of French Biologists to Darwin's *Origin of Species*," *Journal of the History of Biology* 7, no. 2 (1974): 275–300, at 294–297. For examples of De Quatrefages's more important works on Darwin, anthropology, and the question of evolution see, e.g., A. De Quatrefages, *Unité de l'espèce humaine* (Paris: Hachette, 1861); A. De Quatrefages, *Métamorphoses de l'homme et des animaux* (Paris: Baillière, 1862); and A. De Quatrefages, *Charles Darwin et ses précurseurs français. Étude sur le transformisme* (Paris: Baillière, 1870). Later editions of *Unité de l'espèce humaine* were published as *L'espèce humaine*, which was also translated into English as A. De Quatrefages, *The Human Species* (New York: Appleton, 1879).

108. De Concilio, *Harmony*, 6.
109. De Concilio, *Harmony*, 7. Adele then defined science, a definition the Doctor approved: "By science we understand that knowledge of God which we acquire by the study of His creatures" (8). The Doctor likewise instructs that, "science means the knowledge of that manifestation which God has made of Himself and of His perfections in the universe, and of all which is contained in it" (9).
110. De Concilio, *Harmony*, 9. He proceeded to explain that in cases where science appeared to contradict the faith, it might be that the data had been misinterpreted (or the data had been misperceived/observed incorrectly), or because of a misunderstanding of the matter of faith, or that it might not actually have been an official doctrine of the Church (10).
111. De Concilio, *Harmony*, 35.
112. De Concilio, *Harmony*, 35.
113. De Concilio, *Harmony*, 36. This microevolution did not admit "that the evolution or development may arrive to the point that one species may develop into another" (36).
114. De Concilio, *Harmony*, 71, 82, and 85–87.
115. De Concilio, *Harmony*, e.g., 156–158 and 165–180.
116. De Concilio, *Harmony*, 156.
117. De Concilio, *Harmony*, "Article 25," 165–173 and "Article 26," 173–180.
118. De Concilio, *Harmony*, 157.
119. De Concilio, *Harmony*, 166–169.
120. De Concilio, *Harmony*, 179–180 and 198.
121. J. De Concilio, "Evolution and Dogma. Criticism by Rt. Rev. Mgr. De Concilio. On Rev. Dr. Zahm's Recent Work," *New York Freeman's Journal* (21 March 1896): 5.
122. As an example see Bowler, *Evolution*, 186, 196, and 204.
123. De Concilio, "Evolution and Dogma," 5.
124. Mendelian genetics initially posed a serious challenge to Darwinian natural selection. The period from 1870–1920, which overlapped both Zahm's and De Concilio's writings, witnessed a low point for natural selection within the diverse scientific community, in part because of questions of genetics and inheritance. See Peter J. Bowler, *Theories of Human*

Evolution: A Century of Debate, 1844-1944 (Baltimore: Johns Hopkins University Press, 1986), 44–50; and Bowler, *Eclipse of Darwinism.* Darwin operated on an earlier notion of blended inheritance, which was no longer tenable after the work of Mendel and his successors. See Peter Vorzimmer, "Charles Darwin and Blending Inheritance," *Isis* 54 (1963): 371–390. On the importance of Mendel's work in genetics and its new widespread appeal around 1900, see especially Sloan, "Bringing Evolution," 196; and Peter J. Bowler, *The Mendelian Revolution: The Emergence of Hereditarian Concepts in Modern Science and Society* (Baltimore: Johns Hopkins University Press, 1989).

125. Although it had precursors in the 1920s and 1930s, that synthesis was often dated to the publication of Julian Huxley, *Evolution: The Modern Synthesis* (Cambridge: MIT Press, 1942); and Ernst Mayr, *Systematics and the Origin of Species from the Viewpoint of a Zoologist* (Cambridge: Harvard University Press, 1942). See Bowler, *Evolution*, 331–339; and Ernst Mayr and William B. Provine, "The Evolutionary Synthesis," *Bulletin of the American Academy of Arts and Sciences* 34, no. 8 (1981): 17–32.

126. Stanley L. Jaki, "Evolution as Science and Ideology," in *Pontificia Academiae Scientiarum Acta 20: The Proceedings of the Plenary Session on Scientific Insights into the Evolution of the Universe and of Life, 31 October—4 November 2008*, ed. Werner Arber, Nicola Cabibbo, and Marcelo Sánchez Sorondo, 517–526 (Vatican City: Pontifical Academy of Sciences, 2009), 518–519. Just before this quotation, Fr. Jaki mentioned, "I hope that a reminder of that difference will not result in my being taken for a minimizer of the scientific merit of Darwinism. I hold its mechanism to be the only, I repeat, the only prospect that any reasoned discourse about the vast variety of species can be turned into science." Nineteenth-century Fr. De Concilio would have disagreed, but we do not know what Fr. De Concilio would have thought had he engaged in these studies and questions in the early twenty-first century.

13 NATURE AND SCRIPTURE: GALILEO'S PROPOSED BIBLICAL HERMENEUTIC IN THE CONTEXT OF EARLY MODERN BIBLICAL INTERPRETATION[1]

JEFFREY L. MORROW, PH.D.

Introduction

In his Gifford Lectures, *The Road of Science and the Ways to God*, Fr. Stanley Jaki writes, "A Perusal of Galileo's *Dialogue* should make it clear, except to close-minded Humeans, that the creative science of Galileo was anchored in his belief in the full rationality of the universe as the product of the fully rational Creator, whose finest product was the human mind, which shared in the rationality of the Creator."[2] Galileo and Fr. Jaki both consider the Catholic faith important, and both are interested in science and astronomy.[3] What I hope to do in this paper is explore the intersection of Galileo's scientific work, which lies outside of my competence, with his more amateur but quite sophisticated theological work, namely his biblical hermeneutic or proposed method of biblical interpretation.

In order to do so I focus on his famous 1615 *Letter to the Grand Duchess Christina* (hereafter LGDC),[4] which we read here at Seton Hall University as part of our required sophomore Core Curriculum course, "Christianity and Culture in Dialogue," and which is arguably Galileo's most important work pertaining to biblical exegesis. I begin with a brief discussion of the "Galileo Affair," after which I turn to a detailed look at Galileo's discussion of biblical hermeneutics in LGDC. Whereas some scholars view Galileo as wholly breaking with traditional Catholic biblical interpretation and others see him as firmly within that prior tradition, I think that Galileo is doing something new in his exegesis, but that this is unintentional.

I argue that, in the LGDC, Galileo attempts to ground Copernicanism in traditional Catholic biblical interpretation of the sort that flourished after the Council of Trent, but that he unintentionally changes it in such a way to include speculative sciences and not just demonstrative sciences. My main argument is that whereas scholars often see Galileo as eschewing recourse to the "literal sense" of Scripture, he employs a different language for the sorts of literalistic interpretations that do not accord with Copernicanism, which he treats as demonstrated science. Instead, he thinks Copernicanism fits the literal sense better, and that the readings of Scripture which appear to contradict Copernicanism, are being misinterpreted literalistically.

The Galileo Affair

In 1999, Thomas Lessl published a little-known article entitled, "The Galileo Legend as Scientific Folklore."[5] Perhaps what remains most important about Lessl's article is his examination of forty textbook accounts of the Galileo Affair that get even the non-controversial aspects of the history wrong, from fabricated stories of Galileo's wallowing in prison to his imagined execution. What Lessl accomplishes is a persuasive account of the work the so-called "Galileo legend" does in modernity.

Indeed, by-and-large the entire Galileo Affair has been deeply polarized.[6] On the one hand, there are large numbers of those who mistakenly think that the Catholic Church of Galileo's day refused to allow anyone to entertain Copernicanism, or a heliocentric (sun-centered) model of the solar system. Instead, so we are told, the Church put forward the geocentric (earth-centered) view as a dogma of faith, and thus Galileo was persecuted by the Inquisition, perhaps tortured, imprisoned, maybe even executed, on account of his scientific findings. This story has been used, time and again, to demarcate the lines between rational scientific enquiry and irrational faith commitments.[7] On the other side, we find the minority defenders of the Catholic Church who maintain that the Inquisition was unconcerned with Galileo's Copernican assertions, but instead were bothered by his theology. Obviously, there are more nuances in the various positions;[8] The reality is more complicated.

We would do well to remember the often-forgotten fact that by the time of the official Inquisition's hearings in 1633, when he was "vehemently suspected of heresy," Galileo was officially recognized as a member of the Catholic clergy in his day, even though he was neither a deacon, priest, nor bishop.[9] In the seventeenth century, there were a variety of minor clerical orders, and Pope Urban VIII bestowed on Galileo, not one, but two canonries, an ecclesiastical title that provided Galileo with an annual salary, but with no real responsibilities, other than receiving a tonsure, the shaving of his head representing his new clerical state.[10]

The drama begins, however, long before his 1633 trial. Galileo assiduously studies Aristotelian philosophy and becomes a professor at the University of Pisa and then at the University of Padua, where he teaches mathematics. He advances Copernicus' heliocentric theory, and it is in fact the controversies surrounding Galileo in the early part of the 1600s that bring Copernicus' work under suspicion. Copernicus' work survives 70 years without incident, but once the impassioned Galileo begins being denounced, the Inquisition examines Copernicus' book, which it censors in 1616, the year after LGDC. The version the Inquisition allows published leaves Copernicus' heliocentrism, scientific findings, and mathematics, untouched; it removes the language of certitude.[11] All the concerned parties basically agree, as William Carroll points out, "that Copernican theory was fine and safe as 'hypothetical knowledge.'"[12]

Galileo's main opponents who denounce him to the Inquisition are his academic rivals, Aristotelian philosophers who are bothered by Galileo's Copernicanism.[13] Some of these conversations occur informally.[14] Galileo writes three letters in which he explains and defends Copernicanism in light of the biblical exegesis of certain passages from the Psalms and the Book of Joshua, as well as in light of his proposed biblical hermeneutical principles, the first to Benedetto Castelli,[15] and the second to Piero Dini.[16] St. Robert Bellarmine is a prominent member of the Inquisition at this time. As the controversy heats up, Paolo Antonio Foscarini writes a letter defending Copernicanism using similar points to Galileo, engaging both Scripture and the interpretation of the Church Fathers, especially St. Augustine. Bellarmine responds to Foscarini, challenging him, and Galileo acquires copies both of Foscarini's letter as well as Bellarmine's response, on which he writes his own private notes in the margins.[17]

In response to this controversy, in which he has placed himself firmly in the middle, Galileo writes his LGDC, in some ways a revision of what he has written to Castelli; LGDC, though is not actually a letter intended for Christina, but for the concerned parties to read.[18] He never publishes the letter, and perhaps the controversy over Foscarini is one reason for this reticence. In 1616, Bellarmine meets with Galileo and urges caution, arguing that prudence dictates that Galileo refrain from treating

Copernicanism as if it were a demonstrated truth, and instead think in terms of it as hypothetical.[19] Years later, when his friend and patron Cardinal Maffeo Barberini becomes pope (1623), taking the name Urban VIII, Galileo probably thinks he might get a better hearing, or, more specifically, that Copernicanism might get a better hearing than it had in 1616.

Galileo is again denounced to the Inquisition, but for more than just heliocentrism. His work, *The Assayer* (1623) causes some a concern that Galileo's atomism may pose a challenge to the Church's teaching on the Eucharist. Although some scholars may have pushed an argument that this doctrinal concern colors the debate over Copernicanism more than the evidence permits, it is clear that, even if the Inquisition finds the denunciations unpersuasive, there are people who have concerns over Galileo's atomism.[20] Galileo's publication of his *Dialogue Concerning the Two Chief World Systems*, which includes his robust defense of Copernicanism, lands him before the Inquisition.[21] We should keep in mind, the context of the recent Protestant Reformation. Mauro Pesce reminds us: "It seems to me to be of extraordinary importance to note the fact that the Catholic Church confronted, in the second decade of the seventeenth century, the new science with the same hermeneutical tools that it had developed to face the Protestant Reformation which exploded about a century earlier."[22]

It is not the case that the Catholic Church forbids writing about, thinking about, or performing experimentation concerning Copernicanism. The standard textbook accounts not only oversimply the matter, but they fail to grasp precisely what the concerns were. It is also clear, however, that the concerns of the theologians on the Inquisition, were not simply matters of theology.[23] A large portion of the problem is the question of scientific demonstration. Galileo, Bellarmine, and the other theologians of the Inquisition, share an Aristotelian notion of science. As much as Galileo opposes Aristotelianism and Aristotelian philosophers, he shares the Aristotelian notion of what is demonstrated.[24]

Throughout his LGDC he speaks about "necessary demonstrations,"[25] "demonstrative science,"[26] "demonstrated truth,"[27] and that which is "truly demonstrated."[28] Such Aristotelian notions of demonstration amount to certain logical proofs, like calculating the third angle of a triangle when you already know the other two. In modern times, much of what we consider science is more experimental and is what Aristotelians understand as speculative science, and thus more hypothetical.[29] It is worth bearing in mind that such a scientific demonstration of the sort Galileo seeks does not appear until the famous 1851 pendulum experiment of Léon Foucault.[30]

Bellarmine is open to the Copernican interpretation if it can be demonstrated in this way. It seems, fundamentally, that the theologians of the inquisition see themselves as defending logic and reason, the understanding of which they derive especially from Aristotle and St. Thomas Aquinas.[31] In the end, Galileo never sees the inside of a prison, nor is there any evidence that he undergoes physical torture.[32] He is found guilty of "vehement suspicion of heresy," but not of formal heresy.[33] The result is that he is placed under house arrest in his luxurious villa.[34]

The Literal Sense and Galileo's Subtle Linguistic Distinctions

It is especially in LGDC, that Galileo lays out principles of interpretation for natural matters in the Bible and it is there that we find his biblical hermeneutic.[35] Galileo makes quite clear that he views the Bible as ultimately about salvation and not primarily about natural phenomena.[36] In his hermeneutic, he relies upon the work of significant contemporaries, like Benito Pereira, who are important sources for his theological interlocutors on the Inquisition.[37] Most importantly, however, is his frequent use of and extensive quotations (in Latin) from the Church Fathers, especially St. Augustine.[38]

Some scholars argue that Galileo wants to silence Scripture so that it only speaks on matters pertaining to salvation and not on nature or the physical world at all.[39] As Pesce writes, "the Galilean thesis ... affirms that the Bible does not have a scientific purpose."[40] Though later Pesce admits, "... Galileo also keeps open the possibility of giving a concordist reading of some passages of the Bible to show how the latter could be reconciled with Copernicanism perhaps better than with the Ptolemaic system."[41] It should be clear, however, that Galileo allows the Bible to speak on "scientific" physical matters of nature, which his examples of exegesis from Joshua and the Psalms show; that seems to be the brunt of his argument.[42]

An important but overlooked hermeneutical starting point for Galileo is that there are no errors in the Bible.[43] He writes in LGDC, "Scripture can never lie or err."[44] He likewise explains, "It is most pious [*santissimamente*] to say and most prudent [*prudentissimamente*] to take for granted that Holy Scripture can never [*non poter mai*] lie [*mentire*], as long as its true meaning [*il suo vero sentimento*] has been grasped [*sia penetrato*]."[45] In this context, Galileo articulates his notion of accommodation, wherein Scripture adapts itself to the mind and comprehensive abilities of its common readers.[46] He explicitly affirms that he feels "reverence for these authorities [Scripture (*Scrittura*), sacred theologians (*i sacri teologi*), and Councils (*Concilii*)] and hold them supreme."[47]

A large part of Galileo's argument about the truths of Scripture and nature cohering has to do with Galileo's use of the traditional medieval Catholic notion that God is the author of both the book of Scripture and the book of nature.[48] In Galileo's words, "... Sacred Scripture and nature proceed equally from the divine Word ..."[49] Thus Galileo likewise argues that the truths of Scripture cannot contradict the truths of nature,[50] since, in his words, "two truths cannot contradict one another."[51] Some scholars argue that Galileo means something different than the traditional Catholic understanding of truth not being able to contradict truth and that he was using this as a sort of Trojan horse to sneak in modern speculative science.[52] This is not clear, at least not from LGDC, but if it is the case, if Galileo is engaging in religious dissembling, then we might situate him within the Latin Averroist tradition. In this paper, I assume that Galileo is not a Latin Averroist, but this question merits further exploration, namely whether Galileo is aware of what he is doing. What is the degree to which the Latin Averroist tradition,[53] in particular the notion of "two truths,"[54] influences Galileo; Latin Averroism was quite prominent at the University of Padua where Galileo taught. More than his use of language concerning "two truths," which he affirms are not in conflict, Galileo's comments concerning the true meaning of Scripture as only available to the wise, seems more in line with the Latin Averroist tradition. So the influence here of Latin Averroism is possible.

There are several places where Galileo's 1615 LGDC represents a subtle change from what he writes in 1613 in his Letter to Castelli. In some instances, the phrasing is basically identical, but he changes a word or two. These subtle changes show Galileo allowing Scripture to speak more to natural matters than his prior language allows, as when he places Scripture in "last place" for examining nature in 1613, but in 1615 he simply cautions against starting from Scripture as the first place to learn about nature. He likewise changes the phrasing on the matter of the Bible's purpose. Before, in 1613, he writes that its "sole" (*solamente*) purpose has to do with salvation, whereas in his LGDC he changes the word to "principally" (*principalmente*).[55] These are subtle but significant changes.

Carroll makes the important observation that:

> There is something more in Galileo's arguments, more than the traditional affirmation that God is the author of the book of nature and the book of scripture and that the truths of nature and the truths of scripture cannot really be in conflict. In the letter to the Grand Duchess there is an additional argument, not well-developed in his earlier letters on the subject: an argument which concerns the role of science in discovering the true senses of those scriptural texts which address scientific questions.[56]

Franco Motta likewise mentions that:

> At this point, the hierarchy of the … "two truths" [*due verità*] through which the divine work is manifested, nature and Scripture, is inverted: if traditionally it was the latter that provided the clear image of earthly reality … to which the profane sciences necessarily had to conform, now only in the artificial space guaranteed by the scientific method [*metodo scientifico*] is it possible to escape the conflict of interpretations that make the biblical dictation unreliable.[57]

Alfredo Damanti argues that this is what makes Galileo's interpretive method so unique: "Given this premise, Galileo goes on to found an original scriptural hermeneutic: when the Bible speaks of natural phenomena, it must be interpreted according to what experience and a rigorous 'scientific' demonstration suggest."[58] What Galileo argues, however, is not that different from what the theologians of the Inquisition, including Bellarmine,[59] hold, except that they are disinclined to go against sensible observation for natural and physical phenomena, and the tradition of interpreting nature, not just Scripture, without Aristotelian demonstration.[60] Galileo appears to be trying to be consistent with the Council of Trent and with traditional Catholic exegesis.[61] Thus, it does not appear that Galileo is consciously trying to do anything novel in biblical interpretation, hence his frequent citation of prominent Catholic exegetes, especially St. Augustine.

It is in the question of the senses of Scripture that I think most scholars seem to be misinterpreting Galileo. I think Carroll is correct to distinguish most of what Galileo describes in LGDC as different from the traditional notion of the "literal sense" (*sensus literalis*).[62] Most scholars seem to take all of the passages Galileo cites regarding how to read Scripture as equivalent to the literal sense.[63] The literal sense, however, is a technical term in the history of biblical interpretation.[64] As John Boyle explains, "The scholastics are generally agreed upon what the literal sense of Scripture is: the thing signified by the words. In this, Scripture is like any other literary work: words signify things. They may signify directly or through metaphor or any number of other literary figures and devices. All of this is a matter of the literal sense."[65]

Most of the time, what Galileo challenges, is *not* the "literal sense," so much as misinterpreting the literal sense, at least as he understands it. That is, what Galileo challenges in this regard is what we might call an overly literalistic interpretation, which does not necessarily conform to the literal sense; a metaphor should be read metaphorically according to the literal sense. I grant that Galileo could be clearer here by explicitly challenging such readings as misinterpretations according to the literal sense, but what is the literal sense in these instances is precisely what is under debate. English translations seem to obscure Galileo's subtle use of language by translating the rare times Galileo uses "literal sense" (*senso literale*) in the exact same way as the other phrases he employs to distinguish what he discusses from the literal sense. Misunderstanding Galileo's words on this point lead to misunderstanding his argument.[66]

So, for example, in LGDC Galileo demurs against reading the text simply according to "the pure signification of the words" (*il puro significato delle parole*), but in English we find this translated as "literal meaning of the words."[67] That English translation is not a bad translation of the Italian, per se, but I am being overly strict in my awkward translation to underscore the subtle differences in the phrases Galileo employs. Italian *significato* can certainly be translated as "meaning," and it is probably better to do so than the awkward "signification," but later when Galileo employs the word *senso*, which can also mean "meaning," I believe he is using it in the technical sense of "sense," as in the classic "senses of Scripture."[68] So, I take *puro significato* to be different from Galileo's later use of *senso literale*, whereas too many English translators, and even Italian scholars like Pesce, see them as synonymous.

Elsewhere in LGDC Galileo writes of the "naked grammatical sound" (*nudo suono grammaticale*).[69] He more than once references the "naked signification of the words,"[70] the "pure verbal sense" (*puro senso verbale*),[71] and later, the "pure signification" (*puro significato*).[72] Galileo only uses the actual technical term, "literal sense," near the end where he believes the Copernican theory might accord with the "literal sense" (*senso literale*), which English translations like Finocchiaro's translate as "literal meaning," the same as they translate the earlier phrases Galileo employs.[73] Galileo affirms that such a Copernican reading was a "sense all the more accommodated to the letter" (*senso tanto più accomodato alla lettera*).[74]

In the end, this is not all that different from Bellarmine, at least as regards the literal sense.[75] The key for Galileo is using "science" to interpret the parts of Scripture that pertain to the physical realm and to nature.[76] This is where the questions really arise for the theologians of the Inquisition. Galileo seems to mean by "science" (*scienza*) an Aristotelian demonstration. If this is the case, then his biblical hermeneutic is not that different from the traditional Catholic exegesis of his day. Consider Galileo's language. He uses both "clear experience" as well as "necessary demonstrations," and "sensory experience."[77]

When Galileo challenges his critics he writes, "they want to extend [*estendere*], not to say abuse [*per non dire abusare*], its [Scripture's] authority, so that even for purely natural conclusions which are not *de fide* one must totally abandon [*lasciar*] sense [*il senso*] and demonstrative reasons [*le ragioni dimostrative*] in favor of any scriptural passage whose apparent words may contain different indications."[78] He writes further that, "if his doctrine was demonstrated it could not contradict [*non poteva contrariare*] the Scriptures, perfectly understood."[79]

Galileo is concerned about what to do when we encounter "problems" (*problemi*) in nature, or rather what he calls, "natural problems" (*problemi naturali*).[80] This is where he maintains, in light of the demand properly to interpret the Bible, that what occurs in nature "which is placed before our eyes by sensory experience or proved by necessary demonstrations [*necessarie dimostrazioni*] should not be called into question, let alone condemned, on account of scriptural passages whose words appear to have a different meaning."[81] He thus concludes, "after becoming certain of some natural conclusions [*conclusioni naturali*], we should use these as very accommodating [*accomodatissimi*] aids to the true [*vera*] exposition [*esposizione*] of the Scriptures."[82] Galileo states that his concern is for "demonstrative sciences" (*scienze demostrative*), and yet that is precisely what is under debate; Copernicanism is not among demonstrated sciences in the seventeenth century.[83] The one thing that is clear is that Galileo's exegesis is an apologetic to defend Copernicanism to the Magisterium.[84]

Conclusion: Galileo's Hermeneutical Significance

In concluding, I wish to point out an area that merits further investigation, namely, the methodological significance of Galileo's argument for modern biblical interpretation. This is my first scholarly foray into Galileo's work. Much of my own scholarship deals with later seventeenth century biblical interpreters, especially the influential Isaac La Peyrère,[85] Thomas Hobbes,[86] Baruch Spinoza,[87] and Richard Simon.[88] Within this history of the development of modern biblical criticism, two of these figures may have been influenced by Galileo: Hobbes, who visits with Galileo,[89] and Spinoza. Galileo's influence on Hobbes seems likely but the specific influence is not clear, particularly in biblical interpretation.[90]

Likewise, it is unclear how familiar Spinoza is with Galileo, since he never cites Galileo, nor are any of Galileo's works preserved in his library,[91] and yet some scholars argue that Galileo's influence on Spinoza is discernable.[92] As I have argued elsewhere, Spinoza attempts to bring Francis Bacon's method of studying nature into modern biblical studies.[93] Whereas, Galileo urges the interpretation of natural phenomena in Scripture to be done in light of scientific enquiry, Spinoza wants to create a method of interpretation of the Bible patterned on the methods of the natural sciences, like Bacon's method. Unlike Galileo,

Spinoza is explicitly trying to develop a new biblical hermeneutic. It is worthwhile exploring further the relationship between Galileo and what comes later in the decades after him in the realm of biblical hermeneutics.

As we have seen, scholars typically understand many of Galileo's statements in LGDC as pertaining to the "literal sense" of Scripture and thus his arguing against recourse to the "literal sense," at least in some passages. My argument in this paper has been that instead Galileo makes a distinction between the "literal sense" and overly literalistic interpretations. Galileo is not trying to be an innovative biblical interpreter, instead he attempts to defend Copernican theory with traditional Catholic biblical interpretive principles. What he does that is novel, however inadvertent it might be, is use the speculative sciences, which he treats as if they are demonstrative, as the lens through which to interpret natural and physical matters detailed in Scripture.

Notes

1. I am indebted to Victor Velarde, Tim Fortin, Justin Anderson, and Eric Johnston for conversations on this topic. Except where noted, English translations are my own.
2. Stanley L. Jaki, *The Road of Science and the Ways to God* (Chicago: University of Chicago Press, 1978), 105.
3. Fr. Jaki wrote extensively on such matters. Consider the following: Stanley L. Jaki, "The Demarcation Line Between Science and Religion," *Angelicum* 87, no. 1 (2010): 81–89; Stanley L. Jaki, "Christ and Science," *Downside Review* 110 (1992): 110–130; Stanley L. Jaki, "Thomas and the Universe," *The Thomist* 53, no. 4 (1989): 545–572; and Stanley L. Jaki, "Physics and the Ultimate," *Ultimate Reality and Meaning* 11, no. 1 (1988): 61–73.
4. When I cite LGDC I will begin with the English from (or, when noted, modified from) Maurice A. Finocchiaro, ed., *The Galileo Affair: A Documentary History* (Berkeley: University of

California Press, 1989), 87–118 (hereafter simply Finocchiaro followed by the page number), followed by the Italian critical edition Antonio Favaro, ed., *Le Opere di Galileo Galilei Volume V* (Florence: Barbèra, 1895), 309–348 (hereafter simply Favaro followed by the page number), followed by the Italian critical edition Franco Motta, ed., *Galileo Galilei, Lettera a Cristina di Lorena: Sull'uso della Bibbia nelle argomentazioni scientifiche* (Genoa: Casa Editrice Marietti, 2000), 86–147 (hereafter simply Motta followed by the page number). There are significant differences between these editions. On the limits of Favaro see Franco Motta, "Due copie della lettera di Galileo a Cristina di Lorena tra Sei e Settecento," *Annali di storia dell'esegesi* 12, no. 1 (1995): 129–143. Unfortunately, I was unable to consult the most recent critical edition of LGDC, Galileo Galilei, *Lettera a Cristina di Lorena: Edizione critica*, ed. Ottavio Besomi, Daniele Besomi, and Giancarlo Reggi (Rome: Editrice Antenore, 2012).

5. Thomas M. Lessl, "The Galileo Legend as Scientific Folklore," *Quarterly Journal of Speech* 85 (1999): 146–168.
6. Maurice A. Finocchiaro, "Science, Religion, and the Historiography of the Galileo Affair: On the Undesirability of Oversimplification," *Osiris* 16 (2001): 114–132.
7. Maurice A. Finocchiaro, *On Trial for Reason: Science, Religion, and Culture in the Galileo Affair* (Oxford: Oxford University Press, 2019), 12 and 215; Maurice A. Finocchiaro, *Retrying Galileo, 1633–1992* (Berkeley: University of California Press, 2005), 3; and Rivka Feldhay, *Galileo and the Church: Political Inquisition or Critical Dialogue?* (Cambridge: Cambridge University Press, 1995), 13.
8. For later receptions, reactions, and myth-making of the Galileo Affair, see Finocchiaro, *On Trial*, 171–201; William R. Shea and Mariano Artigas, *Galileo Observed: Science and the Politics of Belief* (Sagamore Beach: Science History Publications, 2006); Finocchiaro, *Retrying Galileo*, 108–357; and Maurice Finocchiaro, "Galileo as a 'Bad Theologian': A Formative Myth about Galileo's Trial," *Studies in the History and Philosophy of Science* 33 (2002): 753–791.
9. On Galileo's life see William R. Shea and Mariano Artigas, *Galileo in Rome: The Rise and Fall of a Troublesome Genius* (Oxford: Oxford University Press, 2003).

10. Shea and Artigas, *Galileo in Rome*, 131–132.
11. Finocchiaro, *On Trial*, 47–75; Finocchiaro, *Retrying Galileo*, 16–25; Shea and Artigas, *Galileo in Rome*, 84–87; Ernan McMullin, "Galileo on Science and Scripture," in *The Cambridge Companion to Galileo*, ed. Peter Machamer, 271–347 (Cambridge: Cambridge University Press, 1998), 277–291; Pierre-Noël Mayaud, *La condemnation des livres coperniciens et sa révocation à la lumière de documents inédits des Congrégations de l'Index et de l'Inquisition* (Rome: Editrice Pontificia Università Gregoriana, 1997), esp. 37–79; Annibale Fantoli, *Galileo: For Copernicanism and for the Church* (Vatican City: Vatican Observatory Foundation, 1994); and Richard J. Blackwell, *Galileo, Bellarmine, and the Bible: Including a Translation of Foscarini's Letter on the Motion of the Earth* (Notre Dame: University of Notre Dame Press, 1991), 53–54.
12. William E. Carroll, "Galileo, Science and the Bible," *Acta Philosophica* 6, no. 1 (1997): 26. On the different ways in which "hypothesis" and "hypothetical" were being used at the time see Mariano Artigas, Rafael Martínez, and William R. Shea, "New Light on the Galileo Affair?" in in *The Church and Galileo*, ed. Ernan McMullin, 213–233 (Notre Dame: University of Notre Dame Press, 2005), 215–216.
13. Blackwell, *Galileo*, 59.
14. Blackwell, *Galileo*, 53–54.
15. The most important study on the Letter to Castelli is perhaps Andrea Battistini, *Galileo e I Gesuiti: Miti letterari e retorica della scienza* (Milan: Vita e Pensiero, 2000). On this letter see also Mauro Pesce, *L'ermeneutica biblica di Galileo e le due strade della teologia cristiana* (Rome: Edizioni di storia e letteratura, 2005), esp. 29–46 and 55–70; Fantoli, *Galileo*, 165–173; Mauro Pesce, "Le redazioni originali della *Lettera Copernicana* di G. Galilei a B. Castelli," *Filologia e Critica* 17 (1992): 394–417; Mauro Pesce, "Una nuova versione della lettera di G. Galilei a Benedetto Castelli," *Nouvelles de la Republique des Lettres* 2 (1991): 89–122; and Rinaldo Fabris, *Galileo Galilei e gli orientamenti esegetici del suo tempo* (Vatican City: Pontifical Academy of the Sciences, 1986), 12–13.
16. On his correspondence with Dini see William E. Carroll, "Galileo and the Interpretation of the Bible," *Science & Education* 8 (1999): 168–171.

17. On Foscarini's letter, Bellarmine's response, and Galileo's engagement with both, see Franco Motta, "Nature, Faith and the Judge of Faith: Some Considerations on the Historical-Political Context of Copernicus' Condemnation," in *Copernicus Banned: The Entangled Matter of the Anti-Copernican Decree of 1616*, ed. Natacha Fabbri and Federica Favino, 57–100 (Florence: Leo S. Olschki, 2018), 69–72; Paolo Ponzio, "Teologie e copernicanesimo: Bellarmino, Campanella, Foscarini," in *Il caso Galileo: Un rilettura storica, filosofica, teologica: Convegno internazionale di studi Firenze, 26–30 maggio 2009*, ed. Massimo Bucciantini, Michele Camerota, and Franco Giudice, 91–107 (Florence: Leo S. Olschki, 2011), 96–102; Michel-Pierre Lerner, "The Heliocentric 'Heresy': From Suspicion to Condemnation," in *Church and Galileo*, ed. McMullin, 21–25; Fantoli, *Galileo*, 173–180; Pierre-Noël Mayaud, "Deux textes au coeur du conflit. Entre l'Astronomie Nouvelle et l'Écriture Sainte. La lettre de Bellarmin à Foscarini et la lettre de Galilée à Christine de Lorraine," in *Après Galilée. Science et foi. nouveau dialogue*, ed. Paul Poupard, 19–91 (Paris: Desclée, 1994); and Blackwell, *Galileo*, 87–110. For an English edition of Foscarini's initial letter, see Blackwell, *Galileo*, 217–251.
18. On the background to his LGDC see Ernan McMullin, "Galileo's Theological Venture," in *Church and Galileo*, ed. McMullin, 89 and 105–106; and Fantoli, *Galileo*, 180–192. For careful studies of the text of LGDC see Alfredo Damanti, *Libertas Philosophandi: Teologia e filosofia nella Lettera alla granduchessa Cristina di Lorena di Galileo Galilei* (Rome: Edizioni di Storia e Letteratura, 2010); and Pesce, *L'ermeneutica biblica*, 87–116; William E. Carroll, "Galileo and Biblical Exegesis," in *Largo campo di filosofare: Eurosymposium Galileo 2001*, ed. José Montesinos and Carlos Solís, 677–691 (Orotava: Fundación Canaria Orotava de Historia de la Ciencia, 2001); Mauro Pesce, "La 'Lettera a Cristina': Una proposta per definire ambiti autonomi di sapere e nuovi assetti di potere intelletuale nei paesi cattolici," in *Galileo Galilei*, ed. Motta, 9–66; Carroll, "Galileo, Science and the Bible," 5–37; Mauro Pesce, "Momenti della ricezione dell'ermeneutica biblica galileiana e della *Lettera a Cristina* nel XVII secolo," *Annali di storia dell'esegesi* 8, no. 1 (1991): 55–104; Mauro Pesce, "L'interpretazione della Bibbia nella lettera di Galileo a Cristina di Lorena e la sua ricezione, Storia di una difficoltà nel distinguere ciò che non lo è," *Annali*

di storia dell'esegesi 4 (1987): 239–284; and Jean Dietz Moss, "Galileo's Letter to Christina: Some Rhetorical Considerations," *Renaissance Quarterly* 36 (1983): 547–576.

19. See Blackwell, *Galileo*, 29–51.
20. See Mariano Artigas, "Un nuovo documento sul caso Galileo: EE 291," *Acta Philosophica* 10 (2001): 199–214; Rafael Martínez, "Il manoscrito ACDF, *Index, Protocolli*, vol. EE, f. 291 *r-v*," *Acta Philosophica* 10 (2001): 215–242; Lucas F. Mateo-Seco, "Galileo e l'Eucaristia: La questione teologica dell'ACDF *Index, Protocolli*, EE, fol. 291 *r-v*," *Acta Philosophica* 10 (2001): 243–256; and William R. Shea, "Galileo e l'atomismo," *Acta Philosophica* 10 (2001): 257–272.
21. See Shea and Artigas, *Galileo in Rome*, 123–157; and Francesco Beretta, *Galilée devant le Tribunal de l'Inquisition* (Fribourg: Université de Fribourg, 1998).
22. Pesce, *L'ermeneutica biblica*, 98.
23. For example, Paolo Lombardi, *La Bibbia contesa. Tra Umanesimo e razionalismo* (Florence: La Nuova Italia, 1992).
24. See Irving A. Kelter, "The Refusal to Accommodate: Jesuit Exegetes and the Copernican System," *Sixteenth Century Journal* 26, no. 2 (1995): 273–283; William A. Wallace, *Galileo's Logical Treatises* (Dordrecht: Kluwer, 1992); William A. Wallace, *Galileo's Logic of Discovery and Proof* (Dordrecht: Kluwer, 1992); William A. Wallace, *Galileo and His Sources: The Heritage of the Collegio Romano in Galileo's Science* (Princeton: Princeton University Press, 1984), 99; William A. Wallace, *Prelude to Galileo: Essays on Medieval and Sixteenth-Century Sources of Galileo's Thought* (Dordrecht: D. Reidel, 1981), 129–139 and 315–317; and Ernan McMullin, "The Conception of Science in Galileo's Work," in *New Perspectives on Galileo: Papers Deriving from and Related to a Workshop on Galileo held at Virginia Polytechnic Institute and State University, 1975*, ed. Robert E. Butts and Joseph C. Pitt, 209–257 (Dordrecht: Reidel, 1978). On Aristotelian science and demonstration see Justin M. Anderson, *Virtue and Grace in the Theology of Thomas Aquinas* (Cambridge: Cambridge University Press, 2020), 65–107; Eileen Serene, "Demonstrative Science," in *The Cambridge History of Later Medieval Philosophy: from the Rediscovery of Aristotle to the Disintegration of Scholasticism 1100–1600*, ed. Norman

Kretzmann, Anthony Kenny, and Jan Pinborg, 498–507 (Cambridge: Cambridge University Press, 1982); and James A. Weisheipl, O.P., "The Concept of Nature," *The New Scholasticism* 28, no. 4 (1954): 377–408.

25. Finocchiaro (90, 93, and 111); Favaro (312, 317, and 339); and Motta (92, 97, and 131).
26. Finocchiaro (101); Favaro (326); and Motta (114).
27. Finocchiaro (102); Favaro (339); and Motta (131).
28. Finocchiaro (102); Favaro (327); and Motta (115).
29. Wallace, *Prelude*, 9.
30. Joël Sommeria, "Foucault and the Rotation of the Earth," *Comptes rendus physique* 18 (2017): 520–525.
31. Lerner, "Heliocentric," 19; and Carroll, "Galileo, Science and the Bible," 9–10, 22, 26, and 36.
32. Finocchiaro, *On Trial*, 165; Shea and Artigas, *Galileo Observed*, 102–104; Francesco Beretta, "The Documents of Galileo's Trial: Recent Hypotheses and Historical Criticism," in *Church and Galileo*, ed. McMullin, 193; and Finocchiaro, *Retrying Galileo*, 11.
33. Finocchiaro, *On Trial*, 1 and 203–223; and Finocchiaro, *Retrying Galileo*, 11–12.
34. Finocchiaro, *On Trial*, 219; and Finocchiaro, *Retrying Galileo*, 13.
35. McMullin, "Galileo's Theological Venture," 106–112; and Andre-Marie Dubarle, "Les principes exégètiques et théologiques de Galileo concernant la science de la nature," *Revue des sciences philosophiques et théologiques* 50 (1966): 67–78.
36. Pesce, *L'ermeneutica biblica*, 1–2, 103, 105–106, 108, and 218.
37. Motta, "Nature," 76–77; and Fabris, *Galileo*, 29–33.
38. Pesce, *L'ermeneutica biblica*, 110–113; McMullin, "Galileo's Theological Venture," 90–99; Ernan McMullin, "From Augustine to Galileo," *Modern Schoolman* 76 (1999): 169–194; McMullin, "Galileo on Science and Scripture," 291–299; Kenneth J. Howell, "Galileo and the History of Hermeneutics," in *Facets of Faith and Science Volume 4: Interpreting God's Action in the World*, ed. Jitse van der Meer, 245–260 (Lanham: University Press of America, 1995); and Eileen Reeves, "Augustine and Galileo on Reading the Heavens," *Journal of the History of Ideas* (1991): 563–579.
39. Pesce, *L'ermeneutica biblica*, 103, 105–106, 108, and 218; Ernan

McMullin, "The Church's Ban on Copernicanism," in *Church and Galileo*, ed. McMullin, 183; and Giorgio Stabile, "Linguaggio della natura e linguaggio della scrittura in Galilei dalla Istoria sulle macchie solari alle lettere copernicane," *Nuncius* 9, no. 1 (1994): 37–64.

40. Pesce, *L'ermeneutica biblica*, 2.
41. Pesce, *L'ermeneutica biblica*, 91. Damanti, by contrast, argues that Galileo's views are not concordist. See Damanti, *Libertas Philosophandi*, 343–353.
42. Lerner, "Heliocentric," 21; Carroll, "Galileo and Biblical Exegesis," 680 and 685; and Carroll, "Galileo and the Interpretation," 160–162 and 165. On his exegesis of the Psalms see Paolo Rossi, "Galileo Galilei e il libro dei Salmi," *Rivista di filosofia* 10 (1978): 45–71.
43. Carroll, "Galileo and Biblical Exegesis," 680.
44. Finocchiaro (92); Favaro (315); and Motta (95).
45. Finocchiaro (92); Favaro (315); and Motta (95).
46. McMullin, "Galileo on Science and Scripture," 303–304 and 306; Kelter, "Refusal," 273–283; Blackwell, *Galileo*, 66–67; Carlo M. Martini, "Galileo e la teologia," in *Saggi su Galileo Galilei*, ed. Carlo Maccagni, 441–451 (Florence: Barbèra, 1972); and Heinrich Karpp, "Der Beitrag Keplers und Galileis zum neuzeitlichen Schriftverständnis," *Zeitschrift für Theologie und Kirche* 67 (1970): 53. For an important study on Galileo's notion of accommodation in its context see Michele Camerota, "Galileo e la *accomodatio* copernicana," in *Il caso Galileo*, ed. Bucciantini, Camerota, and Giudice, 129–151.
47. English slightly modified from Finocchiaro (91); Favaro (314) is almost identical to Motta (94).
48. Carroll, "Galileo and Biblical Exegesis," 678. On the use of this notion in the early modern period see Kenneth J. Howell, *God's Two Books: Copernican Cosmology and Biblical Interpretation in Early Modern Science* (Notre Dame: University of Notre Dame Press, 2004), 186–196 on Galileo.
49. The English translation is my own: Favaro (316); and Motta (96).
50. Carroll, "Galileo and Biblical Exegesis," 678 and 686–687.
51. Finocchiaro (96); Favaro (320); and Motta (104–105). The language of "two truths" not contradicting seems to relate to the tradition that "truth cannot contradict truth," which can be

traced at least as far back as Pope Leo X. In his 1613 papal bull, *Apostolici Regiminis,* Leo X writes, "and truth does not contradict truth at all." His papal bull is found in *Sacrorum Conciliorum Nova et Amplissima Collectio Volume 32,* ed. Johannes Dominicus Mansi and Philippe Labbe (Paris: Akademische Druck, 1902). Leo X was responding to thinkers coming out of the Latin Averroist tradition, who became known for the notion of "double truth," or "two truths," using language not dissimilar from Galileo's here. See, e.g., Luca Bianchi, *Pour une histoire de la "double verité"* (Paris: Vrin, 2008), 117–156.

52. Lombardi, *La Bibbia contesa,* 214–216.
53. It is highly contested whether or not Averroes (Ibn Rushd) actually holds the notion of "double truth," but this is clearly the way the Latin Averroist tradition receives and redeploys Averroes. See, e.g., Luca Bianchi, "From Pope Urban VIII to Bishop Étienne Tempier: The Strange History of the 'Doctrine of Double Truth,'" *Freiburger Zeitschrift für Philosophie und Theologie* 64 (2017): 9–26; Richard C. Taylor, "Averroes: God and the Noble Lie," in *Laudemus viro gloriosos: Essays in Honor of Armand Maurer, CSB,* ed. R.E. Houser, 38–59 (Notre Dame: University of Notre Dame Press, 2007); Richard C. Taylor, "'Truth Does Not Contradict Truth': Averroes and the Unity of Truth," *Topoi* 19 (2000): 3–16; and Richard C. Dales, "The Origin of the Doctrine of the Double Truth," *Viator* 15 (1984): 169–179.
54. On this question see, e.g., Iroda Rikhsivayevna Ikramova, "Influence of Ibn Rushd's Concept of 'Two Truths' on Western Philosophers," *Central Asian Journal of Social Sciences and History* 4, no. 4 (2023): 49–59. It is interesting that as Galileo a student at the University of Pisa, Girolamo Borro, a Latin Averroist whose lectures on Aristotle are completely based on Averroes' commentary on Aristotle, is among Galileo's primary teachers. On this intellectual formation, see Michele Camerota and Mario Helbing, "Galileo and Pisan Aristotelianism: Galileo's *De Motu Antiquiora* and the *Quaestiones De Motu Elementorum* of the Pisan Professors," *Early Science and Medicine* 5 (2000): 319–365; and Anna De Pace, "Galileo lettore di Girolamo Borri nel *De motu,*" in *De motu. Studi di storia del pensiero su Galileo, Hegel, Huygens e Gilbert,* ed. University of Milan Department of Philosophy, 3–69 (Milan: Cisalpino, 1990).
55. See the important discussion here in Carroll, "Galileo and

Biblical Exegesis," 682–685 and 683–684n12; Carroll, "Galileo and the Interpretation," 162–164; and Carroll, "Galileo, Science and the Bible," 20–21.

56. Carroll, "Galileo and Biblical Exegesis," 685–686.
57. Motta (105n49).
58. Damanti, *Libertas Philosophandi*, 52.
59. See Joseph Zepeda, "Galileo's *Letter to the Grand Duchess Christina*: Genre, Coherence, and the Structure of Dispute," *Galilaeana* 16 (2019): 61: "Bellarmine concedes the principle that demonstrated truths take precedence over apparently contrary passages of Scripture about nature; he believes that some claims about nature cannot ever be truly demonstrated; he is potentially alert (though alas, not sensitive enough) to the damage done by yoking Scripture's credibility to natural propositions that might turn out to be false."
60. Carroll, "Galileo and Biblical Exegesis," 689; and François Laplanche, "Herméneutique biblique et cosmologie mosaïque," in *Les Églises face aux sciences du Moyen Age au XXe siècle*, ed. Olivier Fatio, 29–51 (Geneva: Droz, 1991).
61. Motta, "Nature," 76; and Carroll, "Galileo, Science and the Bible," 14. On the question of Trent and its importance in this debate see: Olaf Pedersen, *Galileo and the Council of Trent* (Vatican City: Vatican Observatory Foundation, 1991).
62. Carroll, "Galileo and Biblical Exegesis, 680; and Carroll, "Galileo and the Interpretation," 161–162.
63. Pesce, *L'ermeneutica biblica*, 102; and McMullin, "Galileo's Theological Venture," 108–109. Consider Maurice Finocchiaro, "The Methodological Background to Galileo's Trial," in Reinterpreting Galileo, ed. William A. Wallace, 241–272 (Washington, D.C.: The Catholic University of America Press, 1986), 265.
64. John F. Boyle, *The Order and Division of Divine Truth: St Thomas Aquinas as Scholastic Master of the Sacred Page* (Steubenville: Emmaus Academic, 2021), 9–12, 28–33, 77–78, and 83–85; John F. Boyle, "Authorial Intention and the *Divisio textus*," in *Reading John with St. Thomas Aquinas: Theological Exegesis and Speculative Theology*, ed. Michael Dauphinais and Matthew Levering, 3–8 (Washington D.C.: The Catholic University of America Press, 2005); and Robert George

Kennedy, "Thomas Aquinas and the Literal Sense of Sacred Scripture," (Ph.D. diss., University of Notre Dame, 1985).

65. Boyle, *Order and Division*, 28.
66. Consider Finocchiaro, "Methodological Background," 265: "the literal interpretation of the Bible is not binding when we are dealing with physical propositions that are capable of being conclusively proved."
67. Finocchiaro (92); Favaro (315); and Motta (95).
68. On the "senses" of Scripture see the classic studies Gilbert Dahan, *Lire la Bible au moyen-âge: Essais d'herméneutique médiévale* (Geneva: Librairie Droz, 2009); Gilbert Dahan, *L'Exégèse chrétienne de la Bible en Occident médiéval: XIIe-XIVe siècles* (Paris: Cerf, 1999); Beryl Smalley, *The Study of the Bible in the Middle Ages*, 3rd ed. (Oxford: Blackwell, 1983 [1941]); Henri de Lubac, *Exégèse médiévale: les quatre sens de l'Écriture I* (Paris: Aubier, 1959); Henri de Lubac, *Exégèse médiévale: les quatre sens de l'Écriture II* (Paris: Aubier, 1959); Henri de Lubac, *Exégèse médiévale: les quatre sens de l'Écriture III* (Paris: Aubier, 1961); and Henri de Lubac, *Exégèse médiévale: les quatre sens de l'Écriture IV* (Paris: Aubier, 1964). For contemporary Catholic articulation of these sense see especially *Catechism of the Catholic Church*, no. 115–117, online at: https://www.vatican.va/archive/ENG0015/_INDEX.HTM; Pope Benedict XVI, *Verbum Domini* (2010), nos. 37–41, online at: https://www.vatican.va/content/benedict-xvi/en/apost_exhortations/documents/hf_ben-xvi_exh_20100930_verbum-domini.html; and Joseph Cardinal Ratzinger, "Handing on the Faith and the Sources of the Faith," in *Handing on the Faith in an Age of Disbelief: Lectures Given at the Church of Notre-Dame de Fourvière in Lyons, France and at Notre-Dame Cathedral in Paris*, Joseph Cardinal Ratzinger with Archbishop Dermot J. Ryan, Godfried Cardinal Danneels, and Franciszek Cardinal Macharski, with commentaries by Pierre Eyt, Bernard Bro, O.P., Georges Bonnet, and Jacques Guillet, S.J., with a preface by Jean-Marie Cardinal Lustiger and Archbishop Albert Decourtray, trans. by Michael J. Miller, 13–40 (San Francisco: Ignatius Press, 2006 [1983]), 32–35.
69. Motta (95). The text of Favaro (315) differs slightly.
70. Favaro (316 and 330); and Motta (97 and 119).
71. Favaro (330); and Motta (119). Motta includes an editorial note

which shows that he understands this also as just another way of Galileo discussing the literal sense: "The addition of the adjective 'verbal' seems to respond to a need for more clarity, unequivocally revealing Galileo's reference to the literal sense." I think Motta is incorrect here. When Galileo wishes to discuss the literal sense [Motta's "*senso letterale*"] he does so by using "*senso literale*."

72. Favaro (344); and Motta (139).
73. Favaro (346); Motta (141 and 142).
74. Favaro (347); and Motta (143).
75. Carroll, "Galileo, Science and the Bible," 14; and Blackwell, *Galileo*, 166–170.
76. Pietro Redondi, "Natura e Scrittura," in *Il caso Galileo*, ed. Bucciantini, Camerota, and Giudice, 153.
77. Favaro (312 and 316); and Motta (92 and 96).
78. English is modified from Finocchiaro (90); Favaro (313) has minor differences from Motta (92).
79. English is modified from Finocchiaro (91); Favaro (314); and Motta (93).
80. Favaro (316); and Motta (96).
81. Finocchiaro (93); Favaro (317); and Motta (97).
82. English modified from Finocchiaro (93); Favaro (317); and Motta (97).
83. Finocchiaro (101); Favaro (326); and Motta (114).
84. Pesce, *L'ermeneutica biblica*, 116.
85. See John S. Bergsma and Jeffrey L. Morrow, *Murmuring Against Moses: The Contentious History and Contested Future of Pentateuchal Studies* (Steubenville: Emmaus Academic, 2023), 209–213; Jeffrey L. Morrow, *Pretensions of Objectivity: Toward a Criticism of Biblical Criticism* (Eugene: Pickwick, 2019), 35–41; Jeffrey L. Morrow, "The Acid of History: La Peyrère, Hobbes, Spinoza, and the Separation of Faith and Reason in Modern Biblical Studies," *Heythrop Journal* 58, no. 2 (2017): 169–171; Jeffrey L. Morrow, *Three Skeptics and the Bible: La Peyrère, Hobbes, Spinoza, and the Reception of Modern Biblical Criticism* (Eugene: Pickwick, 2016), 54–84; Jeffrey L. Morrow, "Pre-Adamites, Politics and Criticism: Isaac La Peyrère's Contribution to Modern Biblical Studies," *Journal of the Orthodox Center for the Advancement of Biblical Studies* 4, no. 1 (2011): 1–23; Jeffrey L. Morrow, "French Apocalyptic Messianism: Isaac La Peyrère

and Political Biblical Criticism in the Seventeenth Century," *Toronto Journal of Theology* 27, no. 2 (2011): 203–213; and Jeffrey L. Morrow, "The Politics of Biblical Interpretation: A 'Criticism of Criticism,'" *New Blackfriars* 91 (2010): 534–535.

86. See Bergsma and Morrow, *Murmuring Against Moses*, 214–216; Scott W. Hahn and Jeffrey L. Morrow, *Modern Biblical Criticism as a Tool of Statecraft (1700–1900)* (Steubenville: Emmaus Academic, 2020), 14; Morrow, *Pretensions of Objectivity*, 42–48; Jeffrey L. Morrow, *Theology, Politics, and Exegesis: Essays on the History of Modern Biblical Criticism* (Eugene: Pickwick, 2017), 9–10; Morrow, "Acid of History," 171–174; Jeffrey L. Morrow, "Averroism, Nominalism, and Mechanization: Hahn and Wiker's Unmasking of Historical Criticism's Political Agenda by Laying Bare its Philosophical Roots," *Nova et Vetera* 14, no. 4 (2016): 1320–1322; Morrow, *Three Skeptics*, 85–105; Jeffrey L. Morrow, "The Untold History of Modern Biblical Scholarship's Pre-Enlightenment Secular Origins," *Journal of Theological Interpretation* 8, no. 1 (2014): 150–151; Jeffrey L. Morrow, "*Leviathan* and the Swallowing of Scripture: The Politics behind Thomas Hobbes' Early Modern Biblical Criticism," *Christianity & Literature* 61, no. 1 (2011): 33–54; Morrow, "Politics of Biblical Interpretation," 535–536; and Jeffrey L. Morrow, "The Bible in Captivity: Hobbes, Spinoza and the Politics of Defining Religion," *Pro Ecclesia* 19, no. 3 (2010): 294–296.

87. See Bergsma and Morrow, *Murmuring Against Moses*, 216–223; Jeffrey L. Morrow, "Methods of Interpreting Scripture and Nature: The Influence of the Baconian Method on Spinoza's Bible Criticism," in *Studies in the History of Exegesis*, ed. Mark W. Elliott, Raleigh C. Heth, and Angela Zautcke, 157–173 (Tübingen: Mohr Siebeck, 2022); Hahn and Morrow, *Modern Biblical Criticism*, 15–16; Morrow, *Pretensions*, 48–54; Jeffrey L. Morrow, "Spinoza and the Theo-Political Implications of his Freedom to Philosophize," *New Blackfriars* 99 (2018): 374–387; Morrow, *Theology, Politics, and Exegesis*, 16–34; Morrow, "Acid of History," 174–176; Morrow, "Averroism," 1322–1328; Morrow, *Three Skeptics*, 104–138; Jeffrey L. Morrow, "Spinoza's Use of the Psalms in the Context of His Political Project," *Interdisciplinary Journal of Research on Religion* 11 (2015): 1–18; Morrow, "Untold History," 151–152; Jeffrey L. Morrow, "Historical Criticism as Secular Allegorism: The Case of Spinoza," *Letter & Spirit* 8 (2013): 189–221; Jeffrey L. Morrow, "The Early Modern Political

Context to Spinoza's Bible Criticism," *Revista de Filosofía* 66, no. 3 (2010): 7–24; Morrow, "Politics of Biblical Interpretation," 536–538; and Morrow, "Bible in Captivity," 296–299.

88. See Bergsma and Morrow, *Murmuring Against Moses*, 223–225; Hahn and Morrow, *Modern Biblical Criticism*, 16–17; Morrow, *Pretensions*, 67–68; Morrow, *Theology, Politics, and Exegesis*, 11–12; Morrow, "Averroism," 1328–1330; Jeffrey L. Morrow, "Faith, Reason and History in Early Modern Catholic Biblical Interpretation: Fr. Richard Simon and St. Thomas More," *New Blackfriars* 96 (2015): 659–664; Morrow, "Untold Story," 152; and Morrow, "Politics of Biblical Interpretation," 538–539.

89. Franco Giudice, "Echi del caso Galileo nell'Inghilterra del XVII secolo," in *Il caso Galileo*, ed. Bucciantini, Camerota, and Giudice, 280; and Jaki, *Road of Science*, p. 58.

90. Gianni Paganini, "Hobbes's Galilean Project: Its Philosophical and Theological Implications," in *Oxford Studies in Early Modern Philosophy Volume VII*, ed. Daniel Garber and Donald Rutherford, 1–46 (Oxford: Oxford University Press, 2015); Luc Foisneau, "Hobbes's First Philosophy and Galilean Science," *British Journal for the History of Philosophy* 19, no. 4 (2011): 795–809; and Douglas M. Jesseph, "Galileo, Hobbes, and the Book of Nature," *Perspectives on Science* 12 (2004): 191–211.

91. Jacob van Sluis and Tonnis Musschenga, ed., *De boeken van Spinoza* (Groningen: Bibliotheek der Rijksuniversiteit Groningen and Haags Gemeentearcheif, 2009).

92. Franco Biasutti, "Galileo and Spinoza: Historical and Theoretical Perspectives," *Intellectual History Review* 23, no. 1 (2013): 3–23; Herman De Dijn, "Spinoza and Galileo: Nature and Transcendence," *Intellectual History Review* 23, no. 1 (2013): 99–108; Pietro Redondi, "Spinoza, Critic of Galileo," *Intellectual History Review* 23, no. 1 (2013): 109–118; Tamar Rudavsky, "Galileo and Spinoza: The Science of Naturalizing Scripture," *Intellectual History Review* 23, no. 1 (2013): 119–139; Jacob Adler, "Joseph Solomon Delmedigo: Student of Galileo, Teacher of Spinoza," *Intellectual History Review* 23, no. 1 (2013): 141–157; and Tamar Rudavsky, "Galileo and Spinoza: Heroes, Heretics, and Hermeneutics," *Journal of the History of Ideas* 62, no. 4 (2001): 611–631.

93. Morrow, "Methods of Interpreting Scripture and Nature," 157–173.

14 PYTHAGOREAN ARGUMENT OF THE GRAND DESIGN

Alexey Burov, Ph.D. and Alexei Tsvelik, Ph.D.

1. Introduction

The crucial importance of Christianity for the possibility and value of natural science was a central idea of Fr. Jaki, as is clear already from the title of his famous book, *The Savior of Science.*[1] As for the inverse, science supporting religion, Fr. Jaki expressed a rather nuanced view. In his last book, *Questions on Science and Religion*, he wrote,

> As to religionists, they look in vain in science for proofs of God or for anything about God. If believers want solid intellectual foundations for religious tenets, they better find it outside science before they say anything on the relation of science and religion. They merely reveal their insecurity when they resort to science for support, let alone for a lifebelt or lifeline. From science they can obtain only quantities and nothing else.[2]

Within the context of that book, the word 'religionists' relates to rather unphilosophical religious enthusiasts, alien to the spirit of science but still eager to extract support from scientific material. For them, Fr. Jaki suggests a "cold shower":

> Science merely presents a vast range of evidence on behalf of the fact that everything is queerly strange, quantitatively that is. It therefore should enforce the wish to ponder the question: why such and not something else? More than that science cannot do in reference to the topic discussed in this book. Any

> book on science and religion that claims to derive more than that from science, offers only dreams and fantasies.[3]

This is not to say that Fr. Jaki did not see the possibility of a healthy productive dialog between science and religion. Such a dialog, however, requires, according to him, creation of a special philosophical space from which both sides would adequately open up, and through which they could understand each other and themselves:

> One can, however, form a clear idea about assuming that a meaningful dialogue can be carried on only when one's meaning is clear about the platform on which that dialogue is to take place. The platform is not a simple uniform construct.[4]

The present work is aimed precisely at promoting the development of such a philosophical platform. The proposed "Pythagorean Argument of Grand Design" is our version of the teleological or physico-theological argument for the existence of God, which derives from reflecting upon the state of modern physics and its spiritual premises. This argument demonstrates that the creation of the universe by a supreme intelligence is not only in every respect a reasonable hypothesis about the origin of and the status of man in the universe but also the only hypothesis that can withstand rational critique, at least for today. This argument has, in our opinion, a great and yet to be fully appreciated importance not only for theism in general but especially for Christianity.

The argument, as we formulate it, is twofold.

Its first, objective, logical, bottom-up branch is based on the demonstration of the very special character of physical laws, which combine the requirements of both sufficient complexity for the possibility of life and sufficient simplicity for discoverability by emerging living thinking beings. The achievements of modern physics have revealed the tension between these two requirements in all its dramatic majesty, so that a resolution of this contradiction seems to be a truly impossible task. The laws falling into this narrow *minimax of complexity* can be explained only by purposeful choice, which points to a super-intelligent Creator, interested not only in the appearance of living and thinking beings, but also in their ability to discover these very laws in an enormous range of parameters and with incredible accuracy. This objective-logical branch of the argument is discussed in the next section of the paper, concluded with our criticism of Fr. Jaki's views on mathematics and its role in physics.

The second branch of our argument is subjective, historical, top-down. The discoverability of physical laws is just a possibility. The fact that they are being discovered is due to some special mindset or system of beliefs of the scientists involved. This belief system, which we call *the Pythagorean faith*, is a certain synthesis of Platonism and Christianity that promised both the possibility of a deep knowledge of the cosmos through mathematics and the highest value, even sacrality, of this task as eminently worthy of service. Here we echo in part Fr. Jaki's *Savior of science* and supplement the ideas therein. The level of physical knowledge today demonstrates a fantastic success of the Pythagorean faith, whose equal in power and universal significance is hard to find. By its fruits, such a belief system at least deserves the status of the leading working hypothesis about the universe and man. This branch of the argument is discussed in the third section of our paper.

2. Character of the physical law

Physics is known to be the science of matter, based on the belief that the Universe is mathematical, that is, on faith in the existence of an elegant logical-mathematical structure of ideas, physical laws, to which experimental observations can be effectively reduced and using which detailed predictions be made. The idea of atemporal laws, mathematical principles of motion, in the foundation of physics proved to be exceptionally fruitful. The physical principles known to modern science allow not only to describe already known phenomena but also to predict fundamentally new ones in the enormous space-time range of 45 orders of magnitude, and with the accuracy up to 12 decimal digits. The first of these numbers is the ratio of the size of the largest object of science, the visible universe, to the characteristic size of its smallest objects, the top quark and Higgs boson. The second characterizes the degree of agreement between theory and observations, achieved by ultra-precise experiments testing Quantum Electrodynamics and the General Theory of Relativity.

This *cosmic discoverability* implies a special character of the natural regularities expressed by the physical laws. On the one hand, these regularities made possible the emergence of life, with its subsequent unfolding into a rich biosphere up to the emergence of intelligent beings, *sapiences*. On the other hand, the Universe has also turned out to be accessible to the cognition of these intelligent beings, and that not just within the habitat, but on the cosmic scale.

The first of these architectural qualities of laws denotes their friendliness to intelligent life; it can be called *sapienceness*, which we use here. The second architectural quality of laws, cosmic discoverability, combines such of their properties as universality, sufficient mathematical simplicity, symmetries, invariances, correspondences, equivalences, and sufficient richness of solutions. This architectural quality of the laws is inseparable from their aesthetic appeal, mathematical beauty, or elegance. This special aesthetic guides physicists in their search of both compact and property-rich mathematical structures; consciously or latently it is connected with all the great discoveries of physics.

Let us try to outline the qualities of the physical laws known today with the degree of detail that is still compatible with the genre of a philosophical article:[5]

a. Mathematical compactness, reflecting the simplicity and reasonableness of the mathematical principles behind the laws.
b. Aesthetic appeal (elegance): the laws possess symmetries, invariances, correspondences, and equivalences.
c. Universality, the validity within enormous spans of space, time, energies.
d. Sufficient completeness and extreme accuracy of description of material reality; today, up to 12 decimal digits.
e. Asymptotic exactness: classical conclusions about observations coincide with those of relativistic and quantum theories at the corresponding mathematical limits: infinite speed of light and infinitesimal Planck's constant. Neither the new knowledge invalidates the old advances, but it appeared due to them relying on them as on already known limit cases.
f. Syntactic correspondence: some deeper laws were obtained from less deep "classical" laws by merely syntactic transformation.

g. Sapienceness, i.e., compatibility with intelligent life.

h. Gradual increase of conceptual difficulty. The laws of planetary motion are simpler than the laws of electricity and magnetism, which are simpler than non-relativistic quantum physics, etc., and not vice versa. The previous conceptual level of difficulty plays an invaluable role as a steppingstone to the next.[6]

i. Notwithstanding all the simplicity and reasonableness, all the "hints", parallels and correspondences, the laws are barely discoverable; they were found through extreme efforts of the genius pioneers. "Like an excellent tutor, the universe has not been so demanding as to ensure failure but rather has allowed us to succeed while still presenting us with worthy challenges."[7]

Note that items (a)-(g) reflect mainly objective properties of the laws, whereas the last two items are rather associated with human abilities. By virtue of objective properties of laws, especially (a)-(d), the physical theories are fundamentally different from empirical fitting the data by more or less arbitrary formulas—like in the algorithms of Ptolemy, Copernicus, or artificial intelligence. Such fits describe only the facts they already contain, they require updating when more or better data become available, so that complexity of the fits progressively increases with the data. On the contrary, physical theories are fundamentally simple, and they are confirmed by new measurements with better and better accuracy while maintaining the same simplicity. The crucial distinction is the predictive power: such theories predict phenomena not yet observed, frequently not even suspected of by their authors. This predictive power is due to the same properties of physical theories as their

Popperian falsifiability:[8] simplicity, universality, precision, and completeness of their mathematical principles. Contrary to physical theories, fitting algorithms are non-falsifiable: any contradiction with the new data is trivially resolved with one more fitting parameter, one more 'epicycle' or 'different'.

As we have stated above, the properties of sapienceness and aesthetic attractiveness, both necessary for the discoverability of the laws, set seemingly unresolvable opposing demands. Indeed, sapienceness requires sufficient complexity of the laws for the possibility of constructively complex states of matter, the *fine tuning* for complex chemistry.[9,10,11] However, intellectual attractiveness requires that the mathematical forms of the same laws be simple and symmetrical enough to allow their discovery by sapiences, to comply with their mathematical taste and intuition.

This *minimax of complexity*, the conflicting requirement for the laws' complexity to be sufficiently high for life and sufficiently low for successful guesses, so strongly limits the set of laws that the fulfillment seems hardly possible at all. The number of conditions that ensure the possibility of life, and a very diverse life at that, is enormous; it highly exceeds the small number of free parameters of any elegant set of formulae. For example, the number of dimensionless parameters in the equations describing the chemistry relevant for life is just three.[12] If we imagine each of such conditions for life as a mathematical restriction on the free parameters, we get a vastly overdetermined system of equations and inequalities, the solution to which does not generally exist. The situation is like an impossible task to tune a large number of music instruments by means of a few knobs. At best, solutions could be found only for a very special structure of laws, for which this system of conditions would be many times degenerate.[13]

The fact that such a special set of elegant laws not only exists in principle but is materialized seems to be a miracle. Numerous estimates on the fine-tuning of fundamental physical constants show that if the physical laws were simpler than they are, life would likely be impossible regardless of the remaining free constants. If the laws were a little more complex, however, it is unlikely they would have ever been discovered; in their present form they have been discovered already at the limit of ability of genius founding fathers of physics. This unity of the opposite architectural qualities of the laws of nature raises a fundamental question that requires explanation: what is the reason for such a special, *Pythagorean*, character of the universe?[14]

Why did nature turn out to be describable with a high degree of accuracy and at such a level of universality, by formulas selected according to human aesthetic preferences? This special productivity of elegant mathematics is the subject of a famous essay, "The Unreasonable Effectiveness of Mathematics in the Natural Science" by Eugene Wigner,[15] one of the founders of quantum physics. It concludes that "the miracle of the appropriateness of the language of mathematics for the formulation of the laws of physics is a wonderful gift which we neither understand nor deserve. We should be grateful for it and hope that it will remain valid in future research". Let us note that gratitude requires a person to address. We cannot be grateful to material objects or to ideas. Thus, Wigner's commandment implies the transcendent personal author of the universe.

Why do the 'keys' chosen on the human principle of elegance so nicely fit the 'locks' of nature? Could not this consistency between such different entities indicate that beautiful mathematics lies at the logical foundation of the universe itself? Is this not evidence in favor of the assumption of a transcendent mind that specified this elegant and life-friendly architecture of matter in a way that intelligent beings like us could emerge and discover it? Notwithstanding centuries-old efforts, so far, there is only one reasonable idea about that, and it points to the design of the universe by a superior mind. If so, this idea should be assumed as a working hypothesis: we should live and think relying on it; any other attitude would be irrational.

Let us emphasize that such questions and such an assumption are posed in a rational philosophical context, not in the context of this or that religious confession.[16],[17],[18] The argument stresses a certain conformity between elegant mathematics and the observable universe; honoring its roots, we call it *Pythagorean*. The argument is based on the special character of physical laws that allow emergence of living thinking beings, capable of discovering these very laws. In other words, this argument is grounded on the observation of *cognitive self-consistency of the universe*. Since the only explanation at hand, at least so far, is the purpose of a transcendent Architect, it gives this explanation a power of the working hypothesis. Being rational, this argument of *Grand Design*[19] is based on reflection upon physics; it brings us to the frontier of religions but does not force us into any of them.

The leading role of mathematical aesthetics in physical discoveries has been discussed since the beginning of the last century. The first here, it seems, was the great mathematician and physicist Poincaré, followed then by Einstein, Jeans, Heisenberg, Pauli, Dirac, Wigner, Feynman, Weinberg, Penrose, Wilczek, and others, all referring to their own experience and one of the predecessors. Despite this, philosophers, with very few exceptions, have remained deaf to this central aspect of the cosmic cognition. Unfortunately, we did not find any reflection on that in Fr. Jaki's texts. We think that this is not our oversight, but a feature of his misconception about mathematics.

By emphasizing that science, especially physics, is entirely about such a meager entity as 'quantity', Fr. Jaki, in our opinion, makes the mistake of identifying mathematics with knowledge about quantities. Well, this is not the reason why mathematicians, starting from the ancient Greeks, are interested in their subject. Mathematics, as Godfrey Hardy wrote, is the art of considering extremely universal, beautiful structures of ideas.[20] Mathematical proofs are demonstrations of necessary connections between elements of these eternal structures. Understanding mathematics as operations with quantities is just as reasonable as understanding sculpture as operations with solid bodies or music as manipulation with sounds. Conceptions of this kind lose the most important thing—the search for beauty, the aesthetic thread of Ariadne leading science and art. Beauty cannot be defined, but this had stopped the founders of mathematical physics as much as it had sculptors or musicians. Fr. Jaki overlooks this key contribution of physics to philosophy and religion. Were sculptures mere stones, placing some inside temples would make little sense. The wonder of the Universe lies not in the fact that its fabric can be described in terms of quantities, but in that the physical laws are mathematically beautiful structures. Quoting Steven Weinberg,

> It is very strange that mathematicians are led by their sense of mathematical beauty to develop formal structures that physicists only later find useful, even where the mathematician had no such goal in mind. A well-known essay by the physicist Eugene Wigner refers to this

> phenomenon as "The Unreasonable Effectiveness of Mathematics." Physicists generally find the ability of mathematicians to anticipate the mathematics needed in the theories of physicists quite uncanny. It is as if Neil Armstrong in 1969 when he first set foot on the surface of the moon had found in the lunar dust the footsteps of Jules Verne ... We believe that, if we ask why the world is the way it is and then ask why that answer is the way it is, at the end of this chain of explanations we shall find a few simple principles of compelling beauty. We think this in part because our historical experience teaches us that as we look beneath the surface of things, we find more and more beauty. Plato and the neo-Platonists taught that the beauty we see in nature is a reflection of the beauty of the ultimate, the "nous". For us, too, the beauty of present theories is an anticipation, a premonition, of the beauty of the final theory. And in any case, we would not accept any theory as final unless it were beautiful.[21]

3 Pythagorean Faith

The above describes the objective, metaphysical or ascending line of the Pythagorean argument, reflecting the movement from science to metaphysics, theology, and religion. Historically this movement has been, however, two-way. Ideas about the hidden mathematical nature of matter appeared alongside the birth of mathematics in the Pythagorean schools of southern Italy in the sixth to fifth centuries BCE and were religious in nature. Then they were picked up and developed by Platonism, and in due time incorporated into Christian theology and philosophy through Neoplatonic schools. The philosophical and religious visions of mathematics as an element of perfect knowledge, as a communion with the supreme mind, gave rise to the enthusiastic expectation that matter too is structured by these perfect forms. Hence stemmed the natural-philosophical intuitions of Kepler, Galileo, Descartes, Leibniz, and Newton, from which mathematical physics was born and developed.[22],[23],[24],[25],[26]

The Pythagorean argument thus has a two-way character. On the one hand, the elegant mathematical laws discovered by physics, by their very refined character, raise the question, "how is this possible?" and point to the hypothesis of a superior mind as the origin of the laws. Such is the objective side of the argument. Logic of this kind contributed to natural theology; Leibniz called it anagogical or ascending: "I have shown on several occasions that the final analysis of the laws of nature leads us to the most sublime principles of order and perfection, which indicate that the universe is the effect of a universal intelligent power."[27]

On the other hand, there was also a downward path: mathematical physics was born in the space of certain religious and philosophical ideas, namely Renaissance and Modern Christian Platonism. One can look at this ideological synthesis as a "meta-theory" inspiring and guiding the efforts of the founding fathers of physics. The mind-boggling success of this meta-theory is a reason for its acceptance as a working hypothesis.

Galileo's famous words about nature as a book written in the language of mathematics contain much more than is commonly understood. They show that Galileo, first, assumed an ideal logical-mathematical framework of nature, which could be laid down only by its perfect transcendent author. Second, such a belief implied the consistency of this framework with the possibility of life, including the intelligent life of human beings discovering the framework. Third, these Galilean presumptions not only imply the possibility of the discovery, but also stress its special sacred status. The Book of Nature should be read alongside with the Holy Scripture, and only by reading them together can we understand both. Thus, the cognition of the mathematical language of nature is thought by Galileo (and by most of the great physicists afterwards) as a sort of communion with the Supreme Mind. The search for this axiomatic framework constituted the task that Galileo set for science. The goal here was sacral knowledge-communion, not knowledge-power, not power over nature, not worldly utility, and not military might.

> I'm well aware that the Pythagoreans had the highest regard for the science of numbers, and that Plato himself admired the human intellect and regarded it as sharing in divinity simply because it could comprehend the nature of numbers. My own view is not very different,[28]

argued Galileo through Salviati in *Dialogue on the Two Greatest World Systems*. This communion had a special meaning where numbers were also contemplated as the foundations of nature; as Galileo wrote to the Grand Duchess Christina, "God is no less excellently discovered in the effects of nature than in the sacred sayings of the Scriptures."[29]

The outlined set of ideas has a twofold historical origin. First, it is connected with the Biblical ideas about man as the image and likeness of God, or, what is the same, about the Creator as our transcendent Father. Its second source is the Pythagorean doctrine of the sacredness of mathematics and Plato's theory of forms. This complex of Christian-Platonic beliefs deserves a special term; we call it *the Pythagorean faith.*[30] As a special Platonism, this belief further reinforces the emphasis on mathematics. As a special Christianity, it elevates the importance of creative and cognitive possibilities and downgrades ideas of the depravity of man and the distortion of nature by original sin. This belief has never been formalized, its creed has never been discussed in academic or church councils, but nevertheless it has played and continues to play a great role in human history.

The new physics of Galileo, Descartes and those who followed them differed decisively from the traditional empiricism of Aristotle and Francis Bacon. The empirical research of modernity followed the Aristotelian way of categorizing observations: an object is understood if its observed qualities are described and decomposed into some properties and categories. The task of Galilean physics, on the other hand, was: first, to search for and verify the mathematical laws hidden behind the observations; second, to explain new observations with these laws, predicting what has not yet been observed; third, to apply them for technical inventions. For mathematical physics, a phenomenon is considered understood only when it is explained through the universal laws captured in the equations, acting always and everywhere. Before that, the phenomenon is "not comprehended", notwithstanding in how much detail it is described and classified.

By breaking through to the discovery of the laws of nature, theoretical science has been able to accomplish what is inaccessible to empiricism: to predict objects that have not yet been observed or even conceived before: new planets, radio and gravitational waves, antimatter, the early universe, new fundamental particles, and much more.

In its paths of understanding, physics turns not only to real experiments, but, even before that, to mental, imaginary ones These *gedankenexperiments* are speculative considerations of limiting cases, like Galileo's frictionless surface and fluffs falling in vacuum, Newton's infinitely light planets, Langevin's twin flying at the speed of light, Einstein's experiments in a free-falling elevator, Schrödinger's quantum cat and Wigner's quantum friend, etc.

For science of this new, super-empirical type, at first ideal objects, bound by abstract mathematical laws, are assumed, and then real, experimental objects and properties are considered in the light of the former, but not vice versa, as in empirical sciences limited to the classification of observations— "stamp collecting", as Rutherford put it (without belittling). These two opposing yet complementary strategies of cognition are often conflated in the common view of natural science as "empirical" in general. Understanding scientific and technological progress, however, requires a clear distinction between these strategies; it requires awareness of the exceptional role of speculation, imagination, and mathematical guess in the episteme of Modern physics.

Historically, the Pythagorean faith had been gradually formed within the walls of medieval universities, then inherited by the public intellectuals of Renaissance and Modernity. It would be a mistake to take this synthesis of "Athens and Jerusalem" for granted, as a sort of the law of progress. This very special synthesis came against such strong headwinds that one may only wonder how it managed to overcome them.

On the philosophical ground, the Pythagorean faith was opposed by Aristotelianism with its superficially empirical attitude to mathematics, as well as by Hermetism with its passion for magic and sacralization of nature. On the religious ground, it was opposed not only by the mistrust in the cognition of fallen man, not only by the widespread fear and suspicion of science as a tool of diabolical forces, but also by the gnostic-like currents embodied in the dogmas of a sin-damaged nature, of the prince of this world, and of a world lying in evil. On the political and social field it was often opposed by the spirit of absolute obedience and suppression of all initiative. Considering also the uniqueness of the abilities required to break through to the sophisticated Pythagorean mathematics, one can only wonder at how Christian mathematical Platonism has managed to realize itself.

The listed, partly latent, beliefs and intuitions of the Pythagorean faith gave rise to Modern physics. We associate this belief with the name of Pythagoras, not because the father of mathematics himself, whom we know little about, clearly formulated such a creed, but because the fundamental intuition of mathematical physics originates there, in ancient Pythagoreanism.

Explicitly and implicitly proceeding from this creed, Leibniz described the character of the laws of nature even before the first universal physical theory, Newton's Celestial Mechanics, was published:

> God has chosen the most perfect world, that is, the one which is at the same time the simplest in hypotheses and the richest in phenomena, as might be a line in geometry whose construction is easy and whose properties and effects are extremely remarkable and widespread.[31]

Although Leibniz, like all the fathers of the scientific revolution, believed in the discoverability of the logico-mathematical foundations of nature, his faith was, so to speak, naive: the accessibility of laws to the human mind was not perceived by him as something that may not be, even if the laws were simple to God. Such unawareness of the problem of discoverability of the laws of nature was typical for several epochs; it started to subside only in the twentieth century.

In 1988, Freeman Dyson, one of the originators of quantum electrodynamics, proposed the following modification of the Leibnizian principle of "best of all possible worlds":

> When we look at the glory of stars and galaxies in the sky and the glory of forests and flowers in the living world around us, it is evident that God loves diversity. Perhaps the universe is constructed according to a principle of maximum diversity. The principle of maximum diversity says that the laws of nature, and the initial conditions at the beginning of time, are such as to make the universe as interesting as possible. As a result, life is possible but not too easy. Maximum diversity often leads to maximum stress. In the end we survive, but only by the skin of our teeth.[32]

Thus, Dyson not only suggested a solution of the long-standing *problem of evil,* but also deduced the discoverability of laws as a manifestation of the principle of maximum interestingness:[33] "laws of nature are explained if it can be established that they must be as they are in order to allow the existence of theoretical physicists to speculate about them."

In the same year, the same explanation of the laws was independently proposed by Stephen Hawking:

> Einstein once asked the question: "How much choice did God have in constructing the universe?" ... there may well be only one, or a small number, of complete unified theories, ...that are self-consistent and allow the existence of structures as complicated as human beings who can investigate the laws of the universe and ask about the nature of God.[34]

The above demonstrates that the metaphysical foundations of science can remain unexpressed for a long time and not fully grasped even by its founders. Moreover, even what was explicitly stated once could be forgotten in time. Both Leibniz's ascending (anagogical) and descending (Pythagorean-aesthetic) meditations upon laws as forms of the perfect mind were abandoned by the 'normal science'. It is significant, however, that the break with metaphysics got no support from founding fathers of physics, from Newton and Maxwell to Einstein and Heisenberg.

Within 'normal science,' its metaphysical foundations mostly work latently, regardless of how clearly they are perceived by the scientists. They became routine assumptions a long time ago, gave rise to multitudes of specific approaches, ideas and methods and then were safely forgotten by the growing scientific community. The pioneers of the scientific revolution were closer to these foundations than the followers, so they saw and articulated these ground beliefs better. The leading scientific revolutionaries of the twentieth century had often lost the faith in 'personal God', a core element of the meaning of science for previous generations of great physicists. However, the emotionally intense belief in a superior mind responsible for the structure of the universe was common to them all. Here is how Einstein understood this fundamental faith:

> The interpretation of religion, as here advanced, implies a dependence of science on the religious attitude, a relation which, in our predominantly materialistic age, is only too easily overlooked. While it is true that scientific results are entirely independent from religious or moral considerations, those individuals to whom we owe the great creative achievements of science were all of them imbued with the truly religious

> conviction that this universe of ours is something perfect and susceptible to the rational striving for knowledge. If this conviction had not been a strongly emotional one and if those searching for knowledge had not been inspired by Spinoza's Amor Dei Intellectualis, they would hardly have been capable of that untiring devotion which alone enables man to attain his greatest achievements.... This firm belief, a belief bound up with deep feeling, in a superior mind that reveals itself in the world of experience, represents my conception of God.[35]

Among contemporary physicists, especially high-ranked, those who call themselves atheists and agnostics dominate. In fact, it would be more accurate to call them religiously indifferent; they simply do not think at that level. It is reasonable to ask whether this fact cast any shadow on the Pythagorean argument? Does it indicate its weakness? We do not think so. The Pythagorean argument is philosophical; it requires a certain technique of self-reflection, the ability to see the metaphysical and ethical foundations of physics, thinking about its grounding values. This philosophical culture has, for the most part, been lost in the last hundred years or so. What happened is what Ortega y Gasset called *the barbarism of specialization*, an aspect of the *revolt of the masses.*[36] Contemporary physicists do not rebut the Pythagorean argument; as a rule, they simply do not know it, and even if they happen to come across it, they will not hear it and will not be able to discuss it adequately.

This is the general trend, against the background of which glimmer two weak attempts to get rid of the idea of divine design, in the face of the fact of the cosmic discoverability of the universe.

One of these attempts, proposed by physicist Max Tegmark, hypothetically and without any ground, regards our universe as one inside the infinite multiverse where absolutely all laws of nature exist.[37] It just so happened, he suggested, that we were born in such an anthropic universe. The scientific inconsistency of this hypothesis of *chaosogenesis* was shown in Ref.,[38] which won an award of the FQXi, headed, ironically, by Max Tegmark himself. In short, the refutation is that even in a life-friendly universe there is no reason to expect simple physical laws; the measure of universes with simple laws in the assumed subset of life-friendly universes is zero.

Another attempt to avoid the theistic conclusion consists in the proposal to ignore the problem of cosmic discoverability altogether due to its potentially unscientific, incorrect or irrational character. As far as we know, this position was first expressed by Lord B. Russell in his famous debate with the Rev. Fr. F. Copleston;[39] since then it has been periodically reproduced. An objection to Russell and his followers can be formulated as the rationalist presumption of correctness: a question must be presumed correct unless the opposite is proved. Rejection of this presumption would be as disastrous for cognition as rejection of the presumption of innocence for the rule of law. Therefore the question about the reason for the cosmic discoverability must be considered correct. And since the only coherent, at least for today, answer to it is theistic, that answer must be accepted as a working hypothesis until proved incoherent. By virtue of this conclusion, we must live as if there is a God, and as if He cares about us in general and about our cognition of the universe in particular.

Conclusion

On the ground of pure logic, the objective, metaphysical, bottom-up line of the Pythagorean argument of Grand Design leaves largely open the question of the Creator's relation to both man and the ultimate source of being and thinking. The intersubjective, ethical, top-down aspect of the argument, however, allows us to go further, evaluating the truthfulness of the Pythagorean faith by its coherence, by the adequacy of its expectations, by its scientific, technological, and civilizational results, or, biblically speaking, by its fruits. The Pythagorean creed is quite coherent. Very daring notions of the Pythagorean faith, based just on bold intuition, about the exceptional possibilities of mathematical cognition of nature have been justified beyond the wildest dreams of its adherents. As far as we understand, no metaphysical view can compete with the Pythagorean one in that respect. And if so, the Pythagorean creed deserves the status of a metaphysical working hypothesis about the universe, man, and God. Reflections on that, we believe, should be a part of Fr. Jaki's "philosophical platform", where science, religion and their relations can be understood for the common good.

We are grateful to the organizers of the Stanley Jaki International Congress for the chance to express these views. We appreciate the reviewers' recommendations, which were helpful. Our special thanks—to Lev Burov for his valuable suggestions.

Notes

1. Stanley Jaki, *The Savior of Science* (Grand Rapids, Michigan: Wm. B. Eerdmans Publishing, 2000).
2. Stanley Jaki, *Questions on science and religion* (Fort Collins, CO: Gondolin Press, Kindle Edition, 2004), 11.
3. Ibid., 69.
4. Ibid., 78.
5. Alexey Burov and Lev Burov, "Metaphysical Status of Physical Laws," in *Plato in Late Antiquity, the Middle Ages and Modern Times, XVII ISNS Conf., Selected Papers*, ed. John Finamore and Mark Nyvlt (Bream, UK: Prometheus Trust, 2020), 129.
6. Robin Collins, "God and the Laws of Nature," *Philo* 12, no. 2 (2009): 142.
7. Guillermo Gonzalez and Jay Richards, *The Privileged Planet: How Our Place in the Cosmos Is Designed for Discovery* (Washington, DC: Regnery Publishing, 2004).
8. Stephen Thornton, "Karl Popper," in *The Stanford Encyclopedia of Philosophy*, ed. Edward N. Zalta and Uri Nodelman, accessed April 3, 2024, https://plato.stanford.edu/entries/popper/.
9. John Barrow and Frank Tippler, *The Anthropic Cosmological Principle* (Oxford: Clarendon Press; Oxford University Press, 1986).
10. Geraint Lewis and Luke Barnes, *A Fortunate Universe: Life in a Finely Tuned Cosmos* (Cambridge: Cambridge University Press, Kindle edition, 2016).
11. Simon Friederich, "Fine-Tuning," in *The Stanford Encyclopedia of Philosophy*, ed. Edward N. Zalta and Uri Nodelman, accessed April 3, 2024, https://plato.stanford.edu/entries/fine-tuning.
12. Alexei Tsvelik, *Six Days* (Chicago, IL: Bagriy & Co, 2019).
13. Alexei Tsvelik, "Non-Probabilistic Approach to Anthropic Principle and a Fallacy of the Fine-Tuning Argument," in *Plato in Late Antiquity, the Middle Ages and Modern Times*, XVII ISNS Conf., selected papers, ed. John Finamore and Mark Nyvlt (Bream, UK: Prometheus Trust, 2020), 141, https://www.prometheustrust.co.uk/Non-probabilistic_Approach_to_Anthropic_Principle_and_a_Fallacy_of_the_Fine-Tuning_Argument.pdf.

14. Alexey Burov and Lev Burov, "Genesis of a Pythagorean Universe," in *Trick or Truth? The Mysterious Connection Between Physics and Mathematics*, ed. A. Aguirre et al. (Switzerland: Springer, FQXi, The Frontiers Collection, 2016), 157,https://pythagoreanuniverse.com/essays/Genesis-of-a-Pythagorean-Universe_Trick-or-Truth_Springer.pdf.
15. Eugene P. Wigner, "The Unreasonable Effectiveness of Mathematics in the Natural Sciences," *Communications on Pure and Applied Mathematics* 13, no. 1 (1960): 1.
16. Collins, "God and the Laws of Nature."
17. John Lennox, *Cosmic Chemistry: Do God and Science Mix?* (London, UK: Lion Books, Kindle edition, 2021).
18. Mark Steiner, *The Applicability of Mathematics as a Philosophical Problem* (Cambridge, MA: Harvard University Press, 1998).
19. Stephen Hawking and Leonard Mlodinow, *The Grand Design* (New York, NY: Bantam Books, 2010).
20. Godfrey Hardy, *A Mathematician's Apology* (Cambridge, UK: Canto Classics, Cambridge University Press, Kindle Edition, 1940–2012), loc. 769.
21. Steven Weinberg, *Dreams of a Final Theory: The Scientist's Search for the Ultimate Laws of Nature* (New York, NY: Vintage Books, Kindle Edition, 1993), 157, 165.
22. Arthur Koestler, *The Sleepwalkers* (London, UK: Penguin Classics, 1959).
23. Kitty Ferguson, *The Music of Pythagoras* (New York: Walker Publishing, 2008).
24. Melissa C. Travis, *Science and the Mind of the Maker* (Eugene, OR: Harvest House Publishers, 2018).
25. Stephen Barr, *Modern Physics and Ancient Faith* (Notre Dame, IN: University of Notre Dame Press, 2003).
26. Frank Wilczek, *A Beautiful Question: Finding Nature's Deep Design* (New York, NY: Penguin Press, 2015).
27. Gottfried W. Leibniz, "Tentamen Anagogicum," ca. 1696, *in Philosophical Papers and Letters* (Dordrecht: Springer Netherlands, 1989), 477.
28. Galileo Galilei, *Dialogue on the Two Greatest World Systems*, 1632, trans. Mark Davie (Oxford: Oxford University Press, 2022), Kindle edition, loc. 11.

29. Galileo Galilei, "Lettera a Cristina di Lorena, Granduchessa di Toscana," 1615, in *Documentazione Interdisciplinare di Scienza & Fede,* https://disf.org/galileo-lettera-a-madama-cristina-di-lorena.
30. Burov, "Metaphysical Status of Physical Laws."
31. Gottfried W. Leibniz, *Discourse on Metaphysics and Other Essays,* 1686, trans. Daniel Garber and Roger Ariew (Indianapolis, IN: Hackett Classics, 1991), Kindle Edition, loc. 171.
32. Freeman Dyson, *Infinite in All Directions* (New York, NY: Harper & Row Publishers, 1988), 298.
33. Ibid., 296.
34. Stephen Hawking, *A Brief History of Time* (New York, NY: Bantam Books, Kindle edition, 1988), 19.
35. Albert Einstein, *Ideas and Opinions,* 1954 (New York, NY: Three Rivers Press, Kindle edition, 1982), 65, 287.
36. José Ortega y Gasset, *The Revolt of the Masses,* 1930 (London, UK: Routledge, 2021).
37. Max Tegmark, *Our Mathematical Universe: My Quest for the Ultimate Nature of Reality* (New York, NY: Knopf, Borzoi Books, 2014).
38. Burov, "Genesis of a Pythagorean Universe."
39. Bertrand Russell, "The Existence of God: A Debate Between Bertrand Russell and Father F. C. Copleston, SJ," 1948, in *Why I Am Not a Christian and Other Essays on Religion and Related Subjects* (London, UK: Routledge, 2004), 125.

15 ALL THINGS WERE MADE THROUGH HIM

Some thoughts about the teaching of St. Thomas Aquinas on the relationship between matter and God

JORDAN M. MILLER, PH.D.

This paper asks a question about the relationship between matter and God. We might formulate this question in a few ways:

1. Whether in creating material beings out of nothing, God is producing something that is utterly different from what He is, something totally unlike Himself?
2. Whether a material being has similarity to God, not in spite of its materiality, but because of its materiality?
3. Whether in creating matter, God is imparting something of His own, giving something of Himself?

At first glance, any suggestion that God imparts something of Himself to produce matter would seem to be in conflict with the Catholic Faith. On the basis of divine revelation, it has been taught always, everywhere, by every saint, that *God is not the universe,* and *the universe is not God.* Rather, God creates the universe out of nothing.[1] To suggest that God gives something of Himself in creating matter would seem to confuse God with the universe, and to confuse the begetting of the Son, and the procession of the Spirit, with the making of creatures. The thesis would apparently be a departure from Christianity.

Fr. Stanley Jaki repeatedly emphasizes that the Christian doctrine of creation out of nothing provided the "fertile soil" necessary for science to take root and grow.[2] Whereas paganism regards the natural forces of the universe as deities that must be worshipped and propitiated, Christianity proclaims that these natural forces are beautiful and very good, but never to be mistaken for the Creator who brings them into being. Jaki comments:

> If science was to be born, nature had to be de-animized. Animism, which was always an essential feature of *pantheism*, had to retreat in the measure in which the monotheistic doctrine of creation was gaining ground. Animism was no match, in the long run at least, for the impetus of the doctrine of creation when the doctrine was taken in terms of the New Testament. Its doctrine on Christ as the "only begotten Son," in whom the Father created everything, put a damper on *any flirtation with the idea that any other being might be a divine begetting in terms of an emanationism* which always carries an animist touch.[3]

Another statement from Jaki helps to show his thought here:

> [Pierre] Duhem emphatically stated that [Jean] Buridan and [Nicholas] Oresme broke with the debilitating Aristotelian physics of motion by reflecting on what was demanded by the Christian dogma of creation in time and *out of nothing*. Whereas for Aristotle, a *pantheist*, the universe and its motions could have no beginning, for the Christian the world, if it came *out of nothing*, and in time, had to have a beginning even with respect to the quantity of motion contained within it. But since the Creator was *absolutely transcendental to the universe*, the latter's motion could no longer be

> conceived, as was done by Aristotle, as a *continuous participation in the divine*, but as a power given to it once and for all.[4]

Obviously two very brief statements can hardly be said to summarize the position of Fr. Jaki on the question of matter and God. Even so, these statements can at least serve to indicate what is at stake in the question. The universe is created out of nothing; God absolutely transcends all creatures and the universe as a whole; no creature can be compared to God. To say that God creates matter by giving something of Himself, or that matter is similar to God in its materiality, can quickly end up in idolatry, the worship of creatures. The danger of idolatry is that it cuts us off from God; but it also cuts us off from understanding the universe.[5] As Jaki insists, our knowledge of created beings, including the knowledge we acquire via experimental science, depends on our ability to recognize the distinction between creatures and the Creator.

This danger is very real. However, as with many other things, the danger is in two directions. If we confuse God with the universe, we get idolatry. If, however, in our efforts to guard against idolatry, we emphasize the otherness of God and the incomparability of God in a way that is not sufficiently careful, we can end up implying that *the universe is separated from God*, as an *isolated system.*[6] God is not a body, and God is not an object composed of particles; yet if we express this in the wrong way, we can end up thinking of the material, physical universe as something *alien* to God.

Both dangers have a tendency to reduce to atheism, because we can see and touch the universe, but we cannot see and touch God.[7] If we think of God as *indistinguishable* from the universe, the universe soon appears to be the only thing real, and God an anthropomorphic projection. So, too, if we think of the universe as *independent* of God, then God eventually seems superfluous, like a magical entity from some vague realm outside of what we can know through physics and chemistry. The danger of a supposedly independent universe reducing to atheism is not hypothetical; at least the last four centuries have been shaped by this reduction. Many millions of people regard modern science as having rendered the entire question of God hopelessly obsolete. A secularism that seeks to understand all things without God also promises salvation through technological mastery.

In terms of approach, the paper takes some things as axiomatic that cannot be argued for here, such as that God is Other than anything in the universe; that in any likeness of creatures to God there is always a *maior dissimilitudo*; that the creation of the universe is in no way necessary to God, as if God must create to be Himself or to develop Himself; instead, God is completely free to create, and He does not gain anything by creating, nor would He lose anything by not creating. Rather, creatures gain something, their very existence; that God does not change because God is concretely infinite; that the love of the three divine Persons is in no way dependent on creatures, even though creatures have been invited to enter into this love.

All that said, the paper will present a very limited version of the thesis that God creates matter by giving something of Himself. The claim here is that *material beings are analogous to God not in spite of their materiality, but by their materiality.* Therefore, when we say that God is 'immaterial', or affirm that God is not a body, this is true, but we should not take this to mean that God lacks matter, or that matter is something utterly alien to God, as if matter has some aspect of being that God does not have.[8] Rather, God is 'immaterial' in the sense that God is more than matter, transcending matter by including what is transcended. It is only because God transcends matter by including it that He can create matter and constantly sustain matter at every moment.

Further, the eternal generation of the only-begotten Son, and the eternal procession of the Spirit, are utterly different from the making of creatures, and are within God's nature. However, St. John and St. Paul say that creatures are made 'through' (διά) the Son because the making of creatures, which is free and in no sense obligatory for God, is analogous to the eternal gift by which the Father gives His own nature to the Son in the communion of the Spirit.[9] Creatures are loved by God because the Son enables creatures to share in the relation He has to His Father.

In trying to build a case, the paper focuses on St. Thomas Aquinas. At first glance, St. Thomas would seem to rule out any suggestion that God imparts something of Himself in creating matter. However, if we look again, we can see that the position he articulates is that *matter cannot possibly be alien to God*, even while *God is not in any way confused with creatures.*

I. Terminology of matter

Before getting to St. Thomas, a preliminary note on the terminology of 'matter'. For centuries, there has been a narrative that Aristotle's account of physics is fundamentally incompatible with modern physics, and that Aristotle's account is made obsolete by modern physics.[10] Although St. Thomas departs from Aristotle in many critically important places (e.g., teaching that the universe is not eternal, but created), certainly his discussion of physics is profoundly influenced by Aristotle. Hence, if Aristotle is fundamentally incompatible with modern physics, there would be much the same incompatibility between Thomas and modern physics. For this reason, someone may suggest that if we take St. Thomas as a guide in this question of matter and God, a choice is being made to ignore or reject modern physics.

This paper cannot address the enormous topic of whether Aristotle is incompatible with modern science. The paper is not rejecting a modern account of physics; when it asks about the relationship of matter to God, it intends this terminology of 'matter' in the sense used by modern physics (i.e., the Standard Model of particle physics, quantum mechanics, special and general relativity, etc.). However, the paper regards the narrative of a fundamental incompatibility as misleading and overstated, even if it is not without foundation. Although there is no room here to defend it, the paper starts from a presupposition that Aristotle's account of matter-potentiality in composite with form-actuality can be brought into contact with the modern account of energy, mass, particles, wave functions, space-time, etc., even while modern physics works principally in mathematics rather than in words, conducts controlled experiments, and takes extremely precise measurements. When the paper refers to 'matter' it means *That which has energy-mass*, in the conviction that when St. Thomas refers to an object as having matter, he is expressing something along the same lines.

Regarding the terminology of matter in Aristotle, and by extension in St. Thomas, a number of recent authors have argued that there is widespread misunderstanding of what is meant by potentiality/potency. This misunderstanding is that matter is *absolute* or *sheer* potentiality, with no content at all except for the content that it receives from form; therefore, matter is nothing, non-being.[11] But Aristotle's insistence that form cannot exist on its own without matter, and cannot produce a thing without matter, requires that matter must contribute something to the composite, and is not a sheer emptiness or lack of being. If the full reality of the composite is contributed by the form, and the matter contributes nothing, then the form would not need the matter to be real on its own.[12] So matter is not sheer or absolute potential; rather, matter has content and existence, even though it is always in a composite with form.[13] Matter brings properties into the composite; for example, a sword of steel is something quite different from a sword of plastic, even if they are carefully given an identical shape.[14] So potential/potency does not mean 'Not existing, but might exist in the future', but rather 'A thing not as fully developed as it can be'. An alternate way to convey the meaning of potentiality would be "incomplete."[15] Actualization means development to a complete state (for example: a seed develops into a full-grown plant); it does not mean something from nothing.[16] Hence when one encounters the terminology of 'matter' and 'material cause' in Aristotle, the meaning is 'What an object is made of', not 'That which is nothing until it is actualized by form'.

In short, then, the starting point here is that the terminology of matter as it appears in St. Thomas can be brought into engagement with the language of modern physics, even though the paper cannot provide an adequate defense of this approach, as that would require a full treatment of the topic of Aristotle and modern science.

II. Whether matter resembles God in its materiality

In asking the question of whether God creates matter by imparting something from Himself, or whether matter resembles God in its materiality, the initial answer we get from St. Thomas seems to be a very clear and resounding 'No':

> It is impossible that matter should exist in God. First, because matter is in potentiality. But we have shown (ST I, 2, 3) that God is pure act, without any potentiality. Hence it is impossible that God should be composed of matter and form. Secondly, because everything composed of matter and form owes its perfection and goodness to its form; therefore its goodness is participated, inasmuch as matter participates the form. Now the first good and the best—viz. God—is not a participated good, because the essential good is prior to the participated good. Hence it is impossible that God should be composed of matter and form. Thirdly, because every agent acts by its form; hence the manner in which it has its form is the manner in which it is an agent. Therefore whatever is primarily and essentially an agent must be primarily and essentially form. Now God is the first agent, since He is the first efficient cause. He is therefore of His essence a form; and not composed of matter and form.[17]

Thomas likewise argues that God is not a body, and that God does not enter into the composition of things.[18] In continuity with all Scripture and Tradition, Thomas insists that God is not the universe, and the universe cannot be confused for God. In God alone is essence the same as existence; God alone is infinite; God alone is eternal; God alone is everywhere; God is incomparable and not within any genus; God can be known, but no created intellect can comprehend God, even in the beatific vision.[19]

However, right from the outset we can begin to see that Thomas is not teaching that matter is alien to God or separated from God. He is very much aware of the position that would present God and matter as two co-eternal principles of the universe; it may be suggested that this position corresponds to Aristotle's understanding of an eternal universe, and the relationship between the First Mover and primary matter:

> As the first active principle is God, so the first passive principle is matter. Therefore God and primary matter are two principles divided against each other, neither of which is from the other.[20]

In a very significant departure from Aristotle, Thomas rejects the claim of an eternal universe, and argues that God is the Creator of primary/first matter. While matter and God may appear as if they were two co-principles, in truth God is the only principle, because *matter originates from God*:

> Passion is an effect of action. Hence it is reasonable that the first passive principle should be the effect of the first active principle, since every imperfect thing is caused by one perfect.[21]

But if matter is created by God, how should the relationship of matter to the Creator be understood? Is matter utterly unlike God because God is pure actuality, and matter is potentiality? An important clue is provided by Thomas' argument for the creation of primary/first matter:

> Whatever is the cause of things considered as beings, must be the cause of things, not only according as they are 'such' by accidental forms, nor according as they are 'these' by substantial forms, but also according to *all that belongs to their being at all in any way*. And thus it is necessary to say that also primary matter is created by the universal cause of things.[22]

It was indicated very briefly above that several recent authors have sought to show that in Aristotle's account of matter, potentiality is not nothing or non-being, but being that is incomplete or not yet fully developed. Even if Thomas would not necessarily follow these authors in every detail of their reading of Aristotle, we can see that he is in basic agreement at least on this point. Thomas must not understand primary matter as non-being or nothing if it is included in "all that belongs to their being in any way." Non-being neither needs to be created, nor can be created; if primary matter can be created, it is not nothing but something. Thomas presents primary/first matter as having being, albeit in the most imperfect or undeveloped way. So when Thomas says "It is impossible that matter should exist in God", this does not mean that matter is completely unlike God because matter is nothing and God is pure being, but rather that matter is incomplete or undeveloped, and God is utterly complete and in no need of development.[23] This is no small difference; in fact, it is everything in this question, because it means that matter must have a resemblance to God *as* matter, even though matter develops and God does not change.

To see this, we might start by looking at Thomas' argument that even while God transcends the universe absolutely, *He is not outside of the universe as if the universe were on its own*:

> No action of an agent, however powerful it may be, acts at a distance, except through a medium. But it belongs to the great power of God that He acts immediately in all things. Hence nothing is distant from Him, as if it could be without God in itself.[24]

It might be suggested that this is linked to the *prima via*, the argument from motion, which anticipates, or at least significantly overlaps with, what modern physics refers to as the law of conservation (i.e., conservation of energy, of mass, of momentum, of other intrinsic properties such as electric charge, spin, etc.).[25] The argument from motion turns on the fact that *being is conserved,* that energy, mass, momentum do not simply pop into existence from nothing, and therefore that the chain of the transfer of energy, mass, momentum requires a First Mover that does not receive these from any prior event. If the First Mover must be in constant, direct, and immediate contact with the universe to impart to it motion (energy, momentum, mass), then there is direct contact between God and matter, even while God transcends all things.

The position of St. Thomas is further clarified in his well-known argument that in God essence and existence are the same: i.e., God is existence, not something that possesses existence (as all creatures are). We find Thomas saying:

> Just as that which has fire, but is not itself fire, is on fire by participation; so that which has existence but is not existence, is a being by participation. But God is His own essence, as shown above (ST I, 3, 3). If, therefore, He is not His own existence He will be not essential, but participated being. He will not therefore be the first being—which is absurd. Therefore God is His own existence, and not merely His own essence.[26]

This terminology of *participation* in defining the meaning of the word creature is central to the question posed in this paper. When Thomas gives the illustration of fire, he implies that when fire ignites a piece of wood, it communicates *a share in itself* to the wood; incomparably more so, God brings creatures into being that previously did not exist at all by communicating *a participation in Himself* to them.

Thomas reaffirms the constant teaching of the saints that only the eternal Son is begotten. There can be no confusion between begetting and making; the Son is one essence with His Father, sharing eternally in the life of the Father without splitting the divine life into parts. And yet St. Thomas uses this terminology of participation to indicate that there is no being at all without God, that there can be no other source for the being of creatures except God. Thus *the being of a creature is in some way from God's being, and this being is the deepest core of the creature, though the creature is not God.* St. Augustine famously prays, "You are more intimate to me than I am to myself [*interior intimo meo*]," speaking not primarily about a personal experience or an emotional state, but about the being of things: God is the absolute foundation of a thing, of all things; God is subsistent existence, communicating to all things a participation in existence.[27] Hence God is *inside* of everything in the universe, but never in such a way that everything is God.[28]

This is also why the creation of the universe is not only an event in the past, but a continual sustaining; if God is removed from a thing, there is no reality left, because God is existence:

> The being of every creature depends on God, so that not for a moment could it subsist, but would fall into nothingness were it not kept in being by the operation of the Divine power... God alone is Being in virtue of His own Essence, since His Essence is His existence; whereas every creature has being by participation, so that its essence is not its existence.[29]

God is infinite whether or not He creates a universe, and He acquires nothing by creating, but *the universe cannot be isolated from or independent of God,* because there is no universe except insofar as God is inside. An electric light plugged into a wall will immediately shut off if the current flowing through the wires in the wall is removed; incomparably more so, the being of things flows entirely from God. For the same reason, God acts in every action, not in the sense that God is directly controlling all events in the universe, but in the sense that the being of an acting thing is the basis and precondition of its capacity to act as a cause.[30] The primary cause (God) is not outside of secondary causality, but *within* secondary causality.

An especially important aspect of St. Thomas' position is his argument that God must be perfect, containing the perfections of all things.[31] 'Perfections' as Thomas is using the term here principally means forms—i.e., form is that aspect of causality whereby matter (which has being but is incomplete, unfinished, undeveloped) is brought to a state of full development, which is to say perfection. Because God is infinite and unbounded existence, the divine nature/essence contains all possible forms, so that the form of a thing is real in God before it is real in the creature. This is also why Thomas shows elsewhere that God knows all creatures in the act of knowing his own essence/nature.[32] But this passage is worth highlighting:

> God is existence itself, of itself subsistent (ST I, 3, 4). Consequently, He must contain within Himself the whole perfection of being... Since therefore God is subsisting being itself, nothing of the perfection of being can be wanting to Him. Now all created perfections are included in the perfection of being; *for things are perfect, precisely so far as they have being after some fashion.* It follows

> therefore that the perfection of no one thing is wanting to God.[33]

Thomas contrasts matter with perfection, as perfection is actuality, and matter is potential:

> A material principle is most imperfect. For since matter as such is merely potential, the first material principle must be simply potential, and thus most imperfect. Now God is the first principle, not material, but in the order of efficient cause, which must be most perfect. For just as matter, as such, is merely potential, an agent, as such, is in the state of actuality. Hence, the first active principle must needs be most actual, and therefore most perfect; for a thing is perfect in proportion to its state of actuality, because we call that perfect which lacks nothing of the mode of its perfection.[34]

Insofar as matter is potential, and therefore by definition incomplete and not yet fully developed, it is not perfect, and so it would seem at first that "contains all perfections" does not include matter, only all forms. But it is clear from Thomas' argument on the creation of primary/first matter that Thomas does not regard matter as absolute non-being. God's containing of all perfections must necessarily include a containing of matter, even though matter is "most imperfect" of all things. If matter has being at all, even in the least developed and least complete way, then it must have perfection, because "things are perfect, precisely so far as they have being after some fashion."

So if all things exist only by participation in the infinite existence of God, if God is for that reason inside each thing at its very core (even though God is not the thing participating but is Himself), sustaining them in being at every moment from the inside out, and if "all perfections" must include even that which is most imperfect, least developed, should we affirm that *God is giving things their very materiality, their reality as material beings, by the activity of sharing with them a participation in His own being*? And if so, should we affirm that *in containing all perfections, God contains matter and materiality, not in the sense that God is incomplete or undeveloped, but in the sense that material beings have God's own being as the core of their being, not in spite of materiality but as material*?

But if this is what Thomas is saying, how are we supposed to understand his very direct statement that there is no matter in God, and that God does not enter into the composition of things?[35] How can God sit at the core or foundation of the being of things, if God is not the universe and is not confused with any creature? Some clarity is provided here by his teaching on God as the *exemplar cause* of all things.

St. Thomas treats the topic of creation by individually looking at the four causes, or four aspects of causality.[36] Thomas argues that God is the efficient cause of all things (i.e., that which brings them into existence), and God is the final cause of all things (i.e., the goal towards which they are directed).[37] However, his discussion of formal and material causality is somewhat different. We have seen above that Thomas departs from Aristotle by teaching that God is, and must be, the Creator of primary/first matter. But Thomas does not say that God is the material cause of things, as he had said for the efficient and final cause, because to say material cause would mean that God was unfinished or incomplete, thereby denying that God is concretely infinite, that He is eternal, and that He does not change.[38]

However, in asking about formal causality, St. Thomas argues that the forms of all things are patterned after the perfections that are eternally in God's own being.[39]

Thomas does not describe God as the formal cause of things, even though he is saying that the forms of things are present, and real, in God's nature before they are in creatures. Instead, Thomas uses the terminology of *exemplar* cause, rather than saying formal cause, in order to stress that God remains distinct from things and transcends them utterly, even while imparting to them the very form that makes them what they are. As the exemplar cause, God does not enter into the composition of a thing, but its form is still patterned on God and exists in God first before it is in the creature. By means of the terminology of exemplar (model, blueprint, design), *St. Thomas is able to account for the presence of God inside everything, as things only exist by participation in God, and also account for the transcendence of God beyond everything, because God is not the universe, and the universe is not God.*

All of this is closely tied to Thomas' teaching on *analogy*. Having established that God contains the perfections of all creatures, because if God is the source of the entire being of a thing, then all aspects of this thing must already be present in God, Thomas goes on to say that therefore all things must have a *resemblance* to God, not an external resemblance, but a resemblance within their very reality, in the perfections that makes them the particular type of being they are.[40] It is only by this resemblance that humans are able to speak about God at all; without this resemblance to God in the creature, true speech about God would be impossible. All words come from creatures, and when we apply these words to God, the words are neither univocal (same meaning) nor equivocal (utterly different meaning), but analogical (the word applies to God primarily, yet creatures have a true similarity to God).[41] This doctrine of St. Thomas is easily misunderstood as if it were only about human words; but because the resemblance to God is within the creature, because its being is a participation in God's being, it is not principally words or concepts that are analogous. Rather, *it is the creature itself, as a reality, which is an analogy to God,* and *only because the creature is an analogy can human words drawn from the creature be analogous.* So once more: what Thomas is saying about exemplar cause and about analogy indicates both that the being of any creature is in God before it is in the creature, so that God is the core of that creature, but also that God is distinct from the creature and transcends the creature.

If Thomas makes use of the terminology of 'exemplar cause' to designate a singular manner of formal causality unique to God, by which God contains the form of a creature and communicates that form to the creature, without Himself entering into the composition of the creature, is there any basis for saying something similar with regard to the material cause? In Aristotle's account of physics, 'matter' indicates being that is incomplete, unfinished, undeveloped. God is not in any way incomplete, which is why Thomas says that there is no matter in God. Meanwhile, form is actuality, completion, perfection; and Thomas says that God is pure actuality. So someone might say that trying to propose an 'exemplar material cause' is a flat contradiction, a misunderstanding of the point of the word 'matter'.

But again, we are trying to do some 'translating' of St. Thomas' discussion of matter into the terminology of modern physics. If we take full account of everything that Thomas has argued about the way that God communicates a participation in His own being to creatures in order to give them existence, the way that God is in all things because the deepest core of the thing is its existence, and if we have already seen above that Thomas does not understand matter as a sheer potential for being, as if matter was non-being or nothing, then is it possible to affirm (without requiring this paper to suggest a fitting terminology) that there can be something like the exemplar cause, but with regard to material causality?

To be even more specific: By comparison to the way that the terminology of 'exemplar cause' allows Thomas to express simultaneously (1) that the forms of all things are contained in God, so that the forms of creatures are real in God before they are real in creatures, and (2) God is not the universe, and does not enter into the composition of the creature, *can it be argued that the matter of which creatures are made—not in the sense of incompleteness, but in the sense of that remainder of being, however imperfect, which allows matter to contribute something to the composite with form—is in God before it is in creatures, without God entering into the composition of the creature as its material cause*?

III. Is energy-mass present in God?

To put the question of the paper in the terminology of modern physics: Can it be argued that energy-mass is in God before it is in creatures, even though God is not the universe and is not composed of particles? This is dangerous territory, because this attempt to roughly 'translate' Aristotle's account of physics into the account of modern physics, and perhaps vice versa, can very quickly end up in grave theological errors, in addition to scientific absurdities.

What is suggested can be stated as follows: In the terminology used by modern physics, 'energy' and 'mass' (energy the more basic term) designate *That of which things are made*, thus overlapping with Aristotle's use of the terminology of 'matter' and 'material cause'. Starting from an attempt to 'translate' St. Thomas' position into modern physics terminology, we might say that in the account of St. Thomas, energy-mass flows entirely from God, because the being of a creature is a participation in God's being, and thus God is the innermost foundation of the being of a creature, in every aspect. Therefore, while there is nothing incomplete or unfinished in God, nevertheless, energy-mass exists only by participation in God, and therefore must be a resemblance to God, an analogy to God.

If we think about the universe as an isolated system, created out of nothing by God but independent from God in its inner operations, we have missed the dependence of matter on God for its existence *as* matter. If we think about material objects as entirely alien to God, even if we recognize that God has the power to create them, we are thinking about the situation precisely backwards. As stated in the introduction above, God is immaterial in the sense that He is not a body, that He is not an object made of particles; but God is never immaterial in the sense of lacking something that matter has; He is *more than* matter, because He *includes all that is transcended.* The most concrete and solid of material objects are only an analogy to the infinite actuality of the divine essence. St. Gregory of Nazianzus famously responds to those who would deny the human mind of Jesus by reminding them, "That which He has not assumed He has not healed".[42] But a similar principle applies in the creation of the universe: *That which He has not contained He has not created*; yet He contains all the perfections of being, and although matter is the least perfect (i.e., the least complete, the least finished), He contains it as well. He cannot impart energy-mass to the universe if these are not present in Him *incomparably more* than they are in the universe.

Again, if we understand 'non-physical' to mean that God is alien to what we research by physics, we have our thought upside down. His φύσις or nature is the cause of all possible finite natures, and the physical objects that we experience by our senses and research through science (whether stars, animals, plants, rocks, etc.) are merely an analogy to the essence/nature of God, as *God is more physical* than the objects that are from Him and continually sustained by Him. What modern science means when it speaks about 'energy' and 'mass' are in God *first*, and only in creatures because God freely gives things a participation in existence, which is, properly speaking, His alone. But if it makes sense to bring something like St. Thomas' language of exemplar cause into the consideration of material causality, then God communicates energy-mass to things without entering into their composition, much as God communicates form to things without entering into their composition. The created thing has its energy-mass entirely from God, but the created thing is not God, and all created things together as a whole are in no way comparable to God.

In conclusion, the paper has argued, by drawing on St. Thomas, that *matter exists only by a participation in God, and therefore God in a real way creates matter and materiality by giving something of Himself.* More specifically, the argument of the paper depends on St. Thomas' teaching that God creates primary matter (which means that primary matter must have some being, even if it is a very limited and incomplete being), and his teaching that God contains all perfections. If the thesis is true, then God is *more material, more physical* than any created reality. The universe is really distinct from God, but we have badly misunderstood if we think of the universe as an isolated system, operating without God after He creates it. In an effort to proclaim God's absolute transcendence, as

Christianity must, we can convince ourselves that the universe is alien to God's being.

None of this removes the danger of idolatry. If we imagine that everything is God, if we confuse God with the universe, and confuse the begetting of the Son and the procession of the Spirit with the making of creatures, we have mistaken participation in existence for the One who freely gives this participation, and have turned away from Beauty to obtain beautiful finite things. Ultimately, if we mentally collapse God into the universe, we will lose sight of God altogether.

In closing, we might draw out implications of this thesis for the doctrine regarding Jesus Christ. The central claim of the Gospel accounts is that the Son of God has lived with us not only alongside of matter but *as* matter: "And the Logos was made flesh, and dwelt among us".[43] The body of Jesus is not *a garment that the Logos wears*; the body is *Him*. His hands and feet and side are pierced; His own blood is shed, not the blood of a detached instrument that the Logos moves around and controls. The Resurrection, above all, makes clear that *the body is Him*, and not something that belongs to Him externally, because it is the whole Jesus who rises:

> See my hands and my feet, that it is I myself; handle me, and see; for a spirit does not have flesh and bones as you see that I have.[44]

The Logos absolutely transcends the universe that is made through Him, and yet the flesh is truly His flesh, not a puppet dancing on strings but His own true body; and these claims (utter transcendence, and the Incarnation) are presented side-by-side in the same New Testament books.

But the Incarnation cannot possibly mean that the Logos has changed into something He was not, or ceased to be what He always is, because if He changes then he is not concretely infinite. If the Logos does not change, then either the flesh is something that He wields as an instrument, or the flesh is His because it is *Him*. His flesh can only be Him if is not *alien* to Him, not utterly disconnected from what He is. It can only be Him if flesh is created through Him, and therefore if *flesh is His before it belongs to any creature*, because *matter is His before He imparts materiality to any creature*. The flesh that He assumes is able to be truly His without changing Him, because all material things are only real insofar as they are a participation in existence, received from God. The point is not a collapse of the begetting of the Son into the Incarnation, as if the Son must become flesh to be Himself; rather, He freely empties Himself out of love for the human creature. And the point is not that the begetting of the Son is blurred together with creation, as if the entire universe were somehow the Incarnation of the Logos. Jesus alone is the Incarnation; He alone is the Lord dwelling with us, and all things were made through Him.

Notes

1. 2 Macc 7:28: "I beg you, my child, to look at the heaven and the earth and see everything that is in them, and recognize that God did not make them out of things that existed." All quotations from Scripture in the paper are the *Revised Standard Version, Second Catholic Edition.*
2. Stanley L. Jaki, *Science and Creation: From Eternal Cycles to an Oscillating Universe* (Real View Books: New Hope KY, 2016), xiv.
3. Stanley L. Jaki, "The Physics of Impetus and the Impetus of the Koran," in *The Absolute beneath the Relative and Other Essays* (Lanham, MD: University Press of America, 1988), 151 (italics added).
4. Stanley L. Jaki, "Science and Censorship: Hélène Duhem and

the Publication of the Système du monde," in *The Absolute beneath the Relative and Other Essays* (University Press of America, 1988), 185 (italics added).

5. Cf. Pope St. John Paul II, *Evangelium Vitae*, encyclical letter (March 25, 1995), 22: "By living 'as if God does not exist,' man not only loses sight of the mystery of God, but also of the mystery of the world and the mystery of his own being."

6. An 'isolated system' is one in which neither mass nor energy is exchanged with the outside of the system. No perfectly isolated system can be achieved under experimental conditions, but the concept is useful in physics.

7. Cf. 1 Jn 4:20.

8. Several recent Thomist authors have addressed the terminology of 'immaterial.' For example, James Doig reiterates that immateriality means immobility/unchangeability: "Since what does not move is immobile or immaterial, the physician has discovered an immaterial first mover. Given these many and varied texts of the *Commentary*, it would appear correct to understand the birth of metaphysics to be synonymous with the discovery of the immaterial mover" (See James Doig, *Aquinas on Metaphysics: A Historico-Doctrinal Study of the Commentary on the Metaphysics* [The Hague: Martinus Nijhoff, 1972], 246); Daniel Toma emphasizes that immateriality is spirit and mind, and the implication of the hierarchy of being is that the material universe develops to a point of immateriality: "The Thomistic understanding of the hierarchy of nature shows that the material and immaterial are not opposites or irrelevant to each other. In that framework of thought, what is higher on the hierarchy contains or includes what is lower on the hierarchy. Thus, animals with their memory in a small way approximate human beings, although they lack true intelligence. Humans are the material pinnacle of the hierarchy because they share in the material world with other organisms, while they are also able to transcend that world with their immaterial intellects. Life is more than survival or reproduction. Life tends toward intelligence." (See Daniel Toma, *Vestige of Eden, Image of Eternity: Common Experience, the Hierarchy of Being, and Modern Science* [Steubenville, OH: Franciscan University Press, 2019], 122); William A. Wallace argues that modern science will not talk in

terms of immateriality, but will seek to avoid the immateriality it repeatedly discovers by redefining matter: "The concept of immateriality has not been eliminated from scientific thought, even though the word itself may rarely appear in its literature. Now it may be noted that the ways in which immateriality reenters into the thoughts of scientists are not particularly helpful for advancing any of the three moves to transcendence just mentioned [i.e., study of life, study of mind, study of the first Cause of motion or change]" (See William A. Wallace, "Immateriality and Its Surrogates in Modern Science," in *Intelligibility of Nature: A William A. Wallace Reader*, ed. J. Hittinger, M. Tkacz, D. Wagner [Washington, DC: CUA Press, 2023], 234).

9. Jn 1:3, 1:10; 1 Cor 8:6; Col 1:16; Heb 1:2, 2:10.
10. Cf. for example, William A. Wallace, "The Intelligibility of Nature: A Neo-Aristotelian View," in *Intelligibility of Nature: A William A. Wallace Reader*, ed. J. Hittinger, M. Tkacz, D. Wagner (Washington, DC: CUA Press, 2023), 13: "The judgement of history would seem to be that [Aristotle] and his followers failed in their attempts to make nature intelligible to the human mind... Not until Galileo would anyone be found who could rectify these errors and bring about the scientific revolution during which the new physics and the new chemistry would be born."
11. Christopher Byrne, *Aristotle's Science of Matter and Motion* (Toronto: University of Toronto Press, 2018). The entire argument of this book is in response to the widespread misrepresentation that Aristotle regards matter as absolute potentiality or non-being.
12. Byrne, *Aristotle's Science*, 84.
13. A number of recent authors have addressed this, including C. Byrne (cited just above), Devin Henry (*Aristotle on Matter, Form, and Moving Cause: The Hylomorphic Theory of Substantial Generation* [Cambridge, UK: Cambridge University Press, 2019]), and Leo J. Elders, (*Reading Aristotle with Thomas Aquinas: His Commentaries on Aristotle's Major Works*, ed. J. Vijgen [Washington, DC: CUA Press, 2023]).
14. Byrne, *Aristotle's Science*, 85–86.
15. Henry, *Aristotle on Matter, Form, and Moving Cause*, 35: Aristotle responds to Parmenides and the other Eleatics by

"reconceptualizing the concept of *unqualified* non-being in terms of a kind of *incomplete* being" (italics in original).

16. Cf. Edward Feser, *Neo-Scholastic Essays* (South Bend, IN: St. Augustine's Press, 2015), 63: "Change, Parmenides argued, would have to involve being arising from non-being, something coming from nothing. But from nothing, nothing can come. Hence there can be no change. Aristotle identified the error in Parmenides' reasoning... what we have is not a transition to being from sheer non-being or nothingness. What we have is rather the actualization of a potential that was already present... For Aristotle, potentialities are real features of things alongside the ways in which they are actual, and that is why things are capable of changing."
17. All direct quotes from St. Thomas are from: St. Thomas Aquinas, *Summa Theologiae,* 2nd revised edition, trans. Fathers of the English Dominican Province (1920). References to *Summa Theologiae* are given in the paper as ST, part, question, article. This quotation is from ST I, 3, 2 (God not composed of matter and form).
18. ST I, 3, 1 (God is not a body); I, 3, 8 (God does not enter into the composition of things).
19. ST, I, 3, 4 (essence and existence); I, 7, 2 (infinite); I, 10, 3 (eternal); I, 8, 4 (everywhere); I, 3, 5 (incomparable, not in any genus); I, 12, 7 (not comprehend by any finite intellect).
20. ST I, 44, 2 (God creates primary/first matter).
21. Ibid.
22. ST I, 44, 2 (God creates primary matter – italics added).
23. ST I, 9, 1 (God is immutable).
24. ST I, 8, 1 (God is in all things).
25. ST I, 2, 3 (Five ways).
26. ST I, 3, 4 (God is not a composite of essence and existence).
27. St. Augustine, *Confessions,* trans. J.G. Pilkington, *Nicene and Post-Nicene Fathers,* First Series, vol. 1, ed. Philip Schaff (Buffalo, NY: Christian Literature Publishing Co., 1887), III, 6, 11.
28. ST I, 8, 1 (God is in all things).
29. ST I, 104, 1 (creatures continually sustained by God).
30. ST I, 105, 5 (God acts in every agent).

31. ST I, 4, 2 (God contains all perfections).
32. ST I, 14, 5 (God knows all things in knowing His essence) and I, 14, 6 (God has proper knowledge of all things).
33. ST I, 4, 2 (God contains all perfections – italics added).
34. ST I, 4, 1 (God is perfect).
35. ST I, 3, 8 (God does not enter into the composition of things).
36. ST I, 44 (four articles on causality of God).
37. ST I, 44, 1 (efficient cause) and I, 44, 4 (final cause).
38. ST I, 44, 2 (material cause).
39. ST I, 44, 3 (exemplar cause).
40. ST I, 4, 2 (God contains all perfections) and I, 4, 3 (creatures resemble God).
41. ST I, 13, 1 (human words can be used for God) and I, 13, 5 (all words for God are analogy).
42. St. Gregory of Nazianzus, *Letter to Cledonius the Priest Against Apollinarius* (Epistle 101).
43. Jn 1:14.
44. Lk 24:39.

16 EVOLUTION, TELEOLOGY, AND PURPOSE

Transhumanism and the Natural and Supernatural Ends of Being

Braden Molhoek, Ph.D.

From Aristotle to current times, people have debated about the nature of life and purpose in natural processes such as evolution. This paper will explore differing views on the question of teleology in evolution, including Stanley Jaki, O.S.B.'s critique of materialist Darwinism that sought to remove discussions of purpose from nature. Even though contemporary science seeks to present an understanding of evolution that is not teleological, transhumanism as a movement believes that humans are capable of taking control of evolution and should actively use technological means to improve the human condition.[1] Transhumanism is not a monolithic movement so there are some who seek to not only improve or enhance humanity, but to evolve beyond modern humans to one or more post-human species.[2] It is my estimation that Jaki would be quite confused by transhumanists, holding some things in common with them but also rejecting aspects of their thought as well. While not placing teleology or purpose into evolutionary science, transhumanism does appear to make a philosophical or perhaps even theological argument for purpose. If transhumanists are correct in their assertion that humans will be able to control evolution to the point of evolving new post-human species, I believe this raises interesting questions for theological anthropology. In the final section of the paper, I will examine the question of the natural and supernatural ends of humans and potential posthuman species and whether they would differ.

Aristotle's understanding of cosmology and biology, however, is problematic for modern science because "Aristotle's world is finite, unique, eternal, consisting of a finite number of eternally existent species."[3] Marjorie Greene also argues that teleology is another concept in Aristotle's understanding of biology that scientists have tried to reject only to have it continue to appear in discussion of biology and evolution.[4] Two clarifications about Aristotle and *telos* are important for this discussion according to Greene. The first of these is that *telos* does not imply a purpose or plan. What Aristotle is referring to is the natural sequence of events and actions that will occur unless something interrupts or interferes. So, a fertilized robin egg will lead to the hatching on a baby robin, this is the *telos* of the egg. This is mostly descriptive of what nature is and does, and trying to assign agency or purpose to this, such as divine purpose or action, "is to introduce a Judaeo-Christian confusion of which Aristotle must be entirely acquitted."[5] The second clarification is that Aristotle does not view the entire universe as progressing to a single *telos*. This view is again an assertion that comes from Christian views placed on top of Aristotle's view of nature.[6] It is possible, then, to think of Aristotle's notion of *telos* as resonating with the scientific field on ontogenesis, at least to some degree.[7]

Where Aristotle's notion of *telos* becomes untenable for modern science is phylogeny and evolution. Greene states that while it is possible to consider *telos* in embryology or the development of the individual, "it has been repeatedly claimed both that Darwinian evolutionary theory rejects any cosmic *telos* and that it retains the concept of *telos* in some more acceptably 'scientific' sense. Indeed, it has even been argued that it is precisely by virtue of its teleological structure that evolutionary theory, and only evolutionary theory, rescues biology from reduction to physics."[8] This is because evolution involves change, whereas Aristotle's notion of species was fixed. When evolution from one species to another can occur, the only possible *telos* that could remain would be survival, which would be shared by all species.[9]

There have been moves from evolutionary scientists to consider elements of purpose in the discussion of evolution. Ernst Mayr said that "[r]efutation of a finalistic interpretation of evolution or of nature as a whole, however, did not eliminate teleology as a problem in philosophy."[10] Mayr takes a position that I think Jaki would agree with, by saying that deistic teleologists (which I think should also include any theistic understandings of evolution), see *telos* as a cosmic goal.[11] Jaki says the Christian understanding of purpose comes from Moses' last instructions that is built upon in the eschatological view of the prophets. It is in these texts he says, where "the personal God, the Maker of Heaven and Earth, appears even more powerful through his ability to make a new Heaven and Earth. That new set-up is a lasting reward for those who obeyed God's commandments and a lasting punishment for those who disobeyed them."[12] The idea of reward and punishment "becomes an emphatic doctrine in the New Testament. There the personal appearance of God in a visible form puts an even sharper eternal perspective the purpose which free human acts are meant to implement. That purpose is a supernatural and eternal sharing in God's life."[13] Jaki here is agreeing with Aquinas in seeing the Beatific Vision as the supernatural end to humanity, and the *telos* in terms of goal of people and the evolutionary process.

The other use of *telos* is endpoint, and this is what Mayr sees as the proper use in biology. In this case *telos* refers "to the termination of an end-directed process."[14]

Teleology is too strong of a term to describe the endpoint of natural phenomena. Everything in the physical world is capable of changing, but these changes are subject to natural laws. Mayr calls these processes "teleomatic" and argues that even Aristotle made a distinction between teleomatic processes and teleological processes.[15] Teleonomic processes, on the other hand, "one that owes its goal-directedness to the operation of a program. The term teleonomic thus implies goal-direction of a process or activity."[16] This most often applies to the world of living organisms in which goal-directed processes seem ubiquitous. Teleonomic behavior has two parts, Mayr says. It is "guided by "a program" and it depends on the existence of some end point, goal, or terminus which is foreseen in the program which regulates the behavior."[17] Programs, according to Mayr, are material and exist before a teleonomic process starts. Mayr gives a conditional definition of a program "as coded or prearranged information that controls a process (or behavior) leading it toward a goal. The program contains not only the blueprint of the goal but also the instructions of how to use the information of the blueprint."[18] This does not apply to aspects of evolution such as natural selection.

Mayr refutes that there is any cosmic teleology in the science of evolution but says the question that remains is "how can one explain the seemingly upward trend in organic evolution?"[19] Mayr argues that Darwin's understanding of natural selection can still answer that question, even if natural selection itself is not teleonomic or teleological. At first people believed that evolution was teleological, because it led to improvement or progress. Though this perhaps made more sense with Lamarck's understanding of evolution, Mayr argues that Darwinian evolution refutes this idea completely. This is because:

> natural selection is an optimization process, but it has no definite goal, and considering the number of constraints and the frequency of chance events, it would be most misleading to call it teleological. Nor is any improvement in adaptation a teleological process, since it is strictly a post hoc decision whether a given evolutionary change qualifies as a contribution to adaptedness.[20]

Since natural selection works at the generational level, it does not have a sense of a long range goal, but given the lengths of time involved in evolution, looking back through all of the generations, it can appear to have a long range goal of progress or improvement.[21] Another reason that it could appear to directly lead to progress is that it is "a mechanism that would favor the rise of ever better adapted species, it would favor the invasion of new niches and adaptive zones, and as the end-result of competition of species favor developments that are best described as advanced types."[22] It is also possible that the origin of the term "survival of the fittest" and its incorporation into Darwinian evolution has contributed to confusion people have had.

Francisco Ayala argues that it is not survival but adaptability that is central to evolution. Ayala provides an interesting perspective because he was a former Dominican priest as well as a scientist. After studying under Dobzhansky, Ayala's work as geneticist led him in a variety of directions. In 2001 he received the National Medal of Science and in 2010 he won the Templeton Prize from the John Templeton Foundation. He was also a scientific expert in legal cases about the teaching of creationism, intelligent design, and evolution in public schools.[23] He also believed that teleology has a place in biology, in certain and specific ways.

Organisms are adapted to their environments and the parts of organisms are adapted to particular functions. Darwin saw this, Ayala says, and was able to articulate a natural understanding for these observations. In fact, he asserts that for Darwin:

> [o]ne of his greatest accomplishments was to bring the theological aspects of nature into the realm of science. He substituted a scientific teleology for a theological one. The teleology of nature could now be explained, at least in principle, as a result of the natural laws manifested in natural processes, without recourse to an external Creator or to spiritual or non-material forces. At that point biology came into maturity as a science.[24]

Ayala admits that modern science struggles with teleology, but that it need not be the case. Teleology, he says, is "in general disrepute in modern science. The main reason for this discredit is that the notion of teleology is equated with the belief that future events are active agents in their own realization."[25] This does not have to be part of the definition of teleology. In fact, given the importance he places on teleology in the previous block quote, it should not be a surprise that Ayala thinks teleology can be useful and needed in certain areas of science, especially biology.[26]

There are three areas where Ayala believes it is appropriate to discuss teleology. The first of these is when an agent, either a human or another animal, anticipates a goal consciously.[27] The second area is when discussing self-regulating systems, such as the maintenance of body temperature in mammals. The third example is in bodily structures that are designed for a purpose, such as a hand for grasping, or an eye for vision.[28] These are all internal examples of teleology. Evolution as a process does not an external teleology, such as a cosmic teleology.[29]

Philosopher of science Michael Ruse engages with Ayala's understanding to teleology and tries to identify areas of agreement and disagreement when it comes to their views of teleology and evolution. Ruse sees a great deal of overlap in their understandings but argues that human teleology can have a second form. This second form is teleology that humans give to objects or artifacts.[30] Ruse does believe, however, that in terms of design or function in biology, that analogy or metaphor is a better way to think about teleology because he argues we are really thinking of adapted structures as a kind of artifact.[31]These arguments are not meant to be a criticizing of Ayala's perspective, but attempts to clarify some of his themes better and to build on his work.[32]

Soontiëns argues that the "founding fathers" of the modern theory of evolution, which included Ayala's advisor Dobzhansky (as well as Huxley, Mayr, and others) reject teleology because the fossil evidence does not point to a single path in evolutionary history that leads to one progressive goal,[33] what has been referred to previously as a cosmic *telos*. Another reason these thinkers rejected teleology is that chance seems to contradict *telos*. Chance is essential "in mutation, in recombination, in genetic flow, in the pairing of chromosomes and in the mating of gametes and organisms," to the point that Monod attributes all novelty to chance.[34] However, Soontiëns believes that these arguments are insufficient for an outright rejection of teleology. Teleology does not have to mean a single path to a predetermined goal. In fact, searching for a solution often results in a non-linear path.[35] Chance does not have to exclude notions of teleology either. A good example of this, Soontiëns says, is games, which have a goal in terms of winning, but also have ample room and a role for chance.[36]

Stanley Jaki was critical of Darwinism because it was reductive and removed purpose from nature. In *Angels, Apes, and Men* [sic], Jaki says that Darwinism cannot address Huxley's question of whether it is the fittest that survives, or the most ethical "because the only answer—the recognition of a spiritual order in man [sic] who is more than mere matter—cannot come within their ken, a ken constructed to the specifications of strict materialism."[37] Since Darwinism is a materialist position Jaki argued that "the allegiance demanded by Darwinism was simply allegiance to the absence of all norms. It beckoned toward unfathomable whirls in which one was no more than flotsam hurled round and round by the blindest of blind fates."[38] Jaki raised concerns about human nature that are still relevant in today's discussions of emerging technologies. In the final section of this paper I will discuss transhumanism's own understanding of *telos*. It is my estimation that Jaki would be quite confused by transhumanists, holding some things in common with them but also rejecting aspects of their thought as well. Assuming transhumanists are correct in their belief that they could bring about posthuman species, I think the interesting question for Christian anthropology is to explore how we might consider distinctions between the natural and supernatural ends of humans, transhumans, and post-humans.

From the Jaki quote above it is clear that he saw danger in understanding evolution as a process lacking all sense of purpose, comparing people to flotsam completely subjected to absolute randomness. The modern synthesis, he argued, did not take any steps forward regarding purpose compared to the orthodox theory of evolution. Quoting John Dewey, he says "The Origin of Species foreswears inquiry about absolute origins and absolute finalities."[39]

This is a problem Jaki asserts because purpose is inherently important to biologists. Just as "free will should cause nightmares to a consistently materialist determinist...[m]uch the same is true about purpose as far as the Darwinist biologist, orthodox or synthetically reformed, is concerned. As a Darwinist the biologist cannot countenance purpose, but as a biologist he cannot live without it. Consequently, the Darwinist has to live with purpose, or its study, teleology, in a clandestine matter."[40] If evolution is lacking completely in any sense of purpose, Jaki thinks the real question is "How a process void of plan can issue a being, man [sic], whose very nature is to plan, which is always a pursuit of purpose. Much less willing to fathom the question: If there was no absolute finality or purpose, in what sense could purpose be spoken of and with a reliable meaning?"[41]

In my opinion, Richard Dawkins tries to address questions like this in chapter eleven of his book, *The Selfish Gene*. He says that his word choice was intentional to try and include all living beings that have been through the process of evolution in his argument, and that it would take something distinct about a species to be excepted from the control of the selfish gene. He believes that humans are unique in a way that they alone can overcome the selfish gene.[42] Although humans are not the only species with culture, Dawkins does say that "[i]t is our own species that really shows what cultural evolution can do."[43] It is in this argument that even Dawkins refers to progress, both in cultural evolution and biological evolution. Modern science, he argues, is better than older science, and not only does our way of understanding the universe change over time, but that it gets better.[44] While he says that cultural evolution may look like genetic evolution, they are unrelated. However, "[a]s in genetic evolution though, the change may be progressive."[45] He goes on to say that Sir Karl Popper has also highlighted the "analogy between scientific progress and genetic evolution by natural selection."[46] The evolution of life, Dawkins argues, centers around the survival of replicating entities and that for culture he is going to call the replicating entity a meme, the unit of cultural transmission.[47] Though Dawkins is optimistic that humans are capable of true altruism, he does not build an argument for it. Instead, he says that even if we assume the worst, that humans are selfish individuals, "our conscious foresight—our capacity to simulate the future in imagination—could save us from the worst selfish excesses of the blind replicators…[w]e have the power to defy the selfish genes of our birth and, if necessary, the selfish memes of our indoctrination."[48] In this kind of argument I see Dawkins attempting to address the question that Jaki asks. Self-consciousness, creativity, and conscious foresight, it seems are characteristics that are necessary in order to have a being that is capable of planning, and that such a being, while is the product of biological evolution, in some ways is able to transcend that evolution as well. It is this slight optimism that is expanded radically in the transhumanist movement.

It is interesting to me that it appears that some notion of progress in evolution or teleology is a meme that has long managed to survive in human culture, even discussions of evolutionary science. While this may not be a comprehensive cosmic *telos*, the idea that natural processes are leading to some kind of end, even if it is not designed by a creator or agent, still holds some attraction or explanatory power. While Aristotle saw teleology within individuals and within unchanging species, evolutionary biologists also see teleology in the evolution of species. Transhumanism also believes that humans are now capable of directing human evolution in ways that would be beneficial for humanity. Humans can do this in the physical and social world, through actions and culture, but gene editing, computer-brain interfaces (CBIs), pharmaceuticals and other technologies allow humans to do that to their bodies or human nature.[49] While it is possible to compare transhumanist views to a cosmic view of *telos*, I do not think that is the best description. The focus of transhumanist thought is on reshaping humans, but "progress occurs when more people become more able to shape themselves, their lives, and the ways they relate to others, in accordance with their own deepest values."[50] This could allow for a variety of ways of people existing in the world, with some choosing to not change at all on one extreme, to others choosing to alter themselves to the point of being posthuman, that is, "possible future beings whose basic capacities so radically exceed those of present humans as to be no longer unambiguously human by our current standards." The potential existence of posthumans does not mean that humans would no longer exist. In fact, the most likely outcome would be a world with humans, transhumans (humans who enhanced existing traits or introduced new traits into their bodies), and posthumans. Though transhumanists claim that enhancement would be a choice, and choosing to not do so would be respected,[51] I think there are concerns about the coexistence of such beings in a world of finite resources.

At this point I think I want to return to Mayr's discussion of "survival of the fittest" because when combined with Dawkins' perspective on human agency and transhumanism, it raises ethical questions about social pressures to change or conform. Historians trace the term "survival of the fittest" to Herbert Spencer, with the term first appearing in *Principles of Biology*, a book he published in 1864. Spencer's own view of how evolution worked was teleological, and that it "was at work throughout the cosmos and applied to the creation of stars as much as to life on Earth as everything was evolving from simpler forms into more complex ones."[52] Spencer thought that Darwin's notion of natural selection was a useful description of the mechanism that led to improvement. Unlike Darwin, Spencer believed that natural selection applied to a much wider range of fields than biology. In *Principles of Biology,* he argued the concept was also useful for economics[53] but ultimately felt that evolution and natural selection also operated on society as a whole. His thought was distinct from Darwin in that he believed "all societies evolve from more primitive, simpler forms to sophisticated ones, eventually reaching an evolutionary 'end-point'—a final state of perfect equilibrium—at which stage natural selection then stops."[54] It was this perspective that led Spencer to help develop what became known as "social Darwinism,"[55] a position that applied "survival of the fittest" to many facets of human society.

Social Darwinism has been condemned as a form of eugenics, but there has been some disagreement over its origins. In 1944 Richard Hofstadter published *Social Darwinism in American Thought.* In this book, he defined Social Darwinism "as an ideology using the competitive view of nature and Darwin's concept of struggle for existence as a basis for social theory."[56] Scholars see two perspectives from Hofstadter in his book, the first of which saw Social Darwinism as a justification of laissez faire economics. This was more of an individual competition. The latter perspective in the book saw a difference between individual and collectivist competition and raised the issues of racism, eugenics, and imperialism.[57] Although Hofstadter in his book claimed that Social Darwinism was dead in America by 1918, scholars are saying this is not the case. Instead, the argument is, the average person would not have heard of the term much until 1916, and that JSTOR has roughly 50 uses of the term from 1916 to 1943, and 4258 citations from 1944 to 2009.[58]

Generally speaking though, people took a perspective that would be similar to Spencer's thought, that insights from the natural sciences could be applied to the social sciences. However, there is an alternative argument that makes the case that "what was specific about much of Social Darwinism resulted from several shifts in thought in mid-Victorian Britain to which Darwin himself also responded and which therefore also vitally influenced his own development."[59] Gregory Claeys argues that trying to blame Spencer alone for the development of Social Darwinism would be dismissing the development of ideas for at least half a century. A more defensible position would be to say that "Darwin's discoveries occasioned no revolution in social theory, but instead involved remapping, with the assistance of a theory of the biological inheritance of character traits, a preexisting structure of ideas based largely, though not exclusively, upon a Malthusian and economic metaphor of the "struggle for existence."[60] A full explication of Social Darwinism is beyond the scope of this paper, but at least a summary of the core beliefs will remain relevant for the final section of the paper.

There were four assumptions that made up the world view of Social Darwinism. The first is that biological laws govern nature, including humans. The second is that in addition to the biological struggle for existence created by physical finitude, increases in population also make resources scarcer, which creates a struggle for existence. The third assumption is that "physical and mental traits confer an advantage on their possessors in this struggle, or in sexual competition, which advantages can, through inheritance spread through the population."[61] Finally, due to inheritance, the cumulative effect of selection is what leads to the creation of new species and the extinction of others.[62] This allows for a stronger view of purpose or progress in Social Darwinism than biological Darwinism, which makes sense given the discussion thus far, and also supports why transhumanists would still support some of the assumptions of Social Darwinism.

It should not be surprising, then, that many nonexperts continue to misunderstand what is meant by natural selection of survival of fittest. Stanley Fields argues that natural selection:

> is a simple statement of the fact that in dangerous circumstances, only those individuals most adapted to their environment survive—and the world, with its limited food supply, fearsome predators, and devastating diseases is always a dangerous place. An individual born with a physical or cognitive difference that renders it stronger or smarter or faster or sharper-eyed—in short, better adapted than its peers to the conditions it is confronting—will be more likely to live and mate and pass on its advantage to the next generation (through, as we now know, its genes).[63]

It turns out that Darwin included the term "survival of the fittest" in the fifth edition of *On the Origin of Species*, which was published on February 10, 1869. The reason he chose to include the phrase and give credit to Spencer for it is because he believed that the term "natural selection" implied agency, that something must be doing the selecting, whereas "survival of the fittest" did not. Survival of the fittest, however, "What he did not mean was what the phrase has come to mean today, that is 'in the best physical condition.'"[64] For nonexperts, viewing fittest as best or superior seems to hold some vestige of progress or improvement in it, despite how the scientists argue that it not necessary or accurate to think this way.

At this point, I want to quickly return to Jaki's question of how it is possible for humans, who are beings of purpose, to come about through a process of evolution that science claims is utterly devoid of purpose. This is because I think the issues of Social Darwinism, progress, and purposeful beings as the result of evolution all contribute to the views and goals of transhumanists, which will be the focus of the final part of this paper. Transhumanism, according to futurist Nick Bostrom in the *Transhumanist FAQ*, is "the intellectual and cultural movement that affirms the possibility and desirability of fundamentally improving the human condition through applied reason, especially by developing and making widely available technologies to eliminate aging and to greatly enhance human intellectual, physical, and psychological capacities."[65] Autonomy is of utmost importance to transhumanists, allowing people to plan and find purpose as they see fit. While transhumanists typically seek to use technology to enhance human capabilities, they also make the claim that people should be able to choose to not undergo such changes if they so choose.[66] The focus is on technology, which is ultimately an extension of human reason. Jaki, like the theologians of today, raised the question of whether intelligence is the most important thing in terms of being human. In *Apes, Angels, and Men* [sic] he argued that "Intellect is only one chief distinction of being truly human. Another is good will, co-operation, which should seem even more mysterious than intellect in an existence which, within the Darwinian perspective is all 'red in tooth and claw.'"[67] While transhumanism is defined as applying human reason, the goal is to improve the human condition, and for some, to create one or more posthuman species through enhancement. In theory, Jaki and transhumanists would be able to find common ground on the natural end of humans.

Transhumanism, as an intellectual movement or philosophy, is also able to hold a sense of purpose or progress that science alone cannot. I would argue that transhumanism would support at least two of the assumptions of Social Darwinism, in particular that physical and mental traits can provide an advantage in the struggle for existence and that cumulative inherited change drives speciation. While transhumanists might agree that currently, population growth can lead to scarcity, but ultimately technology should be able to address this one way or another, so this would not have to be an inevitability. So not only is it a philosophy, but it is an optimistic philosophy, more so than Christianity when it comes to this world. Bostrom says that:

> While not a religion, transhumanism might serve a few of the same functions that people have traditionally sought in religion. It offers a sense of direction and purpose and suggests a vision that humans can achieve something greater than our present condition. Unlike most religious believers, however, transhumanists seek to make their dreams come true in this world, by relying not on supernatural powers or divine intervention but on rational thinking and empiricism, through continued scientific, technological, economic, and human development.[68]

This quote raises what is likely the most fundamental difference between religious perspectives like that of Jaki and secular transhumanists, where the ultimate end for humans, transhumans, and post-humans lies. For the transhumanists, it is in this world. As materialists, there is no supernatural end for humans or other species because the supernatural does not exist. For theists, on the other hand, the ultimate end for humans is our supernatural end, communion with God. The question, then, is whether transhumans and posthumans would have the same supernatural end as humans.

In this final section I will try and identify possible differences between modern humans, transhumans (modern humans who have undergone intentional changes to enhance existing traits or to incorporate new traits, and posthumans, new species that have evolved beyond humans and transhumans. I will look to the natural end and supernatural ends of such beings to see where there is continuity and where discontinuity could arise. Ultimately I believe there is more overlap between humans and transhumans than there is with posthumans. However, I am not convinced that posthumans necessarily have a distinct supernatural end from the other categories. I admit that this part of the paper is the most speculative and look forward to further discussions on how to further clarify these categories.

I would argue that humans and transhumans still share a common end or *telos*. Transhumanism may enhance human capabilities, perhaps even greatly. It could also introduce new traits, characteristics, or abilities into human nature, but they still share a common purpose. We still consider the humans that exist today to be the same species as *Homo sapiens* from 300,000 years ago. Our abilities have changed drastically in that time, predominately through cultural evolution including technological developments. I would argue that neanderthals, modern humans, and other closely related hominids shared a general sense of what flourishing meant for these related species. Although humans and transhumans share the same end, I would also argue that their list of intellectual and moral virtues do not have to be the same. If Aristotle defines virtues as a mean between an excess and a deficiency, I think it is possible that human enhancement could alter certain human abilities to the point where comparing a transhuman to an unenhanced human would not make sense. Effectively, nonenhanced humans could seen as being completely deficient of something that has been enhanced greatly in transhumans. One possible example of a human virtue that is not a transhuman virtue is courage. For example, if transhumans have computer-brain interfaces that are able to "back-up" memories or consciousness and allow those to be downloaded to a different body, like in the Netflix series *Altered Carbon*, then courage does not really make sense as a virtue anymore. There is no need to fear biological death in almost all circumstances in this case. Life could be seen more like video games, where people can just try again if they fail.

What is harder to imagine is what possible new virtues transhumans could acquire. Elsewhere I have speculated about connectivity being a transhuman virtue.[69] If transhumans have CBIs or other cybernetic implants, connectivity would be the virtue dealing with how organisms interface with one another. Analogously this could build on USB verses Apple's Lightning. The excess of compatibility would be assimilation, in which there is no choice or diversity in how to interface with others, perhaps one company owns the patent on connectivity. The deficiency of connectivity would be isolation, where connectivity would not be possible.

So what kind of distinction would there be between humans and transhumans on the one hand, and posthumans on the other? This is something I have struggled with for a while and my tentative conclusion is that it would take a fundamental change to how a species interacts with the world. Looking to evolutionary history for such an example, I would say that verbal language and bipedalism could be examples of this, but I am more confident in bipedalism and so will focus on that example. Bipedalism allowed organisms to have free use of their arms, for tool making and use, and for social signaling and communication. It likely made them easier to spot by predators because of standing upright, but that also could have allowed them advantages in fighting predators and others of their ingroup for mating opportunities. The number of physiological changes that accompany bipedalism or are possible because of it, I would argue are sufficient to say there is also the possibility of a shift in terms of the natural end or *telos* of an organism or species. It is difficult to predict what kind of future shift might be possible. The most obvious, in my estimation, would be if mind uploading were possible. Existing as a digital entity that could choose to remain disembodied or download temporarily into a body would be a very different way of existing in the world compared to modern humans. Humans would not be able to reproduce with posthumans in these examples, another example of a strict division between the two categories. Other examples could be finding ways to exist in places where modern humans cannot, such as being able to breathe permanently underwater and the physiological changes that would accompany such a change.

Finally, I recognize this is only a concern for theistic understandings of human nature, but what complicates the distinctions I just tried to make is whether humans, transhumans, and posthumans share a common supernatural end. It is difficult for me to envision a distinction supernatural end for these different groups, assuming they are all rational, social beings capable of self-transcendence. For the sake of time in this paper, I would say the two easiest ways to take a position against humans and posthumans sharing a common supernatural end are to deny the supernatural or to take a position that posthumans have no supernatural end. Denying the supernatural would be taking a materialist position where there is no spirit or no spirit apart from matter and that nothing exists beyond the natural end of beings. The same could be said of those who would argue against posthumans having a supernatural end, but this perspective could also include religious thinkers who could take one of several positions. Those who see transhumanism as "playing God" could see the very process of human enhancement as an active turning away from God and denying or saying "no" to the supernatural end of communion with God. It could also include perspectives that deny nonbiological bodies "supernatural" status. For example, if God became flesh, does that mean flesh, and that those who choose to move on from the biological ultimately separate themselves from God. Finally, it could also include perspectives that argue radical life extension or physical immortality replace the supernatural end on the one hand, or at least causes posthumans to actively suppress their supernatural end. While I wish there was more time to try and resolve some of these complications, I believe that this thesis makes some tentative steps in drawing definitional distinctions between humans, transhumans, and posthumans.

Notes

1. Nick Bostrom, "The Transhumanist FAQ," *World Transhumanist Association* 5 (Oxford: Oxford University Press, 2003): 4, https://nickbostrom.com/views/transhumanist.pdf.
2. Ibid., 5.
3. Marjorie Greene, "Aristotle and Modern Biology," *Journal of the History of Ideas* 33, no. 3 (July 1972): 397, https://doi.org/10.2307/2709043.
4. Ibid.
5. Ibid., 398.
6. Ibid.
7. Ibid., 398–399.
8. Ibid., 401.
9. Ibid.
10. Ernst Mayr, "The Idea of Teleology," *Journal of the History of Ideas* 53, no. 1 (January 1992): 120, https://doi.org/10.2307/2709913.
11. Ibid., 123.
12. Stanley L. Jaki, *The Purpose of It All* (Washington, D.C.: Regnery Gateway, 1990), 198.
13. Ibid., 198–199.
14. Mayr, "The Idea of Teleology," 123.
15. Ibid., 125.
16. Ibid., 127.
17. Ibid.
18. Ibid., 127–128.
19. Ibid., 134.
20. Ibid., 132.
21. Ibid., 133.
22. Ibid., 135.
23. Cornelia Dean, "Francisco Ayala, Biologist and Defender of Evolution, Dies at 88," *The New York Times*, March 10, 2023, accessed April 14, 2024, https://www.nytimes.com/2023/03/10/science/francisco-ayala-dead.html.
24. Francisco J. Ayala, "Teleological Explanations in Evolutionary Biology," *Philosophy of Science* 37, no. 1 (March 1970): 2, https:/

/doi.org/10.1086/288276.

25. Ibid., 8.
26. Ibid.
27. Ibid., 9.
28. Ibid.
29. Ibid., 11.
30. Michael Ruse, "Teleology and Biology: Some Thoughts on Ayala's Analysis of Teleology," *History and Philosophy of the Life Sciences* 21, no. 2 (1999): 189.
31. Ibid., 190.
32. Ibid., 193.
33. F. J. Soontiëns, "Evolution: Teleology or Chance?," *Journal for General Philosophy of Science* 22, no. 1 (March 1991): 134–135, https://doi.org/10.1007/bf01801253.
34. Ibid., 135.
35. Ibid., 136.
36. Ibid., 136–137.
37. Stanley L. Jaki, *Angels, Apes, and Men* (La Salle: Sherwood Sugden, 1983), 62.
38. Ibid., 55.
39. Stanley L. Jaki, *The Purpose of It All* (Washington, D.C.: Regnery Gateway, 1990), 47.
40. Ibid.
41. Ibid.
42. Richard Dawkins, *The Selfish Gene* (Oxford: Oxford University Press, 1989), 189.
43. Ibid., 190.
44. Ibid.
45. Ibid.
46. Ibid.
47. Ibid., 192.
48. Ibid., 200.
49. Bostrom, "The Transhumanist FAQ," 4–5.
50. Ibid., 4.
51. Ibid.
52. Justin Pollard, "Did Charles Darwin Coin the Phrase 'Survival

of the Fittest'?," *History Extra*, January 28, 2022, https://www.historyextra.com/period/victorian/survival-of-the-fittest-meaning-origin/.

53. Dan Falk, "The Complicated Legacy of Herbert Spencer, the Man Who Coined 'Survival of the Fittest,'" *Smithsonian.com*, April 29, 2020, https://www.smithsonianmag.com/science-nature/herbert-spencer-survival-of-the-fittest-180974756/.
54. Pollard, "Did Charles Darwin," *History Extra*.
55. Falk, "The Legacy of Herbert Spencer," *Smithsonian.com*.
56. Richard Weikart, "The Origins of Social Darwinism in Germany, 1859–1895," *Journal of the History of Ideas* 54, no. 3 (July 1993): 469, https://doi.org/10.2307/2710024.
57. Thomas C. Leonard, "Origins of the Myth of Social Darwinism: The Ambiguous Legacy of Richard Hofstadter's Social Darwinism in American Thought," *Journal of Economic Behavior & Organization* 71, no. 1 (July 2009): 38, https://doi.org/10.1016/j.jebo.2007.11.004.
58. Ibid., 40.
59. Gregory Claeys, "The 'Survival of the Fittest' and the Origins of Social Darwinism," *Journal of the History of Ideas* 61, no. 2 (April 2000): 225, https://doi.org/10.2307/3654026.
60. Ibid., 228.
61. Ibid.
62. Ibid.
63. Stanley Fields, "The Law of Evolution: Darwin, Wallace, and the Survival of the Fittest," *Genetic Twists of Fate*, 2010, https://www.ncbi.nlm.nih.gov/books/NBK559908/#:~:text=Also%20known%20as%20%E2%80%9Cnatural%20selection,is%20always%20a%20dangerous%20place.
64. Pollard, "Did Charles Darwin," *History Extra*.
65. Bostrom, "The Transhumanist FAQ," 4.
66. Ibid.
67. Jaki, *Angels, Apes, and Men*, 60.
68. Bostrom, "The Transhumanist FAQ," 46.
69. Braden Molhoek, "Moral Enhancement, the Virtues, and Transhumanism: Moving Beyond Gene Editing," essay in *Religious Transhumanism and Its Critics* (London: Lexington Books, 2022), 403.

17 CAN WHAT CANNOT BE MEASURED EXACTLY HAPPEN EXACTLY?

Jaki's Criticism of the Copenhagen Interpretation of Quantum Theory

JOHN M. LONG, PH.D.

Introduction

On accepting the 1987 Templeton prize for progress in religion, Fr. Stanley Jaki stated the four chief goals of his work: First he tried to show that the rise of science occurred within a firmly Christian milieu, and that fact is no mere coincidence. Second, he wanted to make very clear the Christian context of advances in scientific cosmology. Third, following from his firm belief in the philosophy of moderate realism, he wanted to establish the objective reality of the universe, something that not all cosmologists accept. And finally, he wanted to show the differences between real science and philosophical misconceptions about science that have appeared over the centuries.[1]

The proceedings of the first Stanley Jaki Foundation International Congress focussed heavily on his philosophical and historical writings.[2] In addition to these, Fr. Jaki wrote a great deal about physics, its history, almost all prominent contributors to its advance, and how one can make sense of physics in the light of reason and common sense. Of interest here are his extensive commentaries on how one might interpret quantum mechanics, the second of the two most important discoveries of physics in the 20th century.

One of the key outcomes of quantum theory is known as the Heisenberg uncertainty principle, which states that there are limits in how precise one can measure physical interactions on the atomic scale. Many interpreters of this principle extend its scope past the atomic scale, not only to the cosmic scale but also our understanding of reality itself. It has been used as evidence to suggest that talk about cause and effect itself is ultimately meaningless. The most common philosophical interpretation along these lines is called the Copenhagen interpretation, promoted by Neils Bohr and colleagues from the 1930s onward. The Copenhagen interpretation was made popular by numerous scientists, perhaps the most famous being the late Stephen Hawking. Jaki was exceptionally critical of this interpretation, and his comments on the Copenhagen interpretation was the focus of numerous essays. (Table 1 gives a partial listing.) In the proceedings of the previous Jaki congress, this discussion is mentioned three times, but only briefly. Two papers acknowledge how the founders of quantum theory jumped from empirical studies to the nature of things in a move that shook science to its foundations,[3] and a third noted Jaki's tendency to be controversial.[4] Given how important quantum mechanics is to modern science and technology, Jaki's very strong views on how one should interpret it deserve a closer look.

In this paper, I seek to review what he had to say on how quantum mechanics has been interpreted since it was formulated 100 years ago, focussing on the philosophical implications that have found their way into popular literature. First an overview of quantum mechanics will be given as one might learn it in a freshman college physics or chemistry class. Then I discuss the Copenhagen interpretation of quantum theory and how it was applied by Stephen Hawking in his popular books on the universe. Following that will be Jaki's criticism of this interpretation.

Finally I present some lessons that I have learned both from my reading of Jaki and my experience as a practicing materials scientist. The discussion highlights the importance of realism in sensibly interpreting the amazing advances in science, technology, and cosmology, the applications of which our world currently enjoys.

Table 1: Jaki's Discussions on Quantum Theory

Article Title	Book Title	Chapter	Year
The Edge of Precision; Physics and Metaphysics	The Relevance of Physics	6, 8	1966
The Horns of Complementarity	The Road of Science and the Ways to God	13	1978
A Bookish Philosophy	Cosmos and Creator	4	1980
Chance or Reality: Interaction in Nature versus Measurement in Physics	Chance or Reality	1	1986
Turtles and Tunnels; Loaded Dice	God and the Cosmologists	5, 6	1989
In the Twentieth Century	Miracles and Physics	2	1989
The Intelligent Christian's Guide to Scientific Cosmology	Catholic Essays	9	1990
Evicting the Creator; Physics and the Ultimate	The Only Chaos	10, 14	1990
Universes and Universe	Is there a Universe?	3	1993
Determinism and Reality	Patterns or Principles	6	1995
Causality	Means to a Message	6	1999
Computers, Lovable but Unloving; Cosmology: An Empirical Science?	The Limits of a Limitless Science	3, 10	2003
Numbers Decide, or Planck's Constant and Some Constants of Philosophy	Numbers Decide	1	2003
Anthropic Illusion	The Purpose of it All	4	2005

A Brief Overview of Quantum Mechanics and the Uncertainty Principle

One can find a basic treatment of quantum mechanics in a college-level, freshman textbook on physics or chemistry.[5] With relativity, quantum theory was one of the two triumphant theories in physics that led science into the modern age. Quantum mechanics is well established in both theory and experiment, and has countless practical applications: lasers, solar panels, radiography, medical diagnostics and treatment, communications, computers, and materials technology, just to mention a few. Modern chemistry is done by means of the rules established in quantum theory. Without quantum mechanics life for us would be very different, and mostly 19th-century. We certainly would not have a mobile phone in our pockets, and one now wonders how we could live without them.

The field of quantum mechanics began with the proposal by Max Planck in 1900 that the energies emitted (and absorbed) by an ideal thermal radiator come in discrete quantities, called quanta. Einstein cemented this idea with his theory of the photoelectric effect, for which he was awarded a Nobel prize in 1921. In classical (pre-1900) physics, most quantities can take on any value in a continuous range. Post-19th century physics showed that some phenomena occur in a manner that is quantised, only discrete values are allowed. The idea that something we measure can only have discrete values is something found routinely in wave mechanics. Standing sound waves and the notes formed by musical instruments are common examples.[6] For instance, elementary physics books represent a flute as a thin pipe open at both ends. They represent a clarinet as a pipe open at one end and closed at the other end. In a very simplistic sense, musicians know that for a flute, all harmonics are allowed, even as well as odd. Clarinets, on the other hand, can only play harmonics represented by odd numbers, not the even ones. This is one reason why a simple clarinet can play much lower notes (nearly an octave lower), and requires more keys to play the same notes as compared to a simple flute.[7] So the idea of phenomena occurring in discrete packets or multiples of some fundamental value is often found in waves, musical sound being a classic case.

In classical physics, most objects are described as particles. Some phenomena, such as light, splashing water, and music are best represented as waves. It's easy enough to pinpoint a particle — its position and motion. Classical physics tried to apply the particle model to matter at the atomic level, with little success. With better and more precise experiments it became apparent that a particle model of matter does not work at the atomic level. In quantum theory, matter at the atomic level is better understood as a wave. That objects ordinarily seen as particles can indeed have wave-like properties was proposed by Louis de Broglie in 1924, and is now routinely observed and applied in fields such as electron microscopy and neutron diffraction. The wave behavior of particles is most often observed in atomic and sub-atomic particles. This led to the now-accepted quantum model of the atom, with its quantum numbers, and its applications in modern chemistry, physics, and materials science.

The interesting thing about waves is that pinning one down is much less precise than pinning down a particle. A wave is a disturbance moving through something, or in the case of electromagnetic waves, empty space. One of the outcomes of quantum mechanics is Heisenberg's famous uncertainty principle. Thanks to the fact that atomic-sized objects exhibit wave-like properties as well as particle-like properties, it turns out that it is there are fundamental limitations to the precision with which we can simultaneously measure a particle's position and its momentum, or in a similar vein, its energy and the time it takes to determine that energy.

In quantum mechanics, position and momentum are known as conjugate variables. Conjugate means joined or linked, especially in a reciprocal relation. For example, in thermodynamics pressure and volume are conjugate variables. For a gas with a fixed number of particles and constant temperature, when the volume decreases, the pressure increases; if the volume increases, the pressure decreases. Quantum theory states that the product of the uncertainties of particle's position and momentum (mass times velocity) is Planck's constant, a fixed and well-known number. If you increase the precision of a position measurement, the corresponding uncertainty of the momentum measurement must decrease, and vice versa. The general model of the atom shifted from that of a particle, where its properties and motion could be described very accurately and precisely to a wave model. In the wave model, we describe the position and behavior of particles with the laws of probability. We talk not of where an electron is in an atom, but what is the probability that it will be found in such a position and with such behavior.

Interpreting Quantum Mechanics

Expressing atomic-sized objects as both waves and particles is not easy to get your head around. Neils Bohr proposed the Bohr model of the atom and was a pioneer of quantum theory. He proposed his principle of complementarity as how we might think about the particle-wave duality of matter. One popular physics text expresses his principle this way: "The wave and particle aspects of a quantum entity are both necessary for a complete description. However, both aspects cannot be revealed simultaneously in a single experiment. Which aspect is revealed is determined by the nature of the experiment being done".[8] This discovery made many physicists in the early 20th century very uncomfortable, especially Einstein. The reason for this is because most physicists interpreted the innate uncertainty of measuring atomic particles as being a reflection on reality itself. Does modern physics thus imply that somehow reality itself is fuzzy; and by implication is our knowledge of reality also fuzzy?

Many physicists and historians of science make a causal connection between Heisenberg's uncertainty principle and the nature of reality itself. Another popular physics text describes causality as the idea that one event can cause another event.[9] For instance, one throws a stone against a window and the window breaks. We say that the moving stone caused the window to break, and we can describe the pair of events with physics. Cause and effect take on very strong meanings when used in the context of Newton's laws of motion. Newton's second law states that an applied force causes an object to accelerate, where the numerical acceleration is proportional to the applied force and inversely proportional to the object's mass. The motion of the accelerating object can be precisely predicted and is said to be determined by its initial conditions and the applied force. Most students of physics are taught that quantum mechanics forces us to radically rethink scientific determinism. In the view of classical physics, the motion of a particle can be predicted by knowing the initial conditions and the forces applied. The wavy nature of particles in the quantum realm means that knowing the initial conditions exactly is not possible, so predicting the particle's exact final state is also not possible. We can only speak of the final state in terms of probabilities. This has led to the notion that at least on the atomic scale, the physicist must abandon the notion of determinism. One text on quantum mechanics puts it this way: "Quantum theory therefore abandons strict causality in individual events in favor of a fundamentally statistical interpretation of their collective regularity... The law of causation, according to which the course of events in an isolated system is completely determined by the state of the system at [time] $t = 0$, loses its validity in quantum theory."[10]

The most common philosophical interpretation of quantum theory was put forward by Bohr and his colleagues, and came to be known as the Copenhagen interpretation. By no means the only interpretation of quantum mechanics,[11] it is however the among the most widely cited and considered by many to be the "orthodox" viewpoint among physicists and philosophers of science.[12] This interpretation grew out of the philosophy of positivism, which asserts that scientific knowledge is the only true knowledge. Taken to its logical conclusion, however, as far as physics is concerned it leads ultimately to the rejection of objective knowledge of the material world independent of the observer, which includes the scientist who devoted his life in the study of the material world. This is in contrast to realism (Jaki's philosophy), which asserts that one can obtain true knowledge of the material world through science and other means. One author summarizes positivism versus realism this way:

We can identify two distinct philosophical positions—positivism and realism—which scientists tend to adopt in their approach to their work... For the positivist, the theory is merely an instrument which can be used to interrelate observed facts and make new predictions. It described elements of an empirical reality which depends on the observer and the measuring device for its existence... For the realist, the aim of a theory should be to describe an independent reality: It should describe how the world *is*."[13]

Stephen Hawking and an Error in Logic

My interest in Jaki's writings began in 1988, after reading his scathing review[14] of Stephen Hawking's popular book, *A Brief History of Time*.[15] When this book was published, the paperback version filled all my local bookshops in floor-to-ceiling displays. In chapter four, Hawking gives an overview of the uncertainty principle. His description matches what one finds in a standard physics text. However, his philosophy sneaks through when he says that "Heisenberg's uncertainty principle is a fundamental, inescapable property of the world."[16] He says that the uncertainty principle marks the end of a deterministic view of physics and the world, in the sense that future physical events cannot be predicted if "one cannot even measure the present state of the universe precisely."[17] A bit further on, he reiterates his earlier statement: "Yet it seems that the uncertainty principle is a fundamental feature of the universe we live in."[18] As for a deterministic view of the physical world, he concludes that the "hopes of determinism cannot be realized ... The uncertainty principle of quantum mechanics implies that certain pairs of quantities, such as position and velocity of a particle, cannot both be predicted with complete accuracy."[19]

In his review, Jaki has no problems with Hawking's contributions to physics and the science of cosmology. He acknowledges Hawking's important contributions to our understanding of black holes. His criticism is directed to Hawking's philosophical and historical analysis of physics and the advancement of science. "Professor Hawking philosophizes from the start and does it badly. He does not notice the irony when at the end he complains about the comedown of philosophy from its Aristotelian heights to its Kantian shallows and to the wastelands where philosophers only offer talk about talk."[20] Jaki wrote that Hawking's (and many others') interpretation of the uncertainty principle led too quickly from limitations on measurement in physics to limitations on the philosophical concept of what nature is, including the indispensable concept of cause and effect, without which there can be no physics. According to Jaki, most, if not all, the fathers of quantum theory "made ontological causality, that takes place in a reality existing independently of the mind, a function of the possibility of measuring physical interactions with complete exactness."[21] Jaki also includes Einstein in his list of philosophically misguided physicists.

Jaki believed that quantum mechanics, and its uncertainty principle, are good, sound physics. He found nothing wrong with experiments establishing the wavy nature of atomic particles, as long as the context remains in the realm of measurement. In his first major study of the history of physics, he writes:

> As a law of physics, [the uncertainty principle] rested both on consistent theoretical grounds and on indirect observational evidence. It was therefore all the more tempting to erect it into a law of philosophy... Heisenberg emphasized man's inability to predict the future with precision and dismissed as sterile and meaningless all speculations about the existence of a real, causally determined world behind the statistically ruled world of quantum mechanics.[22]

Jaki stated repeatedly that this thinking was the result of an error in logic, a *non-sequitur*. The Copenhagen interpretation boils down to the argument that what cannot be measured exactly, cannot happen exactly. "The fallacy of that inference consists in the two different meanings given in it to the word 'exactly'. In the first case it has a purely operational meaning, whereas in the second case the meaning is decidedly ontological."[23] He points out that this error in the logic was spotted as early as 1930 in a little-noticed letter published in *Nature*: "But every argument that, since some change cannot be 'determined' in the sense of 'ascertained', it is therefore not 'determined' in the absolutely different sense of 'caused', is a fallacy of equivocation."[24]

A classic text on logic places equivocation with the fallacies of ambiguity.[25] Many words have more than one meaning. In an argument, if one uses the same word, such as 'determine', with different meanings of that word, he is guilty of equivocation. The key word here is 'determine'. Heisenberg's uncertainty principle points to a limitation of measurement, specifically in atomic and sub-atomic processes. No scientist denies that, and we physicists are taught it from our freshman year. Jaki's point is that there is a large and illogical jump from a limitation in our ability to measure something to a philosophical break in causality and attributing this break to nature itself.

In the physics sense, determinism implies that if one knows the antecedent physical conditions of an object, one can determine (calculate) a later or resultant state by means of mathematics and appropriate physical laws. Even quantum mechanics, in spite of its use of probabilities instead of exact quantities, follows this sense of the word. In his introduction to Heisenberg's *Physics and Philosophy*, Paul Davies notes: "This unpredictability of quantum systems does not imply anarchy, however. Quantum mechanics still enables the relative *probabilities* of the alternatives to be specified precisely. Where it differs from other statistical theories, such as statistical mechanics, weather forecasting or economics, is that the chance element is inherent in the nature of the quantum system and not merely imposed by our limited grasp of all the variables that affect the system."[26] But Jaki would argue that Davies also falls into the trap of equivocation when he claims that quantum uncertainty is fundamental to nature. Jaki says that physicists have interpreted far too much from the uncertainty principle: "That principle, insofar as it is science and not something else, means nothing more and nothing less than that conjugate variables cannot be measured with full precision as long as one applies the statistical formalism of quantum mechanics."[27]

Consequences of the Copenhagen Interpretation

Is this a big deal, or a storm in a teacup? Practicing physicists, chemists, and engineers might note a paradox here, but get on with the job of applying quantum theory in their daily work. A technician installing solar panels on a house or changing a circuit board on a computer is essentially doing the same thing. Jaki however, noted a danger in taking the Copenhagen interpretation to its logical conclusions. "Once the scientific inability to measure reality with complete exactitude was given a philosophical garb, it led with philosophical exactness to the inability to grasp and hold reality. Reality's place was taken by chance, not the chance that stands for ignorance, but which stands for a philosophical ghost residing in the shadowy realm between being and non-being."[28] He calls it "a mixture of brazen pragmatism and rank idealism." Pragmatism is the source of the idea that what cannot be measured exactly, cannot happen exactly, and idealism is at the root of the claim that the act of observation produces physical reality.[29] Furthermore, "once that non-sequitur is accepted, no further objection can be made to the claim that the physicist can literally create matter, indeed an entire universe, out of nothing. And since the ontological distance between non-being and being is infinite, it is no greater a feat to create an entire universe than to produce a mere fraction of the atomic nucleus out of nothing."[30]

Jaki calls this philosophical development a "sleight of hand that pretends to have the very universe for its object ... It took the relentless immersion of two generations of physicists in the Copenhagen philosophy of quantum mechanics to let the claim be hatched that physicists can conveniently rob the 'nothing' and literally create thereby a universe,"[31] something akin to pulling a rabbit out of a hat.[32] He has in mind here the multiverse theory in cosmology, which suggests that the universe we live in could be only one of a multitude that bubbled out from some prime quantum-mechanical system, thanks ultimately to quantum singularities and the uncertainty principle, each with their own individual laws of physics.[33] One of the main proponents of this theory[34] is famous for the saying, "The universe is the ultimate free lunch,"[35] a statement that Jaki found particularly irksome.[36]

The usual places where one encounters multiple universes are in comic books, on TV and in the movies, such as *Star Trek*, or *Men in Black*. We cannot measure them in a lab, even if some theoreticians claim that we could create one. Positivistic philosophies and multiple universes make for great science fiction, but have little or no empirical evidence to support them. The notion that *it is, therefore I can measure it* has been turned on its head into *I can measure it, therefore it is*. But we have to assume that something *is* before we will pay enough attention to measure it. If reality is intrinsically fuzzy because of the wave nature found in quantum theory, then is the reality of other wave phenomena also in question? Does anyone question the reality of classical waves, be they light waves, sound waves, atoms vibrating in a solid, or the waves rode by surfers at Bells Beach in Australia?

Some positivists try to wriggle out of this by proposing that laboratory measurements represent an *empirical* reality but not a *real* reality.[37] This brings no comfort to doctoral students working night and day to get experiments right, "in laboratories where one looks for real things and not merely for one's thought about them."[38] As Jaki points out, "an ontologically sensitive thinker, physicist or not, must take reality for something really real and not for a mere construct of his intellect. Such a reality cannot have for its contours a grayish margin of transition between being and non-being."[39] Modern thinking puts the cart before the horse.[40]

On the other side of the religion-science divide, Jaki is known for his rigorous defence of Christianity in general, and Catholicism in particular.[41] However, he has no time for theological positions which attempt to use physics as a means to defend Christian doctrine.[42] A case in point here is the existence of miracles. Jaki notes that some theologians refer to quantum theory and the uncertainty principle in an attempt to establish the reality of miracles. If causality can be broken because of quantum effects, then it would be possible for God to work a miracle without violating the laws of physics. "Thus began the farcical justification of miracles on the basis of Heisenberg's principle, as if it had secured for God convenient scientifically approved loopholes within which to perform miracles without coming into conflict with science."[43] He points out in the same essay that the business of miracles is the salvation of souls, a supernatural matter. "Therefore, although a miracle will in principle break the natural flow of events, it will not make a theatrical mockery of nature."[44] He has similar criticisms for those who seek to exploit quantum mechanics as a means to prove the existence of free will.[45]

The Need for a Realist Philosophy

"Of course most physicists just do their good quantum mechanics and ignore philosophy, including its Copenhagen version."[46] To these I would add materials scientists, on whom Jaki looked favourably.[47] Those of us who work in physics or materials laboratories measure things daily. We are happy to accept that when working with atoms, molecules, and electron beams, two aspects (properties) of reality are apparent, one particle-like, and the other wave-like. Indeed, the electron microscopist does both. In obtaining electron-microscopy data from a material one first accelerates a beam of electrons to a desired energy, by means of conventional classical physics, employing the voltage of the electron gun and an electron's mass. Once the electron strikes the test sample, wavy effects become apparent in the resulting interference patterns, images, and X-ray spectra.[48] This is as far as a lab researcher or technician would look into Bohr's principle of complementarity, two aspects of the same reality, like two sides of a coin. But "within the Copenhagen view only two sides, two aspects, of a coin or of anything do exist but not the coin or the thing itself."[49] Thus he warns that if one ignores the logic, reality itself can be impaled on the "horns of complementarity."[50]

One would be hard pressed to find a researcher or student labouring away in the lab who denies or even questions the objective reality of what is being painstakingly measured. If the reality of the experiment is in doubt, why bother with all the hard work, the blood, sweat and tears that go with every gain in scientific discovery? Don't forget that scientific research is also very expensive, especially when experimentation is required. Thus, as a materials scientist who works with many engineers and other materials researchers, I agree with Jaki's assessment of quantum theory. In practice, the vast majority of my engineering colleagues would also agree. For example, it is now routine to obtain images of dislocations in planes of atoms via electron microscopy,[51] and the positions of individual atoms can now be measured by atom probes and other instruments.[52] Those researchers who painstakingly obtain such images are highly unlikely to question the objective existence of the atoms shown in them. In their studies of real things, I would suggest that practical scientists and engineers know when to say, "close enough", and avoid splitting philosophical hairs between empirical reality and real reality. In the rivalry between theoreticians and experimentalists—think Sheldon Cooper versus Leonard Hofstadter[53]—the humble 'lab rat' can rest assured that his feet are "firmly planted in in good old reality."[54]

Belief in and attention to hard, observable facts are central to the philosophy required to establish the validity of science (and miracles). For Jaki, this a firm realist philosophy and corresponds with plain common sense. The first Jaki congress summarized moderate realism as advocating universal concepts, the existence of an independent, meaningful reality and the external world, the unity of being, and simultaneously the value of the scientific method.[55] From his earliest writings Jaki stressed that the business of physics and all the physical sciences had at its foundations a philosophy that assumes:

- The physical universe really exists.
- What we perceive of the physical world is real and not a product of the mind.

- We can use our senses to find out real, trustworthy information about physical reality.
- Physical reality and the universe are both rational and consistent.
- One can use logic and mathematics to learn how nature works, and this logic is reliable.

These are generally unspoken assumptions that all scientists take on faith, whether they are aware of them or not.[56] The basis of these assumptions is philosophy, not science.

Non-realist philosophies have been creeping into academic fields for at least three generations. They have been present in the field of education since at least the time of John Dewey.[57] Today one of the most common philosophical lenses through which one does education research is social constructivism, whereby the student constructs his own knowledge.[58] Constructivism has also wriggled its way into theoretical treatment of engineering automation and control.[59] In an attempt to teach physics more effectively to people of non-western cultures, there is an increasing criticism of the implied realism found in science and the scientific method. This includes very recent talk about incorporating non-realist 'indigenous ways of knowing' into engineering and physics education.[60] I'm not blaming the rise of non-realist philosophies in academia on a misinterpretation of quantum mechanics. But when the realist foundation of science is openly abandoned, it is easy for non-realism and its successors to find excuses to take its place.

Suggestions for further work

The problem on how to interpret the uncertainty principle in quantum mechanics is certainly a thorny one.[61] This paper certainly does not seek to be the last word on the matter. Nor does it offer any suggestions on the 'right' interpretation, other than to state that Fr. Jaki's arguments make sense to this practical physicist, and are easier to understand than many other realist interpretations of science.[62] It certainly leads to more questions than it answers. Thus, further work on this issue includes finding out what reception Jaki received for his argument among other physicists and historians of science. Has anyone offered rebuttals, or has anyone proposed how the uncertainty principle should be interpreted in the light of realism, and whether any applications from this approach exist? (This author has not seen any, yet.) As Jaki notes, causation is a metaphysical reality that runs much deeper than the mechanistic version discussed in physics books.[63] The same can be said about determinism. Perhaps then, quantum physics and the uncertainty principle should be studied from the point of view of primary and secondary causes, if this has not been done already. The physicist who has a good grip on realism could also review and address classic quantum conundrums such as Bell's theorem and Schrödinger's cat. There is plenty of work yet to do for both the physicist and the philosopher.

It is now more urgent than ever for Fr. Jaki's lessons and his insistence on realism as the basis for philosophy to be studied, and then taught to the next generation. Perhaps youth have had enough of the relativistic and positivistic world views taught in colleges and universities for the past two or three generations. Perhaps they are hankering for something resembling truth, as I was when nearly 40 years ago, I abandoned formal philosophy studies for the certainty of Newton's laws and the practicality of the laboratory. As G.K. Chesterton once prophetically wrote, "We shall be left defending, not only the incredible virtues and sanities of human life, but something more incredible still, this huge impossible universe that stares us in the face. We shall fight for visible prodigies as if they were invisible. We shall look on the impossible grass and the skies with a strange courage. We shall be of those who have seen and yet have believed."[64] Physicists (especially Christian ones) and students of Jaki's thought ought to be among those leading the charge.

Notes

1. S. L. Jaki, "Address on Receiving the Templeton Prize," in *The Only Chaos and Other Essays* (Lanham, MD: University Press of America, 1990).
2. P. Haffner and J. Laracy, eds., *Stanley Jaki Foundation International Congress 2015* (Leominster: Gracewing, 2020).
3. P. Haffner, "Christology and the Cosmos in Stanley Jaki," in *Stanley Jaki Foundation International Congress 2015*; J.R. Laracy, "Creation, Revelation, and the Emergence of Empirical Science." In *Stanley Jaki Foundation International Congress 2015*.
4. R. M. Liddy, "Jaki and Lonergan: Confrontation or Encounter?" in *Stanley Jaki Foundation International Congress 2015*.
5. See for instance, D. Halliday, R. Resnick, and J. Walker, *Fundamentals of Physics* 11th ed. (New York: Wiley, 2018), chaps. 43–44.
6. I. Johnston, *Measured Tones: The Interplay of Physics and Music* (Bristol, UK: Adam Hilger, 1989).
7. For a brief discussion on the physics of wind instruments, see A. Giambattista, *Physics*. 5th ed. (New York: McGraw Hill, 2020), chap. 12.
8. D. Halliday, R. Resnick, and J. Walker. *Fundamentals of Physics Extended, with Modern Physics*, 4th ed. (Wiley, 1993), 1171.
9. D. C. Giancoli, *Physics for Scientists and Engineers with Modern Physics*. 3rd ed. (Upper Saddle River, NJ: Prentice Hall, 2000), chap. 6.
10. R. Resnick, *Basic Concepts in Relativity and Early Quantum Theory* (Wiley, 1972), 190.
11. F. Laloe, "Quantum Mechanics is Routinely Used in Laboratories with Great Success, but no Consensus on its Interpretation has Emerged," in *The Oxford Handbook on the History of Quantum Interpretations*, edited by O. Freire (New York: Oxford University Press, 2022).
12. W. Myrvold, "Philosophical Issues in Quantum Theory," in *The Stanford Encyclopedia of Philosophy*, edited by E.N. Zalta and U. Nodelman. (Stanford University, Fall 2022).
13. J. Baggott, "What Does It Mean?" in *The Meaning of Quantum*

Theory: A Guide for Students in Chemistry and Physics (Oxford: Oxford University Press, 1992), 81.

14. Jaki, "Evicting the Creator," in *The Only Chaos and Other Essays.*
15. S. Hawking, *A Brief History of Time: From the Big Bang to Black Holes* (New York: Bantam, 1988).
16. Hawking, *A Brief History of Time,* 55.
17. Hawking, *A Brief History of Time,* 55.
18. Hawking, *A Brief History of Time,* 155–156.
19. Hawking, *A Brief History of Time,* 172.
20. Jaki, "Evicting the Creator," 153.
21. Jaki, "Evicting the Creator," 154.
22. Jaki, *The Relevance of Physics,* 362.
23. Jaki, *Miracles and Physics,* 54.
24. J. E. Turner, "Determinism," letter to the editor, *Nature* 126, no. 3191 (1930): 995. https://www.nature.com/articles/126995c0.
25. I. M. Copi, *Introduction to Logic.* 5th ed. (New York: Macmillan, 1978).
26. W. Heisenberg, *Physics and Philosophy* (London: Penguin, 1989), 5.
27. Jaki, *Numbers Decide and Other Essays,* 49.
28. Jaki, *Chance or Reality and Other Essays,* 7.
29. Jaki, *The Purpose of It All,* 113.
30. Jaki, *Is There a Universe?* 48.
31. Jaki, *Patterns or Principles and Other Essays,* 118–119.
32. Jaki, *Patterns or Principles and Other Essays,* 141.
33. See for example, P. Davies, "A Brief History of the Multiverse," *The New York Times.* 12 April, 2003, https://www.nytimes.com/2003/04/12/opinion/a-brief-history-of-the-multiverse.html; F. R. Ellis, "Does the Multiverse Really Exist?," *Scientific American* 305, no. 2 (01 August 2011), 38–43, https://www.scientificamerican.com/article/does-the-multiverse-really-exist; and especially S. Hawking, *The Universe in a Nutshell* (New York: Bantam, 2001).
34. A. H. Guth and P. J. Steinhardt, "The Inflationary Universe," *Scientific American* 250, no. 5 (May 1984), 116–29. https://www.jstor.org/stable/10.2307/24969372.

35. Quoted in Hawking, *A Brief History of Time*, 129.
36. See for instance, Jaki, *God and the Cosmologists*, 138, and *The Only Chaos*, 269.
37. Baggott, "What Does It Mean?" in *The Meaning of Quantum Theory: A Guide for Students in Chemistry and Physics.*
38. Jaki, *Cosmos and Creator*, 102.
39. Jaki, *Patterns or Principles*, 127.
40. Jaki, *Miracles and Physics*, 41–42.
41. P. Haffner, *Creation and Scientific Creativity: A Study in the Thought of S. L. Jaki* (Christendom Press, 1991); S. Trasancos, *Science Was Born of Christianity* (Titusville, FL: Habitation of Chimham Publishing, 2014).
42. Jaki, *Miracles and Physics.*
43. S. L. Jaki, *Bible and Science* (Christendom Press, 1996), 152.
44. Jaki, *Bible and Science*, 153.
45. Jaki, *Patterns or Principles*, 136–137.
46. Jaki, *Patterns or Principles*, 135.
47. S. L. Jaki, "The Relevance of Materials Science," *Ferroelectrics* 267, no. 1 (2002): 77–89. https://doi.org/10.1080/713715905.
48. See for instance, Halliday, Resnick, and Walker, *Fundamentals of Physics* 11th ed., section 38–5.
49. Jaki, *Patterns or Principles*, 140.
50. Jaki, "The Horns of Complementarity," in *The Road of Science and the Ways to God.*
51. See for instance P. Cizek, S. R. Kada, N. Armstrong, R. A. Antoniou, S. Slater, and P. A. Lynch. "Dislocation Structure Evolution During Room Temperature Dwell Loading of a Ti-6Al-4 V Alloy." *Materials Characterization* 209, 113757 (2024). https://doi.org/10.1016/j.matchar.2024.113757.
52. See for instance O. Golim, V. Vuorinen, G. Ross, S. Suihkonen, and M. Paulasto-Krockel. "Investigative Characterization of Delamination at TiW-Cu Interface in Low-Temperature Bonded Interconnects." *Materials Characterization* 209, 113772 (2024). https://doi.org/10.1016/j.matchar.2024.113772.
53. C. Lorre, B. Prady, and D. Goetch. "The Staircase Implementation." The Big Bang Theory. Warner Bros. Television. 2010. Accessed 04 April 2024: https://www.youtube.com/watch?v=L7LiFqugKCU.

54. Jaki, *Patterns or Principles*, 141.
55. J. R. Laracy, "Creation, Revelation, and the Emergence of Empirical Science," in *Stanley Jaki Foundation International Congress 2015*.
56. Jaki, *The Relevance of Physics*, chap. 8.
57. F. R. McKenna, *Philosophical Theories of Education*. (University Press of America, 1995).
58. K. Illeris, ed. *Contemporary Theories of Learning: Learning Theorists ... in Their Own Words*, 2nd ed. (London: Routledge, 2018).
59. J. T. Laracy and J. R. Laracy. "Reconceiving Cybernetics in Light of Thomistic Realism." *Systemics, Cybernetics and Informatics* 19, no. 4 (2021): 24–39.
60. N. Govender and E. Mudzamiri. "Incorporating Indigenous Artefacts in Developing an Integrated Indigenous-Pedagogical Model in High School Physics Curriculum: Views of Elders, Teachers and Learners," *Cultural Studies of Science Education* 17, no. 3 (2022): 827–50. https://dx.doi.org/10.1007/s11422–021–10076–2; K. Peden and S.J. Cassidy. "Excavating Bridges for Future Streams." *Australian Physics* 59, no. 1 (2022): 17–23.
61. W. Greiner, *Quantum Mechanics: An Introduction*. 3rd ed. (Berlin: Springer, 1994), chap.17.
62. J. Leplin, ed. *Scientific Realism* (Los Angeles: University of California Press, 1984).
63. See note 14, p. 123.
64. G. K. Chesterton, *Heretics*, edited by D. Dooley (San Francisco: Ignatius, 1986).

18 THEOLOGY AS SCIENCE AND THE DIALOGUE WITH OTHER SCIENCES FROM A THOMISTIC PERSPECTIVE

FR. JOÃO PAULO SANTIAGO, CSJ

Introduction

In the 13th century, around 1265–67, Aquinas began the process of writing what would later be considered his masterpiece, the *Summa Theologiae*. From the very beginning of this work, Aquinas analyzes the object of his study: the *sacra doctrina*—which translates as "sacred doctrine."[1] One of the important conclusions Aquinas makes is that this doctrine is not only a science but the highest science.[2] Now, such a statement would surprise most average Americans. In 2015, 59% of Americans said that science and religion conflict, which represents a modest increase from 55% in 2009.[3] Obviously, anachronisms must be avoided since Aquinas' conception of science differs from ours as children of the 20th and 21st centuries. The goal of this article, however, is to provide some reflections on theology as a science and to suggest how a dialogue with other sciences could be possible. This study is based on Aquinas' conception of *sacra doctrina* as science and, consequently, theology, as explained below.[4] I do not intend to limit my reflection to a pure exegesis of Aquinas' works; rather, my goal is to show how a Thomistic understanding of theology as a science provides a solid foundation for such a dialogue.

Before detailing the outline of this article, it is worth noting the existence of an important debate around the concept of *sacra doctrina* in Aquinas, whether *sacra doc-*

trina is the equivalent of theology or represents a distinct concept. Since this debate is beyond the scope of this article but important enough to merit acknowledgment, I will align with the opinions of notable Thomists such as Yves Congar, Henry Donneaud, and James Weisheipl and consider, with a high degree of certainty, that *sacra doctrina* means "the entirety of revelation," distinct from our rational effort to understand it, namely theology.[5] Therefore, I will first explain the concept of *sacra doctrina* in Aquinas and its necessity, and secondly, elaborate on what theology is. Third, I will show how Aquinas conceives *sacra doctrina* and, therefore, theology as a science, and, finally, explain the relationship between theology and other sciences from a Thomistic point of view. This article will conclude by drawing some consequences from this study for both believers and non-believers. The *Summa Theologiae* will be the main source of Aquinas's thoughts on this matter though other sources will be used to complement the analysis.

Sacra Doctrina: *Definition and Necessity*

Aquinas never defined *sacra doctrina*. Commentators have questioned if *sacra doctrina* is the equivalent of theology in Aquinas' thought. In the first prologue of the *Summa*, Aquinas gives some indications of what he means by *sacra doctrina*. He introduces his work by saying what a "master of Catholic truth" (*catholicae veritatis doctor*) should do. This means that this work is obviously about the "Catholic truth." A few lines later, he affirms that the purpose of this work is to "instruct beginners" in matters of "whatever belongs to the Christian religion." In short, a master of Catholic truth must teach whatever belongs to the Christian religion. Finally, Aquinas ends his prologue by saying that he wants to try to "set forth whatever is included in this sacred doctrine."[6] Which doctrine? The same as whatever belongs to Christian revelation. Thus, *sacra doctrina* is a teaching as broad as whatever belongs to the Christian religion.[7] Therefore, *sacra doctrina* can be defined as the "entirety of the revelation," what is revealed from God to the believer, and that he must assert as the truth.[8] Such an understanding is confirmed in the prologue of the second question when Aquinas asserts that his writings are "to expound this science," namely, the *sacra doctrina*. In other words, Aquinas writes about the *sacra doctrina* but not the *sacra doctrina* itself.[9] Additionally, Aquinas uses the expression *sacra doctrina* flexibly, sometimes referring to theology, the sacred scriptures, the preaching of the Church fathers, and the teaching of the Church. This flexibility leads us to conclude that the *Summa* is a theological analysis distinct from the *sacred doctrina* itself.[10]

When St. Thomas Aquinas begins to analyze the nature of this doctrine, he chooses to approach the question from the perspective of necessity, focusing on the final cause. God does not reveal himself without a reason. Aquinas asserts that this revelation was necessary for the salvation of mankind, and he provides two reasons for this necessity. First, because in order to act, man needs to know his end. Now, such an end—salvation—exceeds human reason; therefore, this doctrine needed to be revealed so that man can be "led by the hand toward that contemplation" of God's essence.[11] Second, even if some truths could be contemplated by natural reason, they are often difficult to discover and only a few would know it. Thus, to make salvation available for everyone, *sacra doctrina* is necessary.[12]

This point is crucial for our purposes as I maintain that theology qualifies as a science inasmuch as it reflects on and is part of this *sacra doctrina.* In order words, there is a distinction between the rational study of the revelation, namely theology, and the sacred doctrine itself, the "entirety of the revelation." Another way to look at this distinction is through the necessity of the sacred doctrine for salvation, which reiterates why we should maintain such a distinction. Otherwise, only those who study theology would have access to salvation, thus creating a sort of gnostic religion. Could one say that theology—the academic or scholastic effort to understand God's revelation—is necessary for salvation? It would be difficult to maintain such a position when an important number of Catholic saints were not "theologians" in the academic sense. However, if *sacra doctrina* is the broad entirety of revelation, what is theology?

Theology as the Rational Understanding of Sacra Doctrina

Similarly to *sacra doctrina*, there is a challenge in defining what theology means to Aquinas, as "nowhere in his work can be found a fully developed exposition of what he has extensively practiced."[13] In the first question of the *Summa*, the term *theologia* appears only twice: first in article 1, *ad* 2, where Aquinas compares theology belonging to sacred doctrine to theology as part of philosophy; second, in article 7, *sed contra*, where he defines theology simply based on its etymology as "treating of God," and seems not to make a clear distinction between *theologia* and *sacra doctrina.* One conclusion is that Aquinas' usage of *theologia* cannot be confined solely to the academic context or regarded merely as a study conducted by professionals in the field.

Even though Aquinas does not spend time trying to define "theology," it is possible to come up with a definition based on his thought. Yves Congar seems to have the best explanation of what theology means in Thomistic thought:[14]

- Theology, for Aquinas, is the rational construction of Christian doctrine itself.[15]
- Theology, as understood and practiced by Saint Thomas, appears to us as a reflection upon the revealed given [*donné révélé*][16] in a rational and scientific manner, aiming to provide the believing mind with a certain understanding of this given.[17]
- Thus, theology emerges as an endeavor on the part of the rational believer to reconsider reality as God conceives it, not solely at the level of mere faith adherence but rather at the level of discursive and rational knowledge, utilizing its resources and pathways.[18]

These three excerpts highlight different aspects of what theology is. First, it is a "rational construction" of the Christian doctrine, the *sacra doctrina*. The term "construction" cannot mean that theology is based on pure human creation. Congar means that theology is the ability to grasp the divine revelation with our human concepts, which is possible because of the fact that God spoke to us in human language. This understanding is only possible if we translate this revelation into a body of doctrine. Second, this rational endeavor is based on the "revealed given," given that the *sacra doctrina* is the "entirety of revelation." These two points are finally unified by the third, which is the fact that theology is not an irrational discourse, but instead a rational effort to understand what is revealed by God. Obviously, the rational and scientific nature of theology presupposes faith in the revelation from God, since the *sacra doctrina* "proceeds from the principles of faith," and not from any pure human creation.[19] It is a rational systematization and understanding of what God has revealed.

In another significant quote, Congar points out the connection between sacred doctrine and theology:

> Theology is, within this immensely rich framework that is *sacra doctrina*, namely the teaching derived from revelation, the scientific and rational function or modality, the domain where Christian truths are constructed into a body of doctrine in which some play the role of principles and others that of conclusions, as befits reason when it obeys its own laws, which are also a gift and a reflection of God. Theology is the rational and scientific construction of divine teaching.[20]

Thus, theology is the rational and scientific effort to understand sacred doctrine. It is the effort of our human mind to receive the highest level of knowledge, the divine knowledge that has been communicated to us in human words. Therefore, because this communication was given in human words and also derives from God, who is not irrational, human beings can have a rational understanding of this doctrine.

Before moving to the next point, it is important to emphasize that faith as a precondition for theology is not an objection to its rational dimension. Aquinas makes a clear distinction between articles of faith and the preambles of these articles.[21] The first ones are received through revelation and cannot be proven by pure natural reason, such as the belief in the Trinity or the divinity of Christ. Nevertheless, these articles are not irrational. One of the motivations of theology is to expose the possibility of those articles to be true, even though they remain directly, in their source, inaccessible to human reason.[22] The preambles of faith are those truths that are revealed but can be discovered by human knowledge, such as God's existence and the soul's existence. Highlighting the compatibility of sacred doctrine, theology, and reason is crucial to perceiving theology as the scientific effort to understand divine revelation.

Sacra Doctrina and Theology as Science

Now that the concepts of *sacra doctrina*—the entirety of the revelation—and theology—the rational study of such a doctrine—have been explained, distinguished, and established as related to each other, one can better understand how *sacra doctrina* can be considered a science. From an Aristotelian-Thomistic point of view, science is any type of knowledge that can "manifest causal relationships."[23] Indeed, Aquinas follows Aristotle and defines science as "the knowledge of the thing through its cause."[24] That is why *sacra doctrina*, and consequently theology, can be considered science. For Aquinas, "sacred doctrine is a science because it proceeds from principles established by the light of a higher science, namely, the science of God and the blessed."[25] It is a science because it proceeds from a cause which is the knowledge of God and the blessed saints in heaven who contemplate God. Now, if sacred doctrine is science, so is theology, the rational understanding of divine revelation. In other words, theology is science as it reflects on the science of God and the blessed in heaven that is revealed to us.[26]

Obviously, today's concept of science differs from Aquinas.' The problem is that even modern philosophers of science have felt challenged in defining what science is. In her article "Religion and Science," Helen de Cruz tries to distinguish both concepts saying "that science concerns the natural world, whereas religion concerns both the natural and the supernatural." Yet, as she later affirms, this does not define science: "Science and religion defy definition."[27] Moreover, the variety of "sciences" shows the complexity of the challenge: how could one formulate a unique definition of a concept that can comprise several fields, from "hard" and formal sciences to social and religious sciences? Is it even possible to come up with a single definition?

In order to make possible the dialogue between the idea of theology as science and the current conception of science, I will use a broad definition of science formulated by Yves Congar: "Science is any discipline that can justify its own object and method, leading to certainties of a certain type that are communicable to other minds. Thus, history and sociology, for example, will be considered as sciences."[28] The merit of such a broad definition is that theology and any other science can fit into it. In this sense, *sacra doctrina* is science: it has its own object and method and reaches a type of certainty (based on divine revelation) that can be communicable to other minds.[29]

This certainty is not measurable, at least not in the same way that "hard" sciences are. It presupposes faith, but it remains rational and coherent. It comes from a place of trust in someone—God—who has a higher knowledge that man could not reach without having it revealed to him. This is how Aquinas answered one of the objections: "The principles of any science are either in themselves self-evident, or reducible to the conclusions of a higher science; and such, as we have said, are the principles of sacred doctrine."[30] If we perceive this knowledge with uncertainty, it is because of the high level of light that it entails, the most perfect one, the divine revelation itself, compared to our natural knowledge, which is at a lower level: "The fact that some happen to doubt about articles of faith is not due to the uncertain nature of the truths, but to the weakness of human intelligence."[31] As human beings, we need to reason on the basis of faith that enables us to receive God's revelation and understand its content in a rational structure.[32]

Sacra Doctrina *is the Noblest Science: Theology in Dialogue with Other Sciences*

For Aquinas, *sacra doctrina* is the highest science. To prove his point, he first argues that this doctrine is simultaneously speculative (the doctrine of the Trinity, of the Incarnation, etc.) and practical (the practical teachings of moral life).[33] From the speculative side, sacred doctrine is nobler than any other science because of its certitude since it is revealed directly by God, who cannot be wrong, and because of its subject matter, things that are sublime and transcend human reason. From the practical side, this doctrine is nobler than any other science as well since it directs us towards our ultimate end, eternal beatitude.[34] Aquinas uses the traditional expression *ancilla*—handmaid—to explain how other sciences are related to *sacra doctrina*.[35] In other words, the other sciences are servants of such a doctrine, helping the believer to have a better understanding of divine revelation. A theologian should dialogue with other sciences in theological research in order to better manifest the science that is revealed by God himself.[36]

The fact that other sciences should be considered handmaids of theology does not mean that they do not have their autonomy. Since Aquinas, sciences—especially the "hard" sciences—have made progress, and today we have come to a better understanding of their own methods. Any science, including philosophy, must be autonomous and have clear boundaries. This is a crucial condition for a healthy dialogue. Aquinas' conception of other sciences being handmaids of theology is still valid as long as one understands that sacred doctrine is the highest science and dialogue with them but does not undermine their autonomy. Rather, any truth leads necessarily to the one and only truth: God himself.[37]

A clear distinction among sciences is made by their formal object of study. Aquinas asserts that "if we consider, in faith, the formal aspect of the object, it is nothing else than the First Truth."[38] In other words, *sacra doctrina* has God himself—the First Truth—as its formal object.[39] The principle behind theology is faith in God, who reveals the truth about himself. Now, contemporary "hard" sciences can only speak about what is measurable and, consequently, what is material. The fundamental question of God's existence, for example, is beyond the scope of "hard" sciences since the very concept of God is that he is immaterial and not measurable. For Aquinas, it is a preamble of faith that can be discovered by natural reason but through a philosophical method. Contemporary "hard" sciences cannot prove nor disprove God's existence; yet, they can help to add some elements to the question. Also, a theologian can better understand what he believes to be God's creation and the structure of the universe, through scientific data. This is the intention of Pope Francis in his recent apostolic letter *Ad Theologiam Promovendam*. The Holy Father wants to promote a theology that is not self-referential, but open to dialogue. He asks theologians to be attentive to the transdisciplinarity of theology, meaning a strong interdisciplinary approach that opens theology to better integration with the other sciences.[40]

Conclusion

The Thomistic perspective of theology as science gives us the foundations for a better understanding of how faith and the other sciences can enter into dialogue. Theology emerges as a rational grasp of God's revelation, a scientific reflection upon revealed truths aiming to provide a rational understanding of divine knowledge. While Aquinas emphasizes the nobility of *sacra doctrina* as the highest science, this study pointed out that the autonomy and boundaries of each science must be respected, developing a nuanced understanding of their roles. This short analysis shows us how theology and other sciences can coexist and interact with one another in the search for truth.

This conclusion is especially fitting in the context of the Stanley Jaki International Congress. Jaki was a prolific writer who dedicated himself to understanding how theology and "hard" sciences should dialogue. He was an admirer of Aquinas and knew that Aquinas would agree with his way of referring to religion as a "two-way street": "whatever moves along that street is a move from God to man and a move from man to God."[41] Theology makes human beings understand themselves in the light of God and opens space for other sciences to explain the measurable material component of their nature. Instead of contradiction, theology and science are invited to live in harmony.

Two key points could be particularly helpful for believers. First, such an understanding of theology as a science helps them understand the certitude of their faith. Theology provides a rational understanding of God's revelation who cannot be wrong. The "science of God and the blessed" is the most sublime knowledge that invites our intelligence to make an act of trust. Second, faith cannot lead to a fundamentalism that would present "hard" sciences as an evil to be eradicated. If it ever happens that faith and "hard" sciences would seem to contradict, it is because there is either a mistake made somewhere or methodological confusion in the research. True scientific research should not prevent the theologian from fully living out his faith.

This study creates possibilities for a better understanding of the interactions between Catholic theology and science for non-believers as well. The respect for autonomy and boundaries of every field, as Catholic theology professes, helps those who do not have faith to avoid any kind of scientism, which is the tendency to interpret reality exclusively through the lens of "hard" sciences. [42] Since God is necessarily beyond any material measure, the perspective of scientism denies God's existence and blocks any possibility of dialogue between faith and the "hard" sciences. However, if the formal objects and methods of every field are respected, non-believers will be able to respect the theological point of view. Even more, they will be able to attest to the rational nature of theology, which is based on divine revelation but structured around a rational basis.

Notes

1. Torrell established the chronology of Aquinas' works placing the *Summa* among his works of maturity (Jean-Pierre Torrell, *Initiation à saint Thomas d'Aquin*, Nouvelle éd. profondément remaniée et enrichie d'une bibliographie mise à jour (Paris: les Éditions du Cerf, 2015), 436.
2. *Summa Theologiae* I, q. 1, a. 5 (Thomas Aquinas, *Summa Theologiae*, trans. Laurence Shapcote, John Mortensen, and Enrique Alarcón, Latin/English Edition of the Works of St. Thomas Aquinas, Volume 13-20 (Lander, Wyoming: The Aquinas Institute for the Study of Sacred Doctrine, 2012). Unless I explicitly mention otherwise, I will be using this translation throughout the entire article.
3. "Religion and Science." Pew Research Center. Accessed March 03, 2024. https://www.pewresearch.org/science/2015/10/22/science-and-religion/
4. It is interesting to note that James Weisheipl used the expression "scientific theology" referring to theology in the context of Aquinas' thought (James Weisheipl, "The Meaning of Sacra Doctrina in Summa Theologiae I, q. 1," *The Thomist: A Speculative Quarterly Review* 38, no. 1 (January 1974): 49.).
5. Two authors have well developed the history of this debate: Gerald F. Van Ackeren, S.J., *Sacra Doctrina: The Subject of the First Question of the Summa Theologica of St. Thomas Aquinas*

(Romae: Officium Libri Catholici–Catholic Book Agency, 1952), 19–52.; James Weisheipl, "The Meaning of Sacra Doctrina in Summa Theologiae I, q. 1." In order to better understand my position in this debate and the opinion of the authors I mentioned, see below the endnote number 8.

6. *Summa*, prologue. I corrected this translation to be closer to the original Latin, which says *sacra doctrina*.
7. Weisheipl, "The Meaning of Sacra Doctrina in Summa Theologiae I, q. 1," 54.
8. The expression "entirety of revelation" does not come from Aquinas but is a translation from the French expression *ensemble de la Révélation* (Henry Donneaud, "M.-D. Chenu et L'Exégèse de 'Sacra Doctrina,'" *Revue Des Sciences Philosophiques et Théologiques* 81, no. 3 (Juillet 1997): 416. Henry Donneaud is neither the only one nor the first to define *sacra doctrina* on those lines. Other commentators have explained the meaning of sacred doctrine according to Aquinas in similar terms. Yves Congar speaks of *sacra doctrina* as the "revelation, that is, the *doctrina fidei*" (Yves Marie-Joseph Congar, "Théologie," *Dictionnaire de Théologie Catholique* XV (n.d.): 379.). Weisheipl states that *sacra doctrina* is the same as "the doctrine possessed by every Christian and studied by every theologian" (Weisheipl, "The Meaning of Sacra Doctrina in Summa Theologiae I, q. 1," 54.). Weisheipl sums up part of question 1 of the *Summa* by defining *sacra doctrina*: "*Sacra doctrina* is simply wisdom (art. 6) about God (art. 7) in faith, derived from divine revelation (art. 1)" (Weisheipl, 74.). Albert Patfoort reinforces that "*sacra doctrina* does not directly refer to 'theology' but primarily denotes the Holy Scripture, the biblical corpus, and, more broadly, the entirety of the teachings imparted in Christianity based on the revelations made to the prophets" (Albert Patfoort, *Saint Thomas d'Aquin. Les Clefs d'une Théologie.* (Paris: FAC-éditions, 1983), 28.). Brian Davies highlights the authority of this revelation: "*Sacra doctrina* is indeed 'teaching we agree with.' Or more precisely, teaching we ought to agree with ... *sacra doctrina* is nothing else than what we need to know to be saved." (Brian Davies, "Is 'Sacra Doctrina' Theology?," *New Blackfriars* 71, no. 836 (March 1990): 141–42.). He also gives a more thorough definition of *sacra doctrina*: "*Sacra doctrina* is (a) revealed, (b) a science (*scientia*) more noble (*dignior*) than other sciences and

especially to be called wisdom (*sapientia*), (c) a science who subject is God, and (d) a matter in which proof can be said to be involved" (Davies, 142.).

9. *Summa Theologiae*, q. 1, a. 2, prologue.

10. Patfoort, *Saint Thomas d'Aquin. Les Clefs d'une Théologie.*, 27–28. Additionally, Patfoort also states that "such is the true content, it seems, of this first question of the *Summa Theologiae*: to outline the exact form of Christian teaching in all its generality, teaching that the *Summa* aims to provide an *expositio* for the use of beginners" (Patfoort, 46.). In the "Commentary on the Sentences" by Aquinas, the terms *sacra doctrina* and *theologia* do not bear such a clear distinction (Super Sent., q. 1 a. 1 co., Accessed March 3, 2024. https://www.corpusthomisticum.org/snp0001.html), which proves that the term *theologia* is very broad for Aquinas. Another possible reason is that Aquinas' thoughts may have evolved, since the Commentary on the Sentences was the first major work of Aquinas, written 10 to 15 years before the *Summa* (Torrell, *Initiation à saint Thomas d'Aquin*, 432–34.).

11. *Super Sent.*, q. 1 a. 1 co.

12. *Summa Theologiae*, I, q. 1, a.1.

13. Patfoort, *Saint Thomas d'Aquin. Les Clefs d'une Théologie.*, 13. Congar affirms that Aquinas did write about the theological method in three different passages: "In the prologue of the Commentary on the Sentences (1254), in the commentary on Boethius's *De Trinitate*, q. II; and finally, in the Summa Theologica, I, q, 1 (around 1265)." (Congar, "Théologie," 370.9). The question remains if Aquinas is defining the theological method in these passages or rather talking about the *sacra doctrina*. It seems that Patfoort's statement is more accurate.

14. "Thomistic thought" differs from a pure exegetical analysis of Aquinas' works. Again, the term in Aquinas does not have a meaning as defined as we do today. For Aquinas, "it means, roughly, 'discourse about God', including natural theology as practiced by people such as Aristotle" (Davies, "Is 'Sacra Doctrina' Theology?," 141.). Theology can also mean a "discourse about God" based on the *sacra doctrina* (Summa q. I, a. 1, ad 2), but it does not bear a restricted sense of "what theologians teach" (Davies, 146.). Weisheipl says that *sacra doctrina* "is not to be identified with scholastic theology but

with the original revelation of God to man, and can be called 'theology' only in the etymological sense of the term as *Sermo de Deo*" (Weisheipl, "The Meaning of Sacra Doctrina in Summa Theologiae I, q. 1," 79–80.).

15. Congar, "Théologie," 379..
16. French expression referring to the revelation as bestowed by God.
17. Congar, "Théologie," 385.
18. Congar, 460.
19. *Super Sent.*, I, q. 1, a. 1, *sed contra.*
20. Yves Marie-Joseph Congar, *Bulleting Thomiste* V, no. 8 (December 1938): 498–99.
21. *Summa Theologiae*, I, q. 2, a. 2, ad 1.
22. *Summa Theologiae*, I, q. 1, a. 5, ad 2. Patfoort lists three motivations of theology: it assimilates the revealed truth, defends the truth of this revelation, and loves this truth (Patfoort, *Saint Thomas d'Aquin. Les Clefs d'une Théologie.*, 20–25.). Congar speaks about the three roles of reason in theology: establish the preambles of faith, defend the article of faith, and a role of deduction (Congar, "Théologie," 382.).
23. Weisheipl, "The Meaning of Sacra Doctrina in Summa Theologiae I, q. 1," 71.
24. Summa Contra Gentiles, I, c. 94 (Saint Thomas Aquinas, *Summa Contra Gentiles* (Notre Dame [Ind.]: University of Notre Dame Press, 1975). Congar explains this definition by saying that "there is science... when we know a thing by the cause for which it is and cannot be otherwise than it is." (Congar, "Théologie," 459.)
25. Summa Theologiae, I, q. 1, a. 2.
26. There is an important difference between both sciences. S*acra doctrina* is contemplated by the blessed in heaven and gives them the capacity to see God's mystery with their intelligence, illuminated by God himself. This is the *lumen gloriae* which allows them to see God's essence (Summa Theologiae, I, q. 12, a. 2). A theologian knows God only through the light of faith, the *lumen fidei*, which makes them see what they believe (Summa Theologiae, II-II, q. 1, a. 4, ad 3).
27. De Cruz, Helen, "Religion and Science," The Stanford Encyclopedia of Philosophy (Fall 2022 Edition), Edward N.

Zalta & Uri Nodelman (eds.) Accessed March 3, 2024. www.plato.stanford.edu/archives/fall2022/entries/religion-science. She is not the only one who struggles with definitions, Eric Martin presents the same difficulty (Martin, Eric C., "Science and Ideology," Internet Encyclopedia of Philosophy. Accessed March 3, 2024. www.iep.utm.edu/sci-ideo/ .

28. Congar, "Théologie," 460–61. Jacques Maritain says that "science is a knowledge of a perfect mode, more precisely a knowledge in which, under the constraints of evidence, the mind assigns the reasons for the existence of things. The mind is satisfied only when it not only grasps a given entity but also comprehends those factors that ground this entity in being and intelligibility" (Jacques Maritain, *Distinguer Pour Unir Ou Les Degrés Du Savoir* (Paris: Desclée de Brouwer et Cie, 1932), 46.7).
29. Congar, "Théologie," 461.
30. *Summa Theologiae*, I, q. 1, a. 2, ad 1.
31. *Summa Theologiae*, I, q. 1, a 5, ad1.
32. Congar says that "our science is discursive and proceeds through reasoning; however, based on what God has communicated to us about His knowledge of Himself, which we adhere to through faith, we strive to connect the *ignotiora* to the *notis* and ultimately all things, hierarchically, to the unique mystery and the only primal light of God." (Congar, "Théologie," 460.)
33. *Summa Theologiae*, I, q. 1, a. 4.
34. *Summa Theologiae*, I, q. 1, a. 5.
35. *Summa Theologiae*, I, q. 1, a 5, *sed contra*.
36. In *Fides et Ratio*, Pope John Paul II emphasized the necessity and primacy of philosophy for theology without excluding the importance of other sciences. See John Paul II, *Fides et Ratio* (Faith and Reason), 69.
37. "The title [*ancilla*] was not intended to indicate philosophy's servile submission or purely functional role with regard to theology. Rather, it was used in the sense in which Aristotle had spoken of the experimental sciences as "ancillary" to "prima philosophia." The term can scarcely be used today, given the principle of autonomy to which we have referred, but it has served throughout history to indicate the necessity of the link between the two sciences and the impossibility of their separation." (*Fides et Ratio*, 77)

38. *Summa Theologiae*, II-II, q. 1, a. 1.
39. *Summa Theologiae*, I, q. 16, a. 5.
40. Francis, *Ad Theologiam Promovendam*, 5.
41. Stanley L. Jaki, "The Demarcation Line between Science and Religion," *Angelicum* 87, no. 1, Science, Faith and Complexity (2010): 82.
42. *Fides et Ratio*, 88.

19 THE MYSTERY OF CREATION, VARIOUS ONTOLOGIES, AND EFFECTIVE COMMUNICATION

Brian Humphrey, S.T.L.

Introduction

They conversed for two hours. Two geniuses on a podcast—one, a famous philosopher who assumed he could know reality, and another, an influential social activist who intended to change it—faced off to talk about the controversial topic of race.[1] Their language was eloquent; their speech was dispassionate, and their arguments seemed sound enough. Yet something was missing. There was no effective communication. Like Socrates and Thrasymachus in Plato's *Republic*,[2] the two went on talking past each other *ad nauseam*. How could this be, two geniuses unable to communicate? The baffled listener, we argue, would do well to investigate not only their epistemologies, but more so, their ontologies. By ontology we mean "the project of finding out what kinds of things make up reality, what exists, or what there is."[3] It may be distinguished from metaphysics because it deals not just with metaphysical being, but with physical and metaphysical, created, and Uncreated being. Colloquially, the ontologist askes, "What is real?" Therefore, regarding the podcast example, one might ask if both parties presuppose reality is even "there" to be known, and if so, which categories should be used to describe it? Perhaps exploring these questions would shed light on how such a miscommunication could occur.

Stanley Jaki says the admission of ontology is a prerequisite for epistemology and effective communication. He mentions ontology especially when cautioning against the misinterpretation of the Heisenberg principle.[4] According to Jaki, "Heisenberg could not distinguish between mechanical causality and ontological causality, or rather between the measurability of an interaction and its ontological reality."[5] In other words, just because a particle's exact position and exact momentum cannot be measured simultaneously, it doesn't mean they don't have an exact ontological position and exact ontological momentum. In the *Miracle of Physics,* Jaki describes the problematic assumption that underlies this philosophical misstep. He says, "The possibility of perfectly accurate measurements became quickly taken as the justification of ontological causality in physical interactions."[6] Jaki recognized that before the uncertainty principle was discovered, scientists tended to admit ontological causality only *because* they could measure it, as if the cause wouldn't exist without the measurer. He saw that scientists could thus be misled and try to disregard ontology altogether (in this case, ontological causality). Jaki says this would make such a person the "victim of illusion in talking to others." He then says, "For any communication is a realist means whereby one acknowledges something objectively external to one's mind."[7] Thus, Jaki defends the admission of ontology as necessary for epistemology and effective communication. Yet questions remain. Can various ontologies exist? If so, what happens if two parties describe being differently? Will effective communication still be possible? Again, we turn to Jaki who points the way to our answer. When speaking about mysterious hidden variables in physics, he makes the astonishing claim:

> Insofar as hidden variables (or any parameter of mathematical physics, for that matter) need an ontological substratum, the biblical God, *whose self-revealed name I AM WHO I AM is ontology incarnate*, is bound to provide it in one way or another.[8]

Not only does Jaki emphasize the necessity of admitting an ontological substratum, but he also, by introducing the Divine Name, unlocks ontology's most profound dimension. With the revelation of the biblical God as the Ground of all being, the mysterious nature of reality and the possibility of varying ontologies emerge. Created reality is not absolutely self-sufficient; it is dependent on the creative, sustaining, and personal activity of God. Therefore, the deepest knowledge of creation must somehow be open to the knowledge of faith in God. By including the great "I AM WHO I AM" Jaki introduces a workspace where the ontological mystery of creation may be more thoroughly explored by the light of "ontology incarnate."

Paul O'Callaghan embarked into such a space when he put the Balthasarian and Thomistic schools of thought into conversation on the theological question of divine enrichment.[9] This question is about how creaturely being may or may not affect Uncreated Being. While in the past it has sometimes proved difficult to compare these two theological giants, O'Callaghan shows effective communication is possible.[10] Nonetheless, from his work, we can see that both the Balthasarian and Thomistic schools tend to operate within distinct ontological frameworks. Thus, it will be the goal of the rest of this paper to build on Stanley Jaki's claim for the necessity of ontology by asking questions regarding ontological pluralism. We will first survey the notion of ontology in general. Then we will offer reasons for why we claim various ontologies are permissible, and finally, we hope to illustrate how differing ontologies make for difficult conversations by comparing the ontologies of Balthasar and Thomas in the context of the conversation on divine enrichment. We will come to see the very fact that opinions differ about which categories best describe "being" points to the mysterious nature of creation and the need for a properly relational ontology for effective communication.

Various Ontologies and the Question of Permissibility

The Thomist Ralph McInerny says, "Ontology focuses its attention on such seminal notions as substance, the difference between potency and act, between essence and existence, and the various less than substantial ways that being declares itself."[11] Notice how the notion of substance is given primacy of place. Likewise, Jacques Maritain appropriates the Aristotelian framework to point "beneath the accidents" to the substance, which he defines as "a thing or nature whose property it is to exist by itself."[12] These two may be considered spokesmen for a Thomistic view that we may call "substance ontology," a view that picks up and carries on Greek philosophical categories.[13] This understanding of ontology gets re-examined by some contemporary thinkers not only due to a renaissance in Trinitarian theology, but also due to a renewed engagement in the dialogue between science and religion. Joseph Ratzinger, speaks on the topic of epistemology in the context of contemporary physics to say,

> There is no such thing as pure objectivity [when it comes to the result of an experiment]... In the answer, there is always a bit of the question, and a bit of the questioner himself; it reflects not only nature in itself, in its pure objectivity, but also gives back something of man... This too, is true of the question of God.[14]

In other words, while Ratzinger admits ontology ("pure objectivity") he recognizes there is no way for a substance to be perceived "in itself," alone, as the aforementioned Thomistic view might lead one to assume.[15] Ratzinger insists this does not lead to relativism.[16] Nonetheless, as pope, in his encyclical *Caritas In Veritate,* he calls for a renewed way of thinking about reality that entails "a deeper critical evaluation of the category of relation." He says the sciences must work together to achieve this goal: "This is a task that cannot be undertaken by the social sciences alone, insofar as the contribution of disciplines such as metaphysics and theology is needed if man's transcendent dignity is to be properly understood."[17] Here we not only see more value being given to the ontological category of relation and the need for effective communication among various disciplines, but also, with the inclusion of theology, we have an affirmation of Jaki's statement regarding the need for interdisciplinary dialogue to include the theological Ground of all being, that is God. Ratzinger says that man's transcendent dignity renders him a special expression of being that comes from God, and since man is making the scientific measurements, one should even more so be able to justify the admittance of God in scientific discussions.[18]

Thus, alongside the more philosophical ontologies that focus on categories such as substance and relation, there emerges those more theological ontologies that appeal to categories like "gift,"[19] "communion,"[20] "person,"[21] and "thinking love,"[22] just to name a few. Here, we highlight the work of the physicist, Giulio Maspero who offers a Trinitarian ontology which he says, "indicates the metaphysical reinterpretation founded in Trinitarian revelation."[23] This is the seed, he says, for a rethinking of ontology which embraces both substance and relation as co-principles,[24] and ultimately finds its source in the One

Trinity as revealed in Christ. Maspero says, "If we really take seriously... that the archê is one Substance *and* three Persons," then we can see how a rethinking of ontology based on the Persons of the Trinity can emerge. Maspero draws largely from the thought of Gregory of Nyssa to posit "co-relativity" as a key ontological category.[25] Early in his project, however, Maspero points to Klaus Hemmerle's letters to von Balthasar as justification for such an approach.[26] He sees Hemmerle's work as seminal in the field of "new ontologies" and Balthasar did too.[27]

Balthasar's Ontology

Having shown the existence and permissibility of various ontologies, we now move, by way of comparison, to the question of effective communication. Do Balthasar and Thomas operate within distinct ontological frameworks, and if so, does this pose a challenge for constructive dialogue? As we just mentioned, Klaus Hemmerle wrote Balthasar a series of letters entitled, "Theses Towards a Trinitarian Ontology" where he said Balthasar's work reaches right down into the mystery of creation and "transforms structures of thinking, modes of seeing, the encounter with reality and the ways in which persons act."[28] This enthusiastic reading of Balthasar, coupled by his reminder that "the answers given at Nicaea and Chalcedon... are not merely applications of Greek philosophy, but transformations of it," lead him to conclude, "In the reciprocal relationship between theology and philosophy, there lies, perhaps, even today, a chance for the reconquest of ontology."[29] Hemmerle then goes on to champion the development of what we may call a Balthasarian view of being "from above."[30] In other words, according to Hemmerle, Balthasar's ontology necessarily involves the perspective that only faith in

Christ can offer. In the preface to his *Theology of History*, Balthasar claims to be "looking at the subject exclusively from above, not explicitly stating but rather taking for granted created things as the content which is given its form by the Christological categories."[31] Matthew Levering takes Balthasar at his word, and says the Swiss theologian employs "a Christological version of Trinitarian ontology."[32] Levering goes on to show how Balthasar's ontological categories are taken especially from the historical dimension, especially the scene of Jesus' abandonment on the cross. These categories are then transposed into the inner life of God. This is followed by Balthasar's next move, which is to bring this kenotic structure back down into the economy as operative in all creation. Balthasar himself even asks the question, "May it not be the case (as Ferdinand Ulrich seeks to show) that the final mystery of the kenosis of God in Christ has an analogous structure in the metaphysical mystery of [created] being?"[33] Leaving aside for the moment any ruling on Balthasar's kenotic ontology, we simply state his position. In the end, we can say with Gerard O'Hanlon that Balthasar works within a Trinitarian ontology conceived as "a scripturally inspired ontology of inter-personal love," or "a relational ontology which goes beyond but includes the categories of substance metaphysics."[34] Now that we have cursorily viewed Balthasar's ontology, we must consider Thomas' and ask if the two can be considered complementary, or even if they even need to be for effective communication.

Thomas' Ontology

In addition to what we already mentioned above regarding Thomas' substance ontology, we can add here what the Common Doctor says in *De Ente et Essentia*:

"We use the term 'a being' (*ens*) absolutely and primarily of substances."[35] Also, in the *Summa Contra Gentiles*, he says "a substance is a thing to which it belongs to be not in a subject ... that which has a quiddity to which it belongs to be not in another."[36] Armand Maurer echoes this reading of Thomas to say, "A substance is a thing or essence that can exist through itself."[37] Without going into more detail on the Thomistic notion of substance, we only wish to say here that it is central to his ontology, and that it is to be conceived as being "in itself."

Advocating for the Thomistic position is Thomas Joseph White, who asks important questions pertaining to Balthasar's employment of "modern ontological terms" in the context of his Christology:

> Should Christian theologians seek to understand the divine nature of Christ not by appeal to pre-modern metaphysical arguments pertaining to the deity, but rather as a mystery of God's self-actualization of inner life revealed in history, specifically in the event of Jesus crucified and resurrected? Can and should Chalcedonian Christology be restated in modern ontological terms and if so, how is it best to do so?[38]

White is speaking here about Balthasar's kenotic Christology, and we may suppose "modern ontological terms" here corresponds to those mentioned above like "gift," "communion," "person," and "thinking love." White goes on to favour an approach to Christological studies based on "pre-modern metaphysical arguments" because, in part, they employ the same language that shaped the Chalcedonian formulas.[39] Paul O'Callaghan agrees with White on the need to maintain continuity with the traditional language and asks another important question: "Is the metaphysics of creaturely existence comparable to the metaphysics that applies to the Divinity?"[40] Here, he hints at a danger in Balthasar's approach, namely that of operating in a seemingly quasi-univocal way when relating ontological categories derived from the historical life of Christ to the imminent life of the Trinity without sufficient analogical purification or proper use of the *communicatio idiomatum*.[41]

Along with these Thomistic advocates, we must remember that while Ratzinger questions a view of "being" that over-emphasizes the notion of substance conceived of as standing in-itself,[42] he also, as pope, continued the Church's long-standing tradition of promoting the study of St. Thomas.[43] Furthermore, we recall here that Stanley Jaki takes more of a Thomistic position. As John Rooney states, "While [Jaki] prefers not to take to himself any particular labels of systems of philosophy, he considers the 'methodical realism' of Étienne Gilson, which is a form of Thomism, to be the most akin to his own ... He holds with St. Thomas that in the knowledge of reality, not only what is sensory is revealed but the intelligible nature of physical entities."[44] Considering these factors, we may claim that while various ontologies are permitted, all ontological innovations should be placed at some point in dialogue with the thought of St. Thomas, as he is an approved representative of the perennial tradition. Here, another important question comes to the fore: What happens if two ontologies differ, as seems to be the case between Balthasar and Thomas on the point of divine enrichment?

Case Study—The Enrichment of God?

Balthasar claims that God can be "enriched" by His creation.[45] At once, the Thomistic view seems challenged. Clearly, Balthasar's claim that something can be "added" to God doesn't fit neatly into a substance ontology that conceptualizes God as pure act. Wouldn't it seem that Balthasar is importing potentiality into the divine essence, thereby making God subject to change and calling into question the traditional teachings on the divine attributes?[46] Here, we find helpful a brief review of the basics of ontology. There are two distinct ontological levels, that of God, and that of creation. The gap between them safeguards God's transcendence and the integrity of creation. All ontologies must be relational, therefore, not only because God is relational, but also because creation is relational (since it is constantly being held in existence by God, and since pure nature is a hypothetical state). In fact, this ontological gap is what accounts for the mysterious nature of created existence. Of course, within this relational framework, substance ontology is still permissible. It just needs to remain open to relation.

This introduces the distinction between "closed" and "open" ontologies. While it is easier to see the error of a closed substance ontology, it is more difficult to imagine a relational ontology that is too open. In fact, if an ontology is open to God, it can never be too open. Therefore, the problematic extreme of a relational ontology must be expressed differently, and for this we offer the idea of a "confused ontology." Thus, we can make a conditional statement and complete our table of various ontologies: if the danger of substance ontology is to become closed, the danger of relational ontology is to become confused. The first would lead to individualism and/or extrinsicism, and the second, to relativism and/or immanentism.

Table of Various Ontologies			
Impermissible	**Permissible**		**Impermissible**
	Substance Ontology	Relational Ontology	
Closed	In—itself *(Not Closed)*	In—the other *(Not Confused)*	Confused
Individualism and/or Extrinsicism			Relativism and/or Immanentism

We recognize it is possible to read Balthasar as conflating the two ontological levels by inserting the natural into the supernatural, and to read Thomas as dividing the two levels such that they work in complete independence from one another, effectively deeming the supernatural a mere addition to the natural. Both ways of viewing the mystery of ontology, whether they are justified readings of their respective authors or not, would be deemed inadmissible. Keeping all this in mind, the Thomist might be willing to recognize the ability of working from an ontological starting point that spotlights the relational dimension, even if he or she still reads Balthasar as working from a confused ontology. Perhaps in this openness, even words like "receptivity" and "enrichment" might be able to develop deeper meanings. In any case, we do not claim to offer a definitive answer to the question of divine enrichment. We only wish to use it as an example of how important the question of ontology is for effective communication.

Conclusion

We began our study by affirming Stanley Jaki's assertion that ontology is indispensable for epistemology and effective communication. We then posited the existence and permissibility of various ontologies. On this point, we recognized the importance of exploring the category of relation and pointed to Ratzinger and Maspero as exemplary. We noted how these "new ontologies" must continually be placed in conversation with those traditionally employed by the Church, especially the substance ontology flowing from the Thomistic tradition. Our exploration of ontology in the conversation between Balthasarians and Thomists led us to review the basics of ontology. We recalled the necessity of maintaining the gap between God and creation and including the category of relation and the theological epistemology of faith which this gap introduces.[47] We proposed the distinction between "closed" and "open" ontologies and said "being" can never be too open to God. Instead, the danger that is at the other end of the spectrum is better described as "confused." We expressed this in a table of various ontologies, which included the extremes. Without making a pronouncement on the question of divine enrichment, we recognized the possibility of reading Balthasar as "confusing" the two ontological orders, or Thomas as "closing" the relational dimension. Both approaches would be deemed inadmissible. Finally, we were led to conclude it is possible to work from various ontological frameworks to expand or reinterpret certain terms from the Church's dogmatic formulations and further illuminate the mysteries of God, creation and the relationship between the two.[48] In doing so, one must always safeguard against the error of anthropomorphizing the Divinity, while at the same time, offer enough freedom to expound upon the inexhaustible mysteries.[49] However one approaches this task, it must be done with reverence to the mysterious nature of creation, which includes differing schools of thought, and a properly relational ontology that can ensure fruitful dialogue among the various disciplines.

Notes

1. "Sam Harris Debates Ezra Klein about what He Said about Sam, Charles Murray, and Race." Scott Smith75. March 1, 2022. Video, https://www.youtube.com/watch?v=Tsr7Rv8XnIk.
2. In book one of Plato's *Republic*, Socrates and Thrasymachus, the Sophist, talked past each other about the topic of justice. Thrasymachus' famous definition, "Justice is the advantage of the stronger," seems to reveal his intention is not to know reality, but to change it through power. See Plato, *The Republic of Plato*, trans. Francis MacDonald Cornford (London New York: Oxford University Press, 1970), 8–30.
3. Thomas Hofweber, *Ontology and the Ambitions of Metaphysics* (Oxford: Oxford University Press, 2018), 1. For a similar definition, see Louis Bouyer, *Dictionary of Theology*, trans. Charles Underhill Quinn (Tourani, Belgium: Desclée & Co., 1965), 330. Furthermore, Rahner and Vorgrimler offer a definition in their theological dictionary: "[Ontology:] The philosophical inquiry into that understanding of being and every entity which lies at the basis of any free and thoughtful concern with concrete things, interpreting them in the light of being as a whole and itself interpreted in its light. Ontology is at work in all theology, to which it is an indispensable aid and also a threat. For if reflex philosophical exposition of this original understanding of being is imperfect, yet it is necessary to the understanding of any theological proposition, indeed is present in every theological proposition: it can explain as well as obscure. The rejection of an analytic ontology, particularly in the theological applications, would not make theology 'purer' or 'more independent' but simply expose it to an uncritical, merely implicit ontology." Karl Rahner and Herbert Vorgrimler, *Dictionary of Theology*, trans. Richard Strachan, 2nd ed. (New York: Crossroad, 1965), 324. From yet another angle, "Ontology, 'the science of being,' used to be commonly regarded as a part of metaphysics (q.v.). Today, in an age in which metaphysics is out of favour, ontological knowledge of a non-speculative and non-ratiocinative kind is insisted upon by philosophical theologians of the existentialist school and by certain others. We are directly aware of *being* itself through our own existence-in-the-world (*see Dasein*). Other theologians

speak of a disclosure of being (or of the real, or of God) in certain situations in history and in our own lives. These are not to be dismissed as only emotive or subjective, because they have a cognitive and therefore an ontological content as well as an affective or volitional one." Alan Richardson, ed., *A Dictionary of Christian Theology*, 2nd ed. (London: SCM Press, 1972), 241–42. While we will not adopt either of these last two definitions, we still find them helpful to show how various ontologies exist, and how the nature and purpose of ontology can be perceived differently.

4. The Heisenberg Uncertainty Principle does not imply that an object, at any given moment, is without precise position and precise momentum, but only that the precise position and precise momentum cannot be simultaneously measured with exactitude. In other words, the more precisely one is known, the less precisely the other can be known.
5. Stanley L. Jaki, *The Drama of Quantities* (Port Huron, Mich.: Real View Books, 2005), 40.
6. Stanley L. Jaki, *Miracles and Physics* (Front Royal, VA: Christendom Press, 1999), 41–43.
7. The full quote is helpful for context. In *The Drama of Quantities* Jaki compares Husserl with Kant to show how both are counter examples. He states, "Husserl's principal aim was to part with Kant's program and assure legitimacy to the objective status of the phenomena, though ironically enough, by banishing ontology from the scene. The fact remained that Husserl did not arrive at anything, precisely because, toward the end of his life, he was... still looking for a valid starting point. This could have been found only through an unconditional acceptance of the registering of external reality as the first step in philosophizing. Such a step meant the espousal of ontology, and also the assurance that one was not the victim of illusion in talking to others. For any communication is a realist means whereby one acknowledges something objectively external to one's mind." Jaki, *The Drama of Quantities*, 43–44.
8. Jaki, *Miracles and Physics*, 85 [emphasis added].
9. Paul O'Callaghan, "Can God Be Enriched? On the Metaphysical Underpinnings of Von Balthasar's Theology," *Irish Theological Quarterly* 84, no. 2 (2019): 175–94.
10. Surprisingly, there are not many works that directly engage

these two thinkers. For a primer see Aidan Nichols, *Balthasar for Thomists* (San Francisco: Ignatius Press, 2020). Also there is, Robert Miner, "Thomas Aquinas and Hans Urs von Balthasar: A Dialogue on Love and Charity," *New Blackfriars* 95, no. 1059 (September 2014): 504–24. Michele M. Schumacher, *A Trinitarian Anthropology: Adrienne Von Speyr & Hans Urs Von Balthasar in Dialogue with Thomas Aquinas* (Washington, D.C: The Catholic University of America Press, 2014). Matthew Levering, *The Achievement of Hans Urs von Balthasar: An Introduction to His Trilogy* (Washington, D.C: Catholic University of America Press, 2019). For more on the historical relationship between Thomism and the Nouvelle Théologie see, Romanus Cessario, *A Short History of Thomism* (Washington, D. C: Catholic University of America Press, 2005).

11. Ralph M. McInerny, *The Philosophy of Nature* (Lincoln Nebraska: The Alquin Press, 2001), 13–14.
12. Jacques Maritain, *An Introduction to Philosophy* (Lanham, MD: Rowman & Littlefield Publishers, 2005), 163. In this book, Maritain dedicates three chapters to ontology entitled "Ontology: Essence," "Ontology: Substance and Accident," and "Ontology: Act and Potentiality," respectively.
13. See "Thomistic Ontology," in *The Disfigured Face*, by Luis Cortest (New York: Fordham University Press, 2008), 1–14. Jeffrey E. Brower, *Aquinas's Ontology of the Material World: Change, Hylomorphism, and Material Objects* (Oxford: Oxford University Press, 2014), 22–28.
14. Benedict XVI, *Introduction to Christianity* (San Francisco, CA: Communio Books, 2004), 175–177. In the same place Ratzinger also says, "Even dogmatic formulas such as 'one being in three Persons' include this refraction of the human element." Along with what we already said above regarding Jaki's caution of misinterpreting the Heisenberg principle, we only add here that in this situation Jaki calls for the application of the principle of sufficient reason, which he says "plays a central role in any scientific proof or demonstration." Stanley L. Jaki, *The Relevance of Physics* (Chicago and London: The University of Chicago Press, 1966), 360.
15. It is important to remember that simply saying substance exists "by itself" does not necessarily lead to a closed-off view of substance, but we do recognize the possibility of such an

interpretation. Also, we note here that not all Thomistic ontologies are based on an understanding of substance as being "in itself." For example, William Norris Clarke proposes the category of "being-in-relation." See W. Norris Clarke, *Explorations in Metaphysics: Being, God, Person* (Notre Dame (Ind.) London: University of Notre Dame Press, 1994). W. Norris Clarke, *The Philosophical Approach to God: A New Thomistic Perspective,* 2nd ed (New York: Fordham University Press, 2007). W. Norris Clarke, *The Creative Retrieval of Saint Thomas Aquinas: Essays in Thomistic Philosophy, New and Old* (New York: Fordham University Press, 2009). W. Norris Clarke, *Person and Being,* The Aquinas Lecture 1993 (Milwaukee: Marquette University Press, 1993). Also, see Antonio López who focuses on the category of "gift" in a Thomistic setting. See Antonio López, *Gift and the Unity of Being,* Veritas 11 (Eugene, Oregon: Cascade Books, 2014). Finally, we mention just one more recent study where Giulio Maspero sees a Trinitarian ontology at work in Aquinas. See Giulio Maspero, "Theological Epistemology and Trinitarian Ontology in Aquinas," *Wrocław Theological Review* 28, no. 1 (2020): 85–108.

16. By relativism, we mean, along with Benedict XVI, a theory of knowledge that "does not recognize anything as definitive." Gediminas T. Jankunas, *The Dictatorship of Relativism: Pope Benedict XVI's Response* (New York: St. Pauls, Alba House, 2011), 13. Furthermore, one might even argue the unifying theme of Benedict XVI's theological project as a whole can be set within the framework of his cautioning against the dangers of relativism. For example, see his famous "*Dictatorship of Relativism*" homily which he gave one day before his papal election. Joseph Ratzinger, "Cappella Papale: Mass 'Pro Eligendo Romano Pontifice' Homily of His Eminence Card. Joseph Ratzinger, Dean of the College of Cardinals" (Vatican City, 2005).
17. Benedict XVI, "Encyclical Letter: Caritas In Veritate" (Vatican City, 2009). Also see Joseph Ratzinger, "Concerning the Notion of Person in Theology," *Communio* 17 (Fall 1990): 439–54.
18. Recognizing the contingent nature of human life, and thus, its necessarily relational dimension, Jaki says, "Birth, life, and death as a single, irreversible, and unrepeatable process remains a reminder of the contingent, that is, non-necessary character of existence. This is undoubtedly a potent

psychological preparation, especially in this age of metaphysical insensitivity, for learning from the universe a monumentally deep though stunningly simple lesson in metaphysics." Stanley L. Jaki, *The Savior of Science* (Washington, D.C.: Regnery Gateway, 1988), 109.

19. See López, *Gift and the Unity of Being.*
20. See John D. Zizioulas, *Being as Communion: Studies in Personhood and the Church* (London: Darton, Longman & Todd, 2004).
21. See David L. Schindler, "The Person: Philosophy, Theology, and Receptivity," *Communio* 21, no. 1 (Spring 1994). W. Norris Clarke, "Person, Being, and St. Thomas," *Communio* 19 (Winter 1992): 601–18.
22. See Stefan Oster, "Thinking Love At The Heart of Things: The Metaphysics of Being as Love in the Work of Ferdinand Ulrich," *Communio* 37, no. 4 (Winter 2010): 660–700.
23. See Giulio Maspero, "Trinitarian Ontology and Interdisciplinary Research," in *Social Science, Philosophy and Theology in Dialogue*, ed. Pierpaolo Donati, Antonio Malo, and Giulio Maspero (Abingdon New York: Routledge, 2020), 92–109.
24. Maspero takes a balanced approach saying, "This ontological innovation lies in the fact of having recognized relation to be an original co-principle together with substance... relation has not supplanted substance... instead, the first is placed next to the second, without any superiority of one over the other." Giulio Maspero, *The Cappadocian Reshaping of Metaphysics: Relational Being* (Cambridge, United Kingdom: Cambridge University Press, 2024), 4.
25. See Giulio Maspero, "Patristic Trinitarian Ontology," in *Rethinking Trinitarian Theology: Disputed Questions and Contemporary Issues in Trinitarian Theology*, ed. Robert Józef Wozniak and Giulio Maspero (London: T & T Clark, 2012), 211–29. Giulio Maspero, "Relation and Ontology In Gregory of Nyssa and Augustine of Hippo," *Scripta Theologica* 47 (2015): 607–41.
26. Maspero, "Patristic Trinitarian Ontology." Also see Klaus Hemmerle, *Theses Towards a Trinitarian Ontology*, trans. Stephen Churchyard (Brooklyn, New York: Angelico Press, 2020).
27. Maspero, *The Cappadocian Reshaping of Metaphysics:*

Relational Being, 8. Maspero offers a helpful summary of the history of Trinitarian ontology, tracing two lines that integrate both philosophical and theological approaches. The first moves in the West through Augustine and focuses on relation "building on substance," and "echoing Thomistic-inspired philosophy." In this line of thought, he mentions Ratzinger and others like W. Norris Clarke. The second line moves in the East through the Cappadocians and focuses on "person." Here he mentions Zizioulas and others. In general, Maspero references several thinkers who have contributed to the field of relational ontology. Besides those already named, we may add here Piero Coda, John Milbank, Lubomir Zak, Pavel Florenskij, Sergej Bulgakov, Edith Stein, and Antonio Rosmini. See Maspero, *The Cappadocian Reshaping of Metaphysics: Relational Being*, 8–11.

28. Hemmerle, *Theses Towards a Trinitarian Ontology*, 7–8.
29. Hemmerle, *Theses Towards a Trinitarian Ontology*, 14.
30. Hemmerle, *Theses Towards a Trinitarian Ontology*, 50.
31. Hans Urs von Balthasar, *A Theology of History*, Communio Books (San Francisco: Ignatius Press, 1994), 7.
32. He goes on to say Balthasar does this by "interpreting the Son's 'abandonment' on the Cross by the Father as a revelation of radical intra-divine *kenosis* and finds in this *kenosis* the key to the 'being' of the Trinity." Matthew Levering, *Scripture and Metaphysics: Aquinas and the Renewal of Trinitarian Theology*, Challenges in Contemporary Theology (Malden, MA: Blackwell Pub, 2004), 212.
33. Quoted in O'Callaghan, "Can God Be Enriched? On the Metaphysical Underpinnings of Von Balthasar's Theology," 178. See Hans Urs von Balthasar, *The Glory of the Lord: A Theological Aesthetics: Vol. 5: The Realm of Metaphysics in the Modern Age*, ed. John Riches and Brian McNeill (Edinburgh: T. & T. Clark, 1991), 8. For more on Ulrich's ontology as it relates to Balthasar's view, see Oster, "Thinking Love At The Heart of Things: The Metaphysics of Being as Love in the Work of Ferdinand Ulrich."
34. Gerard Francis O'Hanlon, *The Immutability of God in the Theology of Hans Urs von Balthasar* (Cambridge: Cambridge University Press, 1990), 55, 126, 129, 132, 134.
35. Thomas Aquinas, *On Being and Essence*, trans. Armand Maurer, 2nd ed., Mediaeval Sources in Translation 1 (Toronto:

Pontifical Inst. of Mediaeval Studies, 1983), 32–33.

36. Thomas Aquinas, *Summa Contra Gentiles Book One: God*, trans. Anton C. Pegis, University of Notre Dame Press edition 1975 (Notre Dame, Indiana: Doubleday & Company, Inc., 2014), 128. Quoted in Aquinas, *On Being and Essence*, 33, fn. 15.
37. Aquinas, *On Being and Essence*, 33, fn. 15.
38. Thomas Joseph White, "The Two Natures of Christ in the Crucifixion: The Cross as a Revelation of Divine Love," *The Thomist: A Speculative Quarterly Review* 98, no. 1 (November 2022): 119–49.
39. In a lecture given Dec. 2, 2021, White expresses the necessity of using philosophical language to articulate revelation: "Even if the divine nature of the Son is a mystery of faith... made known to us by way of divine revelation, its reception as mystery into human thought requires an analogue concept drawn from philosophical understanding that allows the human intellect to orient itself towards God." "Chalcedonian Christology and the Concept of Pure Nature." Angelicum Thomistic Institute. December 3, 2021. Video, https://www.youtube.com/watch?v=OWF-fEU-sY0. Balthasar acknowledges this too, but references Hemmerle to say philosophical language can go further: "If mankind is to understand God's Word that is uttered to the world, philosophy is presupposed; conversely, this means that man's finite reason must exhibit an openness beyond itself if it is to be receptive to the divine speech. On the basis of this twofold, reciprocal a priori of philosophy and theology, Hemmerle asserts that, while Christian theology's adoption and adaptation of secular (Greek) philosophy did make significant advances toward a new understanding of Being in its totality, nonetheless there remained a certain 'historical deficit of Christian ontology': 'What is distinctively Christian did not. . . ultimately refashion the anticipatory understanding of the sense of being." Hans Urs von Balthasar, *Theo-Drama: Theological Dramatic Theory: Vol. 5: The Last Act*, trans. Graham Harrison (San Francisco: Ignatius Press, 1998), 39.
40. O'Callaghan, "Can God Be Enriched? On the Metaphysical Underpinnings of Von Balthasar's Theology," 185.
41. White elaborates on the consequences of this danger: "[Christ's] human nature is not indicative univocally of his divine life... [the attempt] to indicate the divinity of Christ

univocally by appeal to his human states and modes of action run the risk of problematically anticipating the eschatological consummation of knowledge." White, "The Two Natures of Christ in the Crucifixion: The Cross as a Revelation of Divine Love," 148. Similarly, O'Callaghan claims it would be better for terms like "receptivity," "donation," and "abandonment" to be employed metaphorically rather than analogically, as Balthasar seems to do. O'Callaghan, "Can God Be Enriched? On the Metaphysical Underpinnings of Von Balthasar's Theology," 193.

42. For example, Ratzinger offers the same reminder as Hemmerle regarding the Trinitarian formula traditionally expressed in classical ontological terminology "one Being – three Persons." He says, "One is not entitled to go too far in the direction of taking these words as the only possible ones and deducing that the matter can be stated only in this way and in no other. That would mean a failure to recognize the negative character of the language of theology, the purely tentative fashion in which it speaks." Benedict XVI, *Introduction to Christianity*, 180–81.

43. In a 2010 Wednesday Audience, he said, "The study of his [Thomas'] thought was explicitly recommended by the Second Vatican Council in two documents, the Decree *Optatam Totius* on the Training of Priests, and the Declaration *Gravissimum Educationis*, which addresses Christian Education. Indeed, already in 1880 Pope Leo XIII… held St Thomas in high esteem as a guide and encouraged Thomistic studies." Benedict XVI, "General Audience Saint Peter's Square: Wednesday, 16 June 2010" (Vatican City, 2010).

44. John C. Rooney, *A Truly Catholic Cosmology: The Perspective of Stanley L. Jaki, O.S.B.* (Yonkers, New York: St. Joseph Seminary, 1989), 111–12. See Stanley L. Jaki, *Cosmos and Creator* (Edinburgh: Scottish Academic Press, 1980), 100, which is also quoted by Rooney.

45. O'Callaghan, "Can God Be Enriched? On the Metaphysical Underpinnings of Von Balthasar's Theology," 175–94.

46. This is one of White's main criticisms of Balthasar, that is, his approach tends to jettison divine attributes such as simplicity, perfection, unity, and, especially apropos, immutability. See White, "The Two Natures of Christ in the Crucifixion: The Cross as a Revelation of Divine Love." Also see Thomas Joseph White, *The Trinity: On the Nature and Mystery of the One God*,

Thomistic Ressourcement Series, Volume 19 (Washington, D.C: The Catholic University of America Press, 2022), 551–687. Furthermore, Guy Mansini claims that this expression of Balthasar's Trinitarian ontology is simply irreconcilable with the approach that underscores the pronouncements of the early and subsequent councils. After offering a refutation of Balthasar's "reinterpretation" of Thomas he says, "[Balthasar's] *Theodrama* itself, so promising in the prospect of a properly theological and indeed Trinitarian reconciliation of modern concerns with the 'older dogmatics,' appears to be yet another demonstration of the impossibility of such a project." Quoted in O'Callaghan, "Can God Be Enriched? On the Metaphysical Underpinnings of Von Balthasar's Theology," 189.

47. Furthermore, we have also seen that this ontological gap between God and creation extends in some way to the relationship between two dialogue partners. Since both are in relationship with God, the gap between God and creation appears somehow in the conversation. This means when we speak about effective communication, the same cautions regarding closed and confused ontologies applies.
48. The precedent is clear when we remember how the Chalcedonian language developed. It was through the human encounter with Trinitarian revelation that words like "substance" and "person" took on deeper meanings.
49. Angela Franz Franks claims Balthasar's laudation of Thomas' real distinction (*analogia entis*) "at the very least, provides a point of departure for further dialogue between followers of Thomas and disciples of von Balthasar." Angela Franz Franks, "Trinitarian Analogia Entis in Hans Urs von Balthasar," *The Thomist: A Speculative Quarterly Review* 62, no. 4 (October 1998): 534.

20 A COMPARISON OF MATHEMATICAL REALISM IN THE WORKS OF STANLEY JAKI AND PLATO

COLLIN DOYLE AND IAIN MORTON

The distinction between the terms "discovery" and "invention" may be logically applied to numerous truths in the physical world. However, the extent to which one is entitled to stamp these classifications on the wide variety of mathematical truths is a source of debate. It is the firm stance of mathematical realism, or mathematical Platonism, that these theorems are fundamental to natural phenomena, and are therefore discovered. It is this inherent reality that will be justified and made obvious in this paper through the examination of various examples and lines of reasoning. The role of proofs within mathematics will be asserted to be to provide logical arguments that ensure theorems reside within the bounds of universal truths and exist in full accordance with laws that are independent of any form of human control or creation. This paper will also discuss the extent to which there is agreement between influential philosophical minds, such as Father Stanley Jaki, and the tenets of mathematical realism. Studies done on the understanding and presence of mathematics in the minds of non-human species will be discussed alongside the role of religion in defining terms that are central to the mathematical realism argument. The overarching point will be made that mathematical realism holds true under most philosophical circumstances; however, it may be observed to fall apart only when certain terms are redefined and the goalposts of meaning and truth are shifted to favor the outlook of the antirealist.

It is the view of many that those who find themselves on opposing sides of the mathematical realism argument "have often appeared to be talking past each other."[1] The burden of proof is often shifted from one to the other, as realists may often assert that they need not prove the reality of mathematical objects until "nominalists can explain how they acquire knowledge that there are some physical things behind their sense-perceptions."[2] It is therefore difficult to begin a conversation, dialogue, or conclusion on the subject of mathematical realism, since the very point at which this exchange begins is hotly contested. When taking into account the tribalism that arises between those who align themselves with mathematical realism and other points of view, it becomes evident that this debate is not one that can be settled objectively. Rather, much like any philosophical debate, the answer is more complex. An allegiance that is formed to either point of view is merely a form of dogma that opposes valuable and honest communication.

Valuable discussion and debate on the topic of mathematical realism has been initiated from various voices with diverse perspectives, however, one figure that left a substantial impact on how this idea would be regarded for years to come is Father Stanley Jaki. It is a frequent theme in his work to regard mathematics as the method through which physical truth is both measured and quantified. More specifically, "integers provide the basis for measurability, which is a kind of universal understanding."[3] Mathematics can therefore be compared to language. Language and speech provide a measurement of human thought and emotion, as they can convey a wide range of concepts that exist within the human brain. Mathematical objects do the same for the truths that exist in the universe. While they exist, they are hard to quantify without the use of mathematics as a set of agreed-upon terms. For instance, the expansion of the universe would be hard to envision were it not for its measurability, which is made possible through the grace of mathematical objects as building blocks of a language of expression. This does not verify the stance of mathematical realism; however, it places into context the view that many philosophers take towards mathematical objects and their stance within reality. Language, after all, may be considered a discovery of the human mind in the same way that it may be perceived to be a human creation.

Mathematical realists regard mathematics not as a language alone, but rather, as its own entity that exists independently of how it is measured. Numbers and sets would therefore exist independently of human measurements and quantification. They are present in the universe in the same way that the planets and their orbits exist.

Jaki's own argumentations also appear to be related to the Fregean Argument for the existence of mathematical objects. Gottlob Frege, a 19th—20th century philosopher and mathematician, argued for the semantics of mathematical arguments. As mathematical statements must necessarily follow the same logical structure in which non-mathematical arguments may be composed, it is thus logical to assume that true statements could lead to mathematical truth. However, this is where Father Jaki and Frege's arguments are divided. While Frege argues for truth-values within language alone, Father Stanley Jaki, while relying on the language of mathematics for the composition of natural laws, advocates for the necessity of mathematical truth in relation to physics.

Thus, the indispensability argument for the existence of mathematical objects takes a heavier context with respect to the views of Father Stanley Jaki. The indispensability argument, proposed as the Quine-Putnam argumentation, provides context to the relation of mathematical truth to natural laws by a classical two-proposition and conclusion argument:

> Proposition 1: It is necessary to have ontological commitment to all indispensable means to explain scientific phenomena.
>
> Proposition 2: Mathematics is indispensable to scientific theorems.
>
> Conclusion: It is necessary to have ontological commitment to mathematics.

Father Jaki seemed to be quite sympathetic to the notion of mathematics being treated as a rather indispensable means to provide a scientific context. As quoted by Pham: "Mathematics, therefore, is extraordinarily helpful in justifying the notion of a universe which is a true totality because of the limitedness of matter to a finite quantity."[4] Jaki, thus, would certainly agree on how mathematics must be treated as an ontological and indispensable means to explain universal phenomena. An antirealist may pose the claim that utilizing the indispensability argument to argue for mathematical realism combines mathematical truths with more substantial and physically tangible scientific truths. To refute this claim, however, one must assert that the first proposition of the argument makes a clear distinction between scientific phenomena and the mathematics used to explain them. Rather than recklessly combining scientific observations with mathematics, the argument instead states that each component relies on the other, and it is the inherent necessity of the role of mathematics in explaining worldly observations that endows it with a unique ontological value. The distinction may be made between mathematics and mathematical objects, and this distinction should be granted, while asserting that the entirety of mathematics is comprised of its parts, which, in their service to the larger discipline of mathematics, have equally commensurate relevance within reality.

However, it would still be difficult to classify Father Jaki as a believer of the Quine-Putnam Indispensability Argumentation, due to the theory's context. While the Quine-Putnam Indispensability Argument follows a naturalist and purely methodological approach to the scientific necessity of mathematics, it is rather important to note that Father Stanley Jaki viewed natural science from a religious perspective. As noted again by Pham: "And it is in appealing to scientific cosmology that they can proclaim the truth of the unique reality of the universe, which in turn is the supreme witness to Truth itself, that Truth which is identical with the one God who has revealed himself to truly be the Father Almighty."[5] To Father Stanley Jaki, the universal laws of matter are not divorced from theological interpretations, but rather used to provide context to religious scripture and phenomena.

Ontologically speaking, the term "commitment" may be confusing to define in terms of mathematics. However, according to Quine: "A theory is committed to those and only those entities to which the bound variables of the theory must be capable of referring in order that the affirmations made in the theory be true."[6] This is rather pivotal to Father Stanley Jaki, who has consistently argued for the basic components of physical theory to "build-up" a universe mandated by natural laws. In such a way, how logical assertions of mathematics are linguistically composed can provide context for ontological commitments in the natural sciences.

Father Jaki's cosmological perspectives, hence, were rather unique, as it can be devised that his stances on mathematical realism, while both Fregean and naturalist, also were combined with his own theology. Mathematics and physics were not treated separately from philosophical notions of Biblical scripture, but rather a means to provide context to a religious whole. It is not difficult to envision Father Stanley Jaki as a Catholic Mathematical Realist.

Jaki's work regarding the perception of the universe as a whole is relevant to an overall understanding of how his philosophies fit into the discussion surrounding mathematical realism. An essential question usually sprouts up in Jaki's writings, which asks "why is the universe what it is and not something else?"[7] It is evident in this question that the basis for the universe being real is the fact that it works a certain way. It has rules that seem to be unbreakable within the bounds of reality. There are clear violations to the ways in which the universe functions, and these violations are well understood. These statements may also be said of mathematics. It is evident that there are rules that comprise the modern and ever-growing understanding of mathematics. These rules are consistently demonstrable, and work under a wide range of circumstances. Mathematics shares this essential trait with the universe, and although the reality of mathematics is contested, the firm foundation of the universe within the bounds of reality is never called into question. When the possession of characteristics such as form, function, and rules is the basis by which something is judged to be real, mathematical realists seem to emerge victorious by way of the comparison of mathematics and mathematical objects to the universe itself.

It is essential to take into account Jaki's theological standpoint when considering his words on the existence and functionality of the universe. While purely scientific discussions may focus on the teleological side of the universe, with respect to the ways in which the universe acts, behaves, and changes, more religious viewpoints may center in on the origination of the universe instead. A focus on a higher power may also serve to set mathematics into a more complex context. The meaning of creation takes a different form when a God is believed to be at the center of all that has been and will be. In essence, the line between invention and discovery may become even more blurred when taking into account Jaki's stance on the existence of God, since all of God's creation may be said to have been invented by God. It is therefore consistent to assume that humans are also invented, which may place mathematics on the same level of other physical phenomena in terms of how they can be considered to have been invented or discovered. Simone Weil's contribution to the understanding of God's place in the modern view on mathematical realism is valuable, and she was quoted as professing that "this very mathematics is above all a kind of mystical poem composed by God himself."[8] The closeness of mathematics to the craftsman of the universe implies that it stood undiscovered long before humans happened upon it, and rather than being discovered, it was instead harnessed. Jaki's firm religious stance therefore predisposes him to being in agreement with the arguments of mathematical realism.

An atheistic standpoint may utilize a philosophy more akin to metaphysical materialism, which seeks to describe and define the world in terms of the matter that makes it up. Rather than focusing on definitions, metaphysical materialism may base something's truth value on how provable it may be in terms of its physical presence in the world. The lack of physical processes behind concepts in mathematics allow for the metaphysical materialist to reduce mathematics to a human invention. One who aligns themself with this philosophy would very well compare the planets and the forces causing their orbits to the abstract and physically elusive nature of concepts such as the Pythagorean Theorem. It is important to understand, however, that this view "denies the existence of spiritual beings or forces."[9] This denial of the existence of a higher power removes the blur between invention and discovery in a way that leaves mathematics squarely on the side of invention, therefore refuting the concept of mathematical realism. This may be countered by asking the antirealist what force caused the universe to be set in motion, if it was not some initiating force. Whether it is referred to as God or some undiscovered force of nature, the argument that everything comprising the universe is essentially harnessed rather than being invented still lends itself to the idea of mathematical realism even in the absence of an omniscient God.

The deciding factor in terms of the classification of mathematics as being invented or discovered rests firmly on the definition and meaning of the words "invented" and "discovered." It is therefore evident that, as long as these terms remain subjective, so too do the arguments for and against the truth value of mathematical realism. It is for this reason that this discussion seems to bring about a sense of tribalism. The true answer lies not in provable facts, but rather, it is reliant on how one views the inception of the universe itself.

Another curse that hangs over the discussion surrounding mathematical realism arises from the fact that mathematics is nestled in the intersection between the hard sciences and the humanities, which some claim to determine truth using vastly different methods. A comparison of the scientific method to philosophical discussions can explain the inability to reach an adequate conclusion between opposing parties. It has been said that scientific thinking deals in a different kind of knowledge than philosophical ponderings. It is the view of some scholars, such as Dr. H. H. Benson, that one should correctly identify "the mathematical method with dialectic (i.e., the method of philosophy)" rather than with dianoetic.[10] The difference between these terms lies in the distinction between thinking about the image of a form, which characterizes dianoetic, and thinking about the actual form itself. This statement can therefore be utilized to promote mathematical realism as a result of its choice to regard mathematics not as an image or projection of the real world, but as a central part of reality. While Benson's words draw a direct connection between the method of acquiring knowledge in mathematics to the method undertaken in philosophy, they also succeed in implying that human knowledge on the subject of mathematics was garnered through the discovery, rather than invention, of one of Plato's forms. The voice of another scholar, Dr. Elaine Landry, uses the concept of "partial homomorphism to express the shared structure between mathematical and physical theories."[11] This assertion that there exists a commonality between practical scientific knowledge and the expanse of supposedly more abstract mathematical knowledge places more emphasis on the truth value of mathematical realism. The antirealist may assume that the weakness in this statement is that these theories, although having a similarity in that they are both provable, are indeed provable by different means. This, however, does not disqualify the argument of the realist. It is true that theories outlined under the discipline of chemical sciences are proven by different means when compared to theories that expand our knowledge of life sciences. They are both provable, and yet they are reached through different methods. The same is true for mathematical theories when placed in comparison to their scientific counterparts. The method by which something is proven to be true does not characterize the extent of its truth value. Rather, it is consistency of results and repeatability that ensures a theory's firm stance in reality.

The inclusion of God at the very center of the discussion behind mathematical realism leads to dispute between those who view knowledge as being gained through observation and experimentation, and those who deem knowledge as being uncovered by thought experiments and extensive lines of reasoning. In essence, mathematics lies somewhere between the hard sciences and the humanities, and neither party is sure how to claim or define it. In a sense, mathematics is a field of study that lends itself to both the observable truths of science and the reasoning of philosophy, and yet defies both fields in terms of being completely pigeonholed.

Father Stanley Jaki, who intertwined both disciplines in his work, was strong in his conviction that modern philosophy had been replaced, either in part or in whole, by the work of cutting-edge physicists. Those who advanced the general understanding of the natural world worked to create an upheaval in what was deemed known, which is similar in many regards to the work of Socrates. In essence, the spirit of the Socratic Method remains within every physicist who asks questions of their surroundings, only to seek the answers through experimentation and gathering data. Science can be described, similarly to the system proposed by Socrates, as "asking continual questions until a contradiction was exposed, thus proving the fallacy of the initial assumption."[12] Einstein, through his thought experiments, was asking pivotal questions that would go on to expose the fallacies and contradictions of how Newtonian and Classical Physics was understood and taught. These questions, such as the thought experiment that led him to the idea of relativity of simultaneity, asked questions of Newtonian physics in the same way that Socrates asked questions of his surroundings.

The work of Plato in the realm of metaphysics influences much of the discussion surrounding mathematical realism, and it is Plato's reliance on the writing of Pythagoras that colors his view on the matter. He shared the view that "we have a priori knowledge of objective, non-physical abstract entities which are constant and unchanging."[13] Plato's emphasis on the reality of immaterial concepts and knowledge is the basis for mathematical realism, as it implies that although the knowledge gained through the constant and unchanging rules of mathematics is separate from the tangible aspects of the physical world, they are no less real and inherent to the world as it exists independent of human influence. Plato deems the ability to perceive a specific phenomenon with the human senses as an unnecessary criterion for proving something's reality. His theory of forms states that forms are not physical and tangible, but rather, they are an understanding of the true nature of an idea that can be attained through routes other than the use of one's human senses such as touch, smell, sight, and hearing. In fact, Plato's emphasis on the timeless nature of mathematics, as well as the way that mathematics may be deemed to be understood through human intellect rather than the senses, all contribute to the foundational arguments of mathematical realism. It is the universality of mathematics that further implies its stance within the bounds of reality, rather than the imperfect furnace that is the human mind.

Plato further breaks down the world and its parts into the realm of being and the realm of becoming. In doing so, he outlines the ways in which parts of the world differ in the forms they take and the ways in which they are uncovered. One specific aspect of mathematics that Plato comments on is the fact that it is unchanging. Its laws do not morph and change under varying circumstances, and its foundations are steady. One does not view mathematics as flexible. Although the understanding of mathematics is one that is constantly expanding and growing in complexity, the rules outlined in mathematics are rigid and unchanging. This is characteristic of something that exists in accordance with the realm of being, as this classification includes the aspects of the world that exist as they are. Humans grappling with mathematics are simply able to learn more about how it works, but do not have the ability to change its fundamental rules. On the other hand, the realm of becoming encompasses all that is changing. To make the distinction between mathematics and the human understanding of mathematics is therefore crucial in this area since the amendments that are constantly made to mathematical knowledge place it within the classification of the realm of becoming rather than the realm of being. Plato's organization of the world into these two separate realms makes clear the fact that the lack of human involvement in the truth and functionality of mathematics is one of the strongest pieces of evidence behind the idea of mathematical realism.

In many senses, the work of Plato allows for the observation that mathematical knowledge aligns itself with knowledge in other disciplines in terms of its foundation in reality. The study of physics allows for the demonstration of mathematical techniques in innumerable applications, and at the heart of physics lies the formulas and rules of mathematics. The language of engineering is rooted in mathematics, and the functionality of chemistry rests on the measurements and quantitative data that can only be predicted, measured, and compared using the rules and doctrine of mathematics. If one were to regard the tangible truths of chemistry as real and substantial, one must also regard the mathematics behind these truths with the same respect. The International Commission on Mathematical Instruction regards mathematics as "having a transversal nature" that is "integral to attempts at understanding the world and ourselves."[14] It is this transversal nature that allows for mathematics not only to serve many purposes as its own area of study, but also as the backbone of other disciplines. It may be said that mathematical realism is supported by the idea that the consequences of mathematics are tangible and real in a physical sense even if the abstractions behind these impacts are not. In the same way that language moves people to action through the communication and discovery of ideas, mathematics provides the ability to discover previously unknown truths about the natural world that prove to be valuable in ways that are oftentimes built on the previously established advancements of those that come before.

The acquisition of mathematical facts is therefore built not on the imaginings of the human mind, but rather, on the wide range of truths that have already been established. The established truths that are agreed upon by mathematicians are, however, based on what can be referred to as assumptions. It is a fact that "proofs have core assumptions on which everything else hinges",[15] and although mathematical realists pose that the existence of certain assumptions for mathematics to build on does not rule out its universality, the antirealist may reason that the assumptions on which mathematics is built is exactly what makes it a creation of the human mind. One may argue that there is a possibility of utilizing different assumptions to create a different kind of mathematics, which therefore challenges the widely held belief that mathematics is truly universal. When recalling the comparison of mathematics to language, however, one may counter that the diverse range of languages that have developed in various parts of the world can be analogous to the different ways of describing and performing mathematics. Just as different sounds may be used to relay a similar meaning, with that meaning being unchanging and representative of the same concept, the history of mathematics proves that different numerical systems lead to the same conclusions. The theorems of Pythagoras would still be true had humans used a different counting protocol.

The ways in which mathematics is presented and framed are inventions of the human mind, however, the truth behind mathematics remains universal. This can be perceived in the various ways that numerals have been represented across cultures and time periods. For instance, the base twenty system created by the Mayan civilization would lead to the same theorems as any other method.

Perhaps one of the most interesting developments in the way that mathematics proliferated among different cultures was the introduction of the idea of zero. This lack of presence of something, or emptiness, traces back to India. Both zero and negative integers were considered more abstract than positive integers, however, their importance grew and with it came a more complex and abstract understanding of mathematics. It has been established that "from early times numbers are words which refer to collections of objects. Certainly the idea of number became more and more abstract and this abstraction then makes possible the consideration of zero and negative numbers."[16] It is this change in the way that mathematics was conceived that led to a long string of advancements. The idea of mathematical realism holds that the introduction of zero and negative numbers into the collective mathematical conscience is not an invention that amends our knowledge, but rather, it is a discovery of a new realm of mathematics that existed, in its entirety, before humans happened upon it. Furthermore, abstract concepts such as infinity seem to fall into this same category. For mathematical realism to hold, it must be asserted that these concepts are not the work of the human imagination, but rather, they are discovered.

The dispute over arithmetic with zero provides some argumentation in favor of mathematical realism, since there is an objective answer when it comes to identities having to do with zero. For example, the work of Indian mathematician Brahmagupta stated that "positive or negative numbers when divided by zero is a fraction with zero as denominator."[17] His work was later disputed and argued against by other mathematicians, who continued to develop a better understanding of arithmetic with zero. The concept of dividing by zero to reach a solution of infinity was one of these developments, however, the mathematical realist would point to the final objective solution, which is that one cannot divide by zero, as proof that mathematics has rules that are discovered rather than invented. If the rules were invented, one may point out, the concept of dividing by zero would not have a solid solution on which to rest. Rather, the solutions of Brahmagupta and other early mathematicians would all have some inherent truth value rather than being subject to the unavoidable fate of being disproven over time. Much like the scientific method, which deals in facts and absolutes, mathematics is not to be argued over. Rather, it is a collection of knowledge that is built through experimentation and proving hypotheses repeatedly. Consensus in mathematics is therefore one of the cornerstones of the mathematical realism argument.

Recalling the previous mention of the applications of mathematics in the sciences, many argue that this application is evidence of the objectivity of mathematics.

Another interesting development in this regard, however, comes from research that was done with life forms who were previously thought to have no knowledge of arithmetic. It has been shown that "arithmetical ability is not only the domain of sufficiently mature human agents" and the ability to harness mathematical truths to form a better and more useful understanding of the world is one that is shared across many species.[18] This seems to hint at the assertion that mathematics is less human and more universal. It is not a unique aspect of the human experience that is inaccessible to other species, but rather, it is a part of the world that is used in many different ways. In this way, mathematical realists may point to the understanding of mathematics among animals as evidence of its commonality. The views of the antirealist may even be characterized by a tone of anthropocentric grandeur. To regard the human mind as uniquely gifted with what is regarded as a human invention in the form of mathematics seems to go against the science surrounding the understanding of arithmetic among other life forms. Just as the geocentric theory was replaced with the significantly less human-centric heliocentric theory, so too has the theory that humans have a unique grasp of mathematics.

Mathematical realism is a concept that has grown over time, and the debate surrounding whether it is fixed within reality is ongoing. It hinges on some of the most fundamental questions, especially with respect to the beginning of the universe and the ways in which humans differ from the rest of the animal kingdom. Voices such as those of Father Stanley Jaki forever shaped the understanding of mathematical realism, and his impact on the field of metaphysics, philosophy, and mathematics, remains exceedingly relevant. Although some aspects of Jaki's work regarding mathematical realism may complicate the simple classification of his viewpoint as being exclusive to either side of the argument, it is clear through his sympathy towards the indispensability argument, as well as his own personal religious beliefs, that his voice was one that overwhelmingly supported the general acceptance of the concept of mathematical realism. Ongoing debate may be bolstered by a growing understanding of the relationship between mathematics and other non-human species, as well as the ways in which religion changes the meaning, and context, of various words used throughout the usual discourse surrounding mathematical realism.

Editors' Note

There are many undergraduate students all over the world interested in the relationship between Religion and Science. They potentially bring to the subject innovative and fresh perspectives, yet most of them have yet to make their mark in the area of science and religion. This Congress offered an opportunity to these two students to share their contribution.

Notes

1. John P. Burgess, "Review of *Platonism and Anti-Platonism in Mathematics,* by Mark Balaguer," *The Philosophical Review* 110, no. 1 (2001): 79–82.
2. *Ibid.* p.181.
3. John-Peter Pham, "The Importance of the Reality of the Universe in the Work of Stanley L. Jaki", *Faith & Reason* 21, no. 4 (Winter 1995), 1–9.
4. *Ibid.*, 5.
5. *Ibid.*, 7.
6. Willard Van Orman Quine, *From a Logical Point of View: Nine Logico-Philosophical Essays* (Cambridge, MA: Harvard University Press, 1961), 12–14.
7. Pham, "The Importance of the Reality of the Universe in the Work of Stanley L. Jaki", 7.
8. Jacques Vauthier, Lecture, Sorbonne (Paris 6, Pierre et Marie Curie), January 2017, accessed March 4, 2024, http://www.sljaki.com/essays/vauthier-metaphysics-and-mathematics.html.
9. "Materialism," *Routledge Encyclopedia of Philosophy,* accessed March 4, 2024, https://www.rep.routledge.com/articles/thematic/materialism/v-1.
10. H. H. Benson, "The Problem is not Mathematics, but Mathematicians: Plato and the Mathematicians Again," *Philosophia Mathematica* 20, no. 2 (2012): 170–199, https://philpapers.org/rec/BENTPI-3.
11. Elaine Landry, "Structural Realism and Category Mistakes," in *Categories for the Working Philosopher* (Oxford Scholarship Online, 2018), 430–450.
12. "The Socratic Method," *The Law School,* University of Chicago, accessed March 4, 2024, https://www.law.uchicago.edu/socratic-method.
13. "Plato's Metaphysics and Epistemology," Florida International University, accessed March 4, 2024, https://faculty.fiu.edu/~harrisk/Notes/Epistemology%20and%20Metaphysics/Plato%20Metaphysics%20and%20Epistemology.htm.
14. "The Role of Mathematics in the Overall Curriculum," International Commission on Mathematical Instruction,

accessed March 4, 2024, https://www.mathunion.org/icmi/role-mathematics-overall-curriculum.

15. Kelsey Houston-Edwards, "Numbers Game: Philosophers Cannot Agree on Whether Mathematical Objects Exist or Are Pure Fictions," *Scientific American*, February 20, 2024, https://www.scientificamerican.com/article/is-the-mathematical-world-real/.
16. "A History of Zero," J. J. O'Connor and E. F. Robertson, University of St. Andrews, accessed March 5, 2024, https://mathshistory.st-andrews.ac.uk/HistTopics/Zero/.
17. Roshani Senior, "The Men Who Invented Zero," *Mathematics Education at the Open University*, August 25, 2022, https://www.open.ac.uk/blogs/MathEd/index.php/2022/08/25/the-men-who-invented-zero/.
18. Markus Pantsar, "Objectivity in Mathematics, without Mathematical Objects," *Philosophia Mathematica* 29, no. 3 (2021): 319, https://academic.oup.com/philmat/article/29/3/318/6291527.

INDEX OF NAMES

www.ingramcontent.com/pod-product-compliance
Lightning Source LLC
LaVergne TN
LVHW090255210926
848480LV00001B/34
* 9 7 8 0 8 5 2 4 4 9 6 0 8 *